Ancient Warfare, Volume II

Ancient Warfare, Volume II:

Introducing Current Research

Edited by

Graham Wrightson
and Jared Kreiner

**Cambridge
Scholars**
Publishing

Ancient Warfare, Volume II: Introducing Current Research

Edited by Graham Wrightson and Jared Kreiner

This book first published 2024

Cambridge Scholars Publishing

Lady Stephenson Library, Newcastle upon Tyne, NE6 2PA, UK

British Library Cataloguing in Publication Data
A catalogue record for this book is available from the British Library

ISBN (10): 1-5275-7035-5
ISBN (13): 978-1-5275-7035-1

CONTENTS

List of Figures.. vii

Introduction 1
The Many Faces of Ancient Warfare
Graham Wrightson

"Dying for 'Doing'": Mythical Athenian Heroines
in Service of the City 7
Ioannis Mitsios

The Appropriation of Women in Classical Literature during
the Great War (1914-18) by the *Girls' Own Paper* 24
Phyllis Brighouse

'Not a Few Hoplites': The Evolution of Illyrian Infantry,
5th to 3rd Century BC.. 40
Florian Feil

A New Interpretation of the Battle of Panion 66
Graham Wrightson

The Unusual Encounter between Demetrius III and Alexander Jannaeus:
Religious Beliefs as a Motivating Factor in Ancient Warfare................. 106
Kenneth Atkinson

Deterrence and Self Representation: The Roman Headquarters
at Syracuse.. 134
Borja Vertedor Ballesteros

*Illa vero pars calliditatis egregia et ab omni reprehensione
procul remota...* Hannibal, the Carthaginians and the
Stratagems in Valerius Maximus 153
Gabriele Brusa

The Suspension of the War-Tax (*Tributum*) in 167 BCE:
A Short-Sighted Decision? 174
François Gauthier

From Battle to Text: A Review of the Word *Actium* in Latin Literature
as a Resource for Identity and Geographical Monumentality 193
Nelson Henrique da Silva Ferreira

The Impact of the Tacfarinas Revolt beyond Africa 212
Jared Kreiner

Neglected Iranian and Syriac Evidence on the Persian
Campaigns of Gordian III and Galerius 227
Byron Waldron

Traumatic Brain Injury in Roman War 254
Brendan McCarthy

With the Courage of an Emperor: *Virtus* and the Empress
in Roman Propaganda 275
Julia Wetzel

Warfare and *Canones Sacrorum Contemptores* 298
Despina Iosif

Conclusion 316
Looking to the future of examining the past
Graham Wrightson

LIST OF FIGURES

Fig. 1.1a. Berlin, Antikenmuseen F 2537. PHOTO: The birth of Erichthonios.
Fig. 1.1b. Berlin, Antikenmuseen F 2537. PHOTO: The Aglaurids (named) in the birth of Erichthonios.
Fig. 1.2. Athens, National Archaeological Museum 17917. Alabastron with the disobedient behavior of Aglauros.
Fig. 1.3 The shrine/cave of Aglauros on the east slope of the Acropolis.
Fig. 1.4: The central scene of the east frieze of the Parthenon.
Fig. 1.5: Delphi 1584. PHOTO: The Akanthus Column in Delphi.
Fig. 4.1: Army dispositions.
Fig. 4.2: The precipitous banks of the river facing north from Snir.
Fig. 4.3: Map of the topography of the battle site with the line representing where Scopas could have drawn up his line to block Antiochus and rest both his flanks on hills.
Fig. 4.4: Map of the plain showing the line from the steep slopes beside Snir to the large rise at the other end of the plain near Ghajar.
Fig. 4.5: Map of the plain from the steep slopes beside Snir to the slight hill at Tel Dan that encompasses the eastern half of the plain.
Fig. 4.6: The two battle lines with the Egyptian line in red and the Seleucid in blue.
Fig. 4.7: Johstono's map of the battlefield from Mount Hermon looking South.
Fig. 4.8: The actions of the left side of the battle.
Fig. 4.9: The end stages of the battle where the Ptolemaic phalanx is surrounded and threatened by the returning cavalry of Antiochus the Younger.
Fig. 4.10: Map showing route from the likely Egyptian camp at Tel Dan to their escape destination at Sidon.
Fig. 10.1: Major Sites and Range of Tacfarinas' Actions c.19 CE.
Fig. 13.1: Silver Denarius, Obv. Laurate bust of Septimius Severus, Rev. Mars advancing right with spear and trophy. *ANS* 1995.11.1490. Images courtesy of the American Numismatic Society. http://numismatics.org/collection/1995.11.1490
Fig. 13.2: Silver Denarius, Obv. Draped bust of Julia Domna, Rev. Vesta seated holding a palladium in right hand and a scepter in the left. *ANS* 1997.10.9. Images courtesy of the American Numismatic Society. http://numismatics.org/collection/1997.10.9

Fig. 13.3: Silver Denarius, Rev. DIANA LVCIFERA, Diana holding a torch. *ANS* 1948.19.1490. Images courtesy of the American Numismatic Society. http://numismatics.org/collection/1948.19.1490

Fig. 13.4: Silver denarius, Obv. IMP CAESAR TRAIAN HADRIANVS AVG with laurate bust of Hadrian, Rev. P M TR P COS III Minerva, helmeted, levelling javelin, and holding spear. *ANS* 1916.192.209. Images courtesy of the American Numismatic Society. http://numismatics.org/collection/1916.192.209

Fig. 13.5: Gold aureus, FAVSTINA AVGVSTA, bust of Faustina the Younger, Rev. VENVS VICTRX, Venus with a victory in one hand and a shield. *ANS* 1955.191.17. Images courtesy of the American Numismatic Society. http://numismatics.org/collection/1955.191.17

Fig. 13.6: Silver cistophorus, Rev. COS II distyle temp of Diana of Perge with a cult statue, and inscribed DIANA PERG. *ANS* 1944.100.44674. Images courtesy of the American Numismatic Society. http://numismatics.org/collection/1944.100.44674

Fig. 13.7: Silver Antioninianus, Rev. VENERI VICTRICI, Venus standing left, holding a Victory and scepter while resting a shield on a helmet. *ANS* 1944.100.51519. Images courtesy of the American Numismatic Society. http://numismatics.org/collection/1944.100.51519

Fig. 13.8: Silver Denarius, Obv. FAVSTINA AVGVSTA - Bust of Faustina Younger, draped, right, head bare, hair waved and elaborately coiled, fastened in a bun on back of head, Rev. VENVS FELIX - Venus, draped, seated, left on throne, holding Victory on extended right hand and vertical scepter in left, *ANS* 1956.127.1020. Images courtesy of the American Numismatic Society. http://numismatics.org/collection/1956.127.1020

INTRODUCTION: THE MANY FACES OF ANCIENT WARFARE

GRAHAM WRIGHTSON

This volume stemmed from the first reiteration of the International Ancient Warfare Conference Series (IAWC) begun and run so successfully before the pandemic by Geoff Lee. During the pandemic we had the idea of a remote conference through Zoom, and I took the initiative in setting that up finally in 2022. That June saw 25 presenters from all over the world connect remotely via zoom to engage in great papers and discussions on topics from Greece and Rome to China.

The first volume of this book series, *Ancient Warfare: Introducing Current Research* resulted from the very first International Ancient Warfare Conference Series (IAWC) meeting Geoff Lee hosted in Aberystwyth in September 2013. That it has taken almost a decade for the second volume to come out is more a testament to the state of academic publishing and the pandemic than the efforts of the editors. Finally, the second volume publishes a great mix of papers on all aspects of Greek and Roman warfare ranging from topics to delight a purist such as military tactics and logistics, through more social aspects of war such as religion and women, to the reception of historical wars in more modern times. It presents papers from graduate students and established professors alike in keeping with the inclusive and welcoming nature of the conferences.

The conference has developed into a hybrid version with two in-person meetings held simultaneously in Europe and the USA both connected remotely through Zoom. The 2023 version saw gatherings at Bonn in Germany and Sioux Falls in South Dakota, as well as online presentations coming from as far afield as Sri Lanka, Australia, and New Zealand. The 2024 conference, under a new name the War in the Ancient World International Conference (WAWIC), will host simultaneous gatherings in Vienna and Chicago with the hope to expand in person offerings in other continents or regions in the future.

We hope to have compiled a volume that demonstrates the wide array of topics currently studied by researchers around the world, while also presenting rigorous scholarship that brings new ideas to the forefront. Rather than having a definitive theme, these papers, just as those in the first volume, represent the broad array of research currently going on around the world.

Arranged chronologically in Greek and Roman history sections, the papers take readers through all manner of current topics on ancient warfare from traditional battle narratives or strategic analyses of campaigns, through the logistical considerations of armies in the field, to the ideology of women in war. As befitting the nature of the conference, the authors featured in this volume blend together current graduate students with both early career and more established academics. The end result, we hope, is a fascinating insight into the many different types of research military historians do and end the stigma of the study of warfare simply being the study of the movements of men with weapons on a battlefield. The study of ancient war deals with a myriad of different topics and deals with themes in all types of history: social, cultural, economic, religious, literary, numismatical, epigraphical, ethnographical, topographical, prosopographical, and mythical as well as the usual political and military. The study of ancient war is a field that is growing in popularity and continues to surprise with many innovative new ideas.

The first two chapters in the book examine women in war and myth from two different perspectives. In the first, "Dying for 'doing': Mythical Athenian heroines in service of the city," Ioannis Mitsios examines the theme of sacrifice for the city or state. He argues that the cases of women sacrificing themselves in war "situated the Athenian virgin women as the ideal examples and paradigms of bravery. The males were expected, given their elevated and superior status, compared to the females, to act with the same (if not with more) courage and bravery during the battle." The power of women to influence their men in war is a fascinating topic in modern studies of warfare both ancient and modern and the second chapter examines how more modern women utilized the actions of ancient women as a model.

Phyllis Brighouse argues in her paper, "The Appropriation of Women in Classical Literature during the Great War (1914-18) by the *Girls' Own Paper*," that women in Britain saw their roles as encouraging the men in war and to do so needed to strengthen their own resolve. These women latched on to the moral behavior exemplified in Greek myths, such as the self-sacrifice described by Mitsios previously, with a specific focus on the Trojan War. She concludes, "A conflation of ancient and modern, it may be seen, was used by the publishers of the *Girls Own Paper* to strengthen the resolve of its female readership at a time of suffering and loss during the Great War." Both these papers confirm the enduring and wide-reaching impact of Greek culture and stories of war.

The next two papers in the volume examine more traditional aspects of warfare studies, namely armament and tactics. Florian Feil in his paper, "'Not a few hoplites': The evolution of Illyrian infantry, 5th to 3rd century BC," examines the evidence for the type of infantry used in Illyria in the Fourth and Third centuries BCE. He argues that contrary to current scholarly opinion though more widely associated with light infantry, the Illyrians actually began relying on a core group of hoplites just like other Greek cities of the era. This changes completely our understanding of warfare in the region and our interpretation of all the battle descriptions that have survived.

Graham Wrightson harks back to more traditional avenues of military history to examine battlefield tactics. He argues in his paper, "A new interpretation of the Battle of Panion," that contrary to popular belief it is possible to reconstruct the events of the Battle of Panion through a comparative analysis of the theory of combined arms and a detailed study of unit deployment within the battlefield topography. His analysis provides an updated detailed examination of a battle that has been largely ignored since Belazel Bar Kochva's influential analysis in 1978. Both these papers demonstrate that innovative recent research can update even more traditional avenues of military history.

The next two papers, though spanning Greek and Roman topics, examine the influence of image, propaganda, and religion in the actual conduct of wars. In his paper, "The Unusual Encounter Between Demetrius III and Alexander Jannaeus: Religious Beliefs as a Motivating Factor in Ancient Warfare," Kenneth Atkinson argues that religious beliefs motivated Jews to wage war and played the key role in the wars of Demetrius III and Jannaeus as well as the associated propaganda. This paper reminds us of the importance of personal motivations for fighting in wars rather than simply the whims of individual statesmen or states.

Borja Vertedor Ballesteros argues in his paper, "Deterrence and Self Representation: the Roman Headquarters at Syracuse," that when in a camp on campaign the Roman army headquarters acted as a conduit for

Roman power abroad and sought simultaneously both to attract allies or peace concessions and to deter enemy action. He concludes that the Roman military "camp was not only a perfect machine for intimidation and violence, but it also acted, systematically, as a point of reference and representation of Rome's political and diplomatic power." The camp itself served as pro-Roman propaganda in presenting the image that furthered Rome's goals in foreign territory and motivated people's actions, similar to the motivation of religion argued by Atkinson above. Both papers show how crucial it is to view military matters through other lenses.

The next two papers both present new interpretations of often discussed aspects about how the Romans waged war. In the first paper, "*Illa vero pars calliditatis egregia et ab omni reprehensione procul* remota... Hannibal, the Carthaginians and the stratagems in Valerius Maximus," Gabriele Brusa examines the long running debate concerning the Roman use of trickery in war specifically in the writings of Valerius Maximus who drew a distinction between treachery and skillful strategies employed by the Romans and the Carthaginians. He argues that Carthaginian treachery is wrong and un-Roman but the use of skill in battle is praiseworthy, yet the Romans unlike the Carthaginians did not need only to rely solely on stratagems because they could win through Roman *virtus* – courage and manliness. The Roman way of war still continues to fascinate scholars and general readers alike.

François Gauthier in his paper, "The Suspension of the War-Tax (*Tributum*) in 167 BCE: A Short-Sighted Decision?," examines how Rome paid for its wars. He argues that "the war-tax paid by Roman citizens eligible for military service" once the mainstay of the Roman military war chest was not canceled in favor of annual income from defeated enemies as the culmination of a determined financial plan. He concludes that "the suspension of *tributum* was not based on a rational long-term plan in which the systematic implementation of provincial taxation was supposed to replace it. Rather, the senate gradually took steps to quickly raise revenues through mining activities instead of creating new provincial fiscal regimes."

The next four papers examine the impact of wars on wider Roman society, even to the soldiers themselves, and how these wars affected other cultures. In the first paper, "From battle to text: a review of the word Actium in Latin literature as a resource for identity and geographical monumentality," Nelson Henrique da Silva Ferreira examines the role that the Battle of Actium assumed in Roman literature as a representation of culture and collective consciousness. He concludes that "in literature, the term seems to be adaptable and, regarding collective memory, although education or propaganda could influence its meaning, it may have been quite open to different approaches." In this paper we see just how impactful even something as simple as the name of a battle could be.

Continuing in this lens of analysis, Jared Kreiner's paper, "The Impact of the Tacfarinas Revolt beyond Africa," assesses just how far reaching was a revolt against Roman rule in northern Africa. He analyzes the grain supply crisis affecting Rome in the year 19 CE, which led to the questioning of Imperial control, and argues that the troubles began with the little-known revolt in Africa. His examination provides a useful assessment of how well the Emperor Tiberius dealt with crises. As a contrast to the prior paper, we see how military setbacks could impact Roman society just as much as victories, especially in the immediate term.

Byron Waldron's paper, "Neglected Iranian and Syriac Evidence on the Persian Campaigns of Gordian III and Galerius," examines an important war in the later Roman Empire through oft-ignored non-Roman sources. His analysis demonstrates the wide-ranging impact Roman wars had on everything around them especially on neighboring cultures and societies. Moreover, Waldron makes the excellent argument for the importance of "seeking to fully employ this array of traditions where appropriate, to challenge existing Romano-centric narratives that have solidified over time." Thus, we are better able to find the truth. "By integrating two non-Roman historical traditions into current reconstructions of Roman campaigns, the case studies in this article argue that we should be reassessing the wars of Gordian and Galerius. Moreover, they suggest that important pieces of information in non-Roman sources might still be awaiting scholarly attention." His excellent paper opens the door for future scholarship to mine more fully other non-Roman sources of information.

The fourth paper in this section, Brendon McCarthy argues that injuries received in war had a greater impact on the everyday experience of Romans than previously understood. Using the evidence of modern studies of Traumatic Brain Injuries and Chronic Traumatic Encephalopathy this paper examines how severe injuries could affect the lives both the veteran soldiers and those around them. He concludes that viewing ancient soldiers, "with a strong understanding of TBI based on modern research allows us to better understand the risks of ancient war." All four of these papers demonstrate the wide-ranging impact of war, and its study, outside of the usual confines of the battlefield and the campaign.

The final two papers examine the ideology of war either in representations on coins or in religious doctrine and serve as a great endpoint for the volume by both examining threads begun in earlier papers. Despina Iosif, in her paper, "Warfare and *Canones Sacrorum Contemptores,"* rounds out the chronological papers with a detailed assessment of the conduct of soldiers under the later Christian emperors and how far religion impacted war. As a counterpoint to Kenneth Atkinson's earlier paper on the importance of Jewish beliefs as an encouragement to wage certain wars, Iosif argues that "Christian involvement in military affairs started creating problems for the Christian conscience from the beginning of the fourth century." Regardless of the era, location or states involved, this paper confirms that religion has always played a crucial role in war.

Julia Wetzel, in her paper, "With the Courage of an Emperor: Virtus and the Empress in Roman Propaganda," connects back to the first papers in the volume concerned with the depictions of women in war. She examines the idea of manly Roman virtue – *virtus* – as a feminine trait when used by empresses in coinage or as Emperors deploying female imagery on their coins. This paper also connects with Gabriele Brusa's examination of the term *virtus* and reveals the importance of considering all aspects of Roman propaganda. She concludes that the Roman empress's "image became an extension of the emperor's image while becoming a tool for him to use, and an example for the masses to emulate." This paper, just as all the others in this volume, demonstrates the enduring appeal of military history in many forms.

There are many intricate areas to examine in the field of ancient warfare and this volume presents just a few. From propaganda to logistics and tactics to religion, this small collection of papers provides insight into the many different themes concerned with ancient warfare and opens numerous doors for other scholars to open to even more analyses. Though an old topic, ancient history can and does still provide us with many new ideas and a myriad of ways to think about the past.

“DYING FOR ‘DOING’”: MYTHICAL ATHENIAN HEROINES IN SERVICE OF THE CITY

IOANNIS MITSIOS

We have no example of historic human sacrifice in ancient Greece. The only exception to this rule may be the sacrifice of three captive Persian princes (nephews of Xerxes) by the Athenian general Themistocles, as attested by Plutarch (*Them.* 13.2-5). On the contrary, we have a number of cases of human sacrifice in Greek mythology. In my paper, by employing an interdisciplinary approach, taking into consideration the literary, epigraphic, iconographic and topographic evidence, along with the historical and ideological context of the classical period, I will examine the phenomenon and context of the voluntary sacrifice of the virgin for the salvation of the city. In another paper I have examined Athenian heroines that sacrificed themselves for the salvation of the city during times of mythical plagues or famines; here I look at self-sacrifice in the context of war.[1]

We have from Greek mythology examples of women (mostly virgin) with unparalleled physical strength that compete against (and sometimes defeat) their male counterparts. Pindar (*Pyth.* 9.26) attests that the heroine Kyrene caught the attention of Apollo because of her fight with lions, while Hesiod (*Ehoiai* fragm. 76 MW), speaks about the heroine Atalanta, who participated in the hunt of the Kalydonian boar and defeated the hero Peleus in wrestling.

Alongside the mythological examples, there are some historical ones (most of them provided by Thucydides) of women who fight at the side of men during times of war and show remarkable strength and bravery, contributing to the defeat of the enemy. Thucydides (3.74.1) in his account on the stasis

---

[1] Ioannis Mitsios, “Ancient pandemics in mythical Athens: the Leokorai and the Hyakinthids,” *Interface* 17 (2022): 85- 108.

of Corcyra speaks of the brave act of women, who joined their men, "fighting from the rooftops" and "standing their ground beyond their nature". In another passage, Thucydides (2.4.2) attests a similar incident, where the women of the Plataeans (along with slaves) joined the fight "pelting their enemies with stones and tiles," contributing to the defeat of the Thebans. Another well-known (and very characteristic) historical example concerns Telesilla, the Argive poetess. Pausanias (2.20.9), in his passage describing the conquest of Argos by the Spartan King Cleomenes, attests in detail her brave acts.

In all these cases, the brave and heroic acts of women, both in the mythological and the historical examples, are active. In the mythological examples, the women show unparalleled strength, something to be admired, while in the historical ones, they compete and fight as active participants in the episodes and contribute to the defeat of the enemy. But what happens if some women show their brave, heroic and patriotic acts through a passive behavior by committing suicide and voluntary sacrifice for the salvation of the city?

The phenomenon of the death of one for the salvation of the many *("unum pro multis dabitur caput")* is known as "*dying for doing*" and the passive behavior of the participant (more specifically the self-killing) is a heroic act, having the same results as fighting in battle, leading to victory and the defeat of the enemy.[2]

My examination will focus on the heroines of Athens: Aglauros, the daughter of Kekrops and the Erechtheids/Hyakinthids, daughters of Erechtheus, who, along with heroines from Boeotia, provide the only known examples of mythical virgins who voluntarily sacrifice themselves for the salvation of the city during times of war.[3] The context of their self-sacrifice has been examined by a variety of scholars.[4] My examination will mostly

---

[2] Emily Kearns, "Saving the City," in *The Greek City: From Homer to Alexander*, eds. Oswyn Murray and Simon Price (Oxford: Oxford University Press, 1990), 329.
[3] Examples from Boeotia include Metioche and Menippe (daughters of Orion), also known as the Koronides, as well as Andokleia and Alkis (daughters of Antipoinos), king of Thebes. Another example may be the daughters of Skedasos, at Leuktra, although they remain nameless. For the issue, see Emily Kearns, *The Heroes of Attica* (London: Institute of Classical Studies, 1989), 59-60.
[4] Kearns, *The Heroes of Attica*, 57-63; Kearns, "Saving the City", 327-32, 338-42; John Wilkins, "The State and the Individual: Euripides' Plays of Voluntary Self-Sacrifice," in *Euripides, Women and Sexuality,* ed. Anton Powell (London:

emphasize: a) the impact of the brave act, b) its relation to the role of men, who die on the battlefield, c) the Funerary Speeches, spoken at Kerameikos, as well as d) the gender and identity parameters.

Human sacrifice is a very common pattern in ancient Greek myth and in some cases the sacrifice is avoided and the victim escapes death. The best-known female sacrificial victim in ancient Greek mythology, and one of the most famous heroines, receiving panhellenic cult in Brauron,[5] Megara,[6] and Achaia,[7] remains Iphigeneia, daughter of King Agamemnon and Queen Clytemnestra. Although Iphigeneia is the most known example of a sacrificial victim, there are several differences between her sacrifice and the sacrifice of the Athenian heroines examined here: a) the decision of her sacrifice is not taken by herself, but by her father, b) she is sacrificed, instead of committing suicide and c) most importantly, she escapes death in the end, being transformed into a deer, by Artemis.[8]

As I have already stated, paradigms of maidens who voluntarily sacrifice themselves for the salvation of the city during times of war exist only in Athens and Boeotia, with Athenian examples being: a) Aglauros, the daughter of King Kekrops and b) the Erechtheids or Hyakinthids, daughters of King Erechtheus and Queen Praxithea.

I begin my examination with Aglauros, daughter of King Kekrops.

Routledge, 1990), 190-91; Christopher Collard, M. J. Cropp and K. H. Lee, *Euripides: Selected fragmentary plays* (Warminster: Aris & Phillips, 1993), 192; Jennifer Larson, *Greek Heroine Cults* (Madison: University of Wisconsin Press, 1995), 101-03; Utah Kron, "Patriotic Heroes" in *Ancient Greek Hero Cult*, ed. Robin. Hägg (Stockholm: Svenska Institutet i Athen, 1999), 78-79; Christiane Sourvinou-Inwood, *Athenian Myths and Festivals: Aglauros, Erechtheus, Plynteria, Panathenaia, Dionysia* (Oxford: Oxford University Press, 2011), 28-9, 78-80, 105-08; Ioannis Mitsios, "The Iconography and Cult of Heroes on the Acropolis of Athens from the Late Archaic to the Classical period: Contribution to the Study of Athenian Autochthony" (Ph.D. diss., University of Athens, 2018), 20-27, 28-35.

[5] For her cult in Brauron, see Euripides (*IT*, 1462-7).

[6] For her cult in Megara, see Pausanias (1.43.1).

[7] For her cult in Achaia, see Pausanias (7.26.5).

[8] Sourvinou-Inwood, *Athenian Myths and Festivals: Aglauros, Erechtheus, Plynteria, Panathenaia, Dionysia*, 31 n. 25, 32. For a comparison between Iphigeneia and the Athenian heroines, especially the Hyakinthids, the daughters of king Erechtheus, see Kearns, *The Heroes of Attica*, 57, 60; Larson, *Greek Heroine Cults*, 104-06.

Aglauros (Daughter of King Kekrops)

The Athenian heroine Aglauros is mostly known from her involvement in the birth of Erichthonios. According to the mythical tradition and the account of Apollodorus (3.14.6), Athena placed her son Erichthonios in a chest, which she entrusted to the daughters of Kekrops, Aglauros, Herse, and Pandrosos (also known as the Kekropids or Aglaurids), though forbidding them to open it. The daughters of Kekrops are often depicted in the episode of the birth of Erichthonios, in the infamous "anodos scenes", where Ge hands the baby Erichthonios to Athena.[9] On a kylix of the Codrus Painter, dating from c.430 BC, now in Berlin, the daughters of Kekrops are securely identified, thanks to the surviving inscriptions (**Fig. 1a+b**).[10]

Fig. 1a. Berlin, Antikenmuseen F 2537. PHOTO: The birth of Erichthonios.

[9] For the depiction of the Kekropids in the birth of Erichthonios, see *LIMC* I (1981) s.v. Aglauros, Herse, Pandrosos no. 6-13, and Mitsios, *The Iconography and Cult of Heroes on the Acropolis of Athens from the Late Archaic to the Classical Period: Contribution to the Study of Athenian Autochthony*, 92 n. 268 (with emphasis on depictions from vases deriving from the Acropolis itself).

[10] Berlin, Antikenmuseen F 2537; BAPD 217211.

Fig. 1b. Berlin, Antikenmuseen F 2537. PHOTO: The Aglaurids (named) in the birth of Erichthonios.

Pandrosos was the only daughter who obeyed the rules of Athena, while Aglauros and Herse disobeyed Athena's orders, resulting in their death by jumping from the hill of the Acropolis. Given the testimonies of Apollodorus, Aglauros is a disobedient heroine and her feminine curiosity causes her death. She serves as a paradigm to avoid and her disobedient behavior is depicted in some vases of the classical period.[11] One example includes an alabastron from the National Archaeological Museum at Athens, attributed to the Karlsruhe Painter, dating from 470/60 BC. (**Fig. 2**).[12]

[11] For the depiction of the disobedient and disloyal behavior of Aglauros and Herse, see Mitsios, *The Iconography and Cult of Heroes on the Acropolis of Athens from the Late Archaic to the Classical Period: Contribution to the Study of Athenian Autochthony*, 81 n.224.

[12] Athens, National Archaeological Museum 17917; BAPD 209099.

Fig. 2. Athens, National Archaeological Museum 17917. Alabastron with the disobedient behavior of Aglauros.

Besides her involvement in the birth of Erichthonios, Aglauros is connected with other mythological motifs, more specifically the war between Athens and Eleusis. In this episode we come across a completely different aspect of Aglauros and the heroine is presented in a very different positive context. According to the testimony of Philochoros (*FGrH 328 F105*), when the Eleusinian King Eumolpos attacked Athens, during the kingship of Erechtheus, the Delphic oracle reported that the city of Athens would be saved only if someone sacrificed themselves. Then, the heroine Aglauros threw herself off the walls of the Acropolis in order to save the city. The act of leaping from a cliff is common in Greek mythology, resulting in heroization and deification, and other examples of such heroines exist, as in the cases of Ino and Molpadia (Hemithea).[13] In terms of Aglauros, her jumping from the cliffs of the Acropolis resulted in the foundation of a sanctuary, where the Athenian ephebes came to take their ephebic oath. The location of the shrine has been securely identified on the east slope of the Acropolis, thanks to the discovery of an inscription found in situ by Dontas, and is associated with a cave (**Fig. 3**).[14] Several literary sources (Demosthenes 19.303; Philoch. *FGrH 328 F 105*; Plut. *Alc.* 15.7-8) attest that the Athenian ephebes took their oath at the sanctuary of Aglauros, invoking her name along with other war deities, like Ares, as a witness for their oath.[15] There is no hesitation that by taking their oath to defend the city of Athens (with their freshly acquired arms) within Aglauros' sanctuary, the eighteen-years-old Athenian ephebes gave Aglauros a warlike character. This specific context is further highlighted by her connection to Ares, the god of war par excellence. According to the testimony of Hellanikos (*FGrH 323 A F1*) and Apollodorus (3.14.2), Aglauros was the wife of Ares and mother of Alcippe, while Porphyry (*Abst.* 2.54) attests that Aglauros was worshiped in Cyprus along with Diomedes, Ares' opponent in the Trojan War. Given the brave act of her self-sacrifice and heroic status, Aglauros was the chief patroness of the ephebes and formed the ideal heroine, a role model for the Athenian ephebes. In fact, the heroic self-sacrifice of Aglauros and her jump from the cliffs of the Acropolis for the salvation of the city of Athens acted as an example during historical times. Herodotus (8.53) attests that some Athenians also jumped from the cliffs of the Acropolis when the Persians invaded the city and were ascending the Acropolis.

---

[13] Gregory Nagy, *The Best of the Achaeans: Concepts of the Hero in Archaic Greek Poetry* (Baltimore: Johns Hopkins University Press, 1979), 203; Gregory Nagy, *Greek Mythology and Poetics (*Ithaca: Cornell University Press, 1990), 223-26; Larson, *Greek Heroine Cults*, 102.

[14] George Dontas, "The True Aglaurion," *Hesperia* 52 (1983): 48-63.

[15] For the ephebic oath, see Reinhold Merkelbach, "Aglauros (Die Religion der Epheben)," *ZPE* 9 (1972): 277-83; Peter Siewert, "The Ephebic Oath in Fifth-Century Athens," *JHS* 97 (1977): 102-11.

Fig. 3 The shrine/cave of Aglauros on the east slope of the Acropolis.

Having examined Aglauros, let us now turn to the examination of the voluntary self-sacrifice of the daughters of King Erechtheus and Queen Praxithea, known as the Erechtheids or the Hyakinthids.

Erechtheids/Hyakinthids (Daughters of King Erechtheus)

The mythological episode of the daughters of Erechtheus, the Erechtheids (or the Hyakinthids), takes place during the war between King Erechtheus of Athens and King Eumolpos of Thrace (instead of Eleusis, as in the case of Aglauros). The context of the mytheme is almost identical to the voluntary self-sacrifice of Aglauros, only with slight differences. King Erechtheus seeks advice from the Delphic oracle, getting the answer that the city of Athens will be saved only if he sacrificed one of his own daughters. Lykourgos (*Leoc.* 98-100), citing Euripides' fragmentary preserved tragedy "*Erechtheus*", states Queen Praxithea's answer, when her husband, Erechtheus, informs her about the answer from the oracle:

> If our household had a harvest of sons in it, rather than daughters, and the flame of war were engulfing the city, would I not send my sons into battle with spears, fearing for their deaths? No, let me have daughters who can

> fight and stand out among men and not be mere figures raised in the city for no purpose. When mothers' tears send their children off to battle, many men become soft. I hate women who choose life for their children rather than the common good, or urge cowardice. Sons, if they die in battle, share a common tomb with many others and equal glory. But my daughter will be awarded one crown all to herself when she dies on behalf of this city, and in doing so she will also save her mother and you (Erechtheus), and her two sisters... are these things not reward in themselves? And do so, I shall give this girl, who is not mine except through birth, to be sacrificed in defense of our land. If this city is destroyed, what share in my children's lives will I then have? Is it not better that the whole be saved by one of us doing our part? And as for the matter which most concerns the people at large, there is no-one, while I live and breathe, who will cast out the ancient holy laws of our fore-fathers. Nor will Eumolpos and his Thracian army ever in place of the olive tree and golden Gorgon head plant the trident on this city's foundations and crown it with a garland, thus dishonoring the worship of Pallas (Athens). Citizens make use of the offspring of my labor pains, save yourselves, be victorious! Not for one life will I refuse to save our city. O fatherland, I wish that all who dwell in you would love you as much as I do! Then we would live in you untroubled and you would never suffer any harm.

The exact number of the daughters of Erechtheus varies, with sources stating either three,[16] four,[17] or six[18] daughters, named Protogeneia, Pandora, Chthonia, Prokris, Kreousa and Oreithyia.[19] In terms of their tragic fate, Euripides (*Ion,* 277-80) states that only one of the daughters, Kreousa, escapes death because of her young age. The brave act of the daughters of Erechtheus is much appreciated in ancient Greek myth and cult. In fact, it is rewarded by the goddess Athena herself.

In the tragedy of Euripides' "*Erechtheus" (*fr. *71-75)*, the goddess Athena declares that "the girls' souls have not gone to Hades and that she (Athena) has translated their spirits (*pneuma*) to the aether" and that "the daughters of Erechtheus will be famous in Greece and will be called goddesses, the

---

[16] Euripides, "*Erechtheus*", talks of three daughters of Erechtheus.

[17] Apollodorus (3.15.1) and Hyginus (*Fab.* 46) attest the following four daughters of Erechtheus: Chthonia, Prokris, Kreousa and Oreithyia.

[18] Photios, s.v. Parthenoi names six daughters of Erechtheus: Protogeneia, Pandora, Chthonia, Prokris, Kreousa and Oreithyia.

[19] For a full account and presentation of the literary sources, see Kearns, *The Heroes of Attica*, 202; Timothy Gantz, *Early Greek Myth: a Guide to Literary and Artistic Sources* (Baltimore: Johns Hopkins University Press, 1993), 242; Douglas Frame, *Hippota Nestor* (Washington, DC/Cambridge: Center For Hellenic Studies, 2009), 448 n. 199.

Hyakinthids." *Catasterism*, or transporting a figure into a star or other heavenly body, is a reward for recognition of superior status and actually is considered the greatest honor of all, given that the shining star becomes one with the cosmos, situating the brave act on a cosmic level.[20]

Additionally, in terms of the cult of the Erechtheids, Athena gives the following instructions: "the girls are to be buried on the spot where they died and no wine will be used in the cult, but water and honey; furthermore, the daughters of Erechtheus will receive annual sacrifice, they will have choruses of girls and their temenos will be abaton" (*Erechtheus*, fr. 75-80). An inscription (*IG II² 1035*), dated to the 1st century BC, informs us about the existence of a Hyakintheion (a cult of the Hyakinthids) but its location is uncertain, since the inscription does not provide further information on its exact spot.[21] The cult and tomb of the Hyakinthids has even been connected with the west side of the Parthenon.[22] Given the importance of the brave act of the Erechtheids and its impact in myth and cult, several iconographic identifications and attributions have been proposed by scholars. The daughters of Erechtheus have been identified with the child figure decorating the central scene of the east frieze of the Parthenon (**Fig. 4**).[23] Additionally, the Erechtheids/Hyakinthids have been identified with the dancing Hyades at the "Akanthos column", in Delphi (**Fig. 5**).[24] Far from being secure on both cases of the identifications, the proposals indicate the

---

[20] Larson, *Greek Heroine Cults*, 103; J.B. Connelly, *The Parthenon Enigma* (New York: Alfred A. Knopf, 2014), 147.

[21] Photius, s.v. Parthenoi, writing on the tragic fate of the Hyakinthids, states that: *«ἐσφαγιάσθησαν δὲ ἐν τωῖ Ὑακίνθωι καλουμένωι πάγωι ὑπὲρ τῶν Σφενδονίων»*. Stephanos Byzantine states that: *«τῶν Ὑακίνθου θυγατέρων ἡ Λουσία ἦν, ἀφ'ἧς ὁ δῆμος τῆς Οἰνηίδος φυλῆς»*, connecting the Hyakintheion with the deme of Lousia. The existence of Hyakintheion in the deme of Lousia has received mixed feedback by scholars: Jacoby, schol. to Phanodemus (*FGrH 325 F4*), page 178, accepts the existence of Hyakintheion at the deme of Lysia. Kearns, *The Heroes of Attica*, 102 does not accept the identification, while Frame, *Hippota Nestor*, 449 remains neutral, stating that the Hyakintheion may have existed on the west part of the city of Athens. For the issue, see also Larson, *Greek Heroine Cults*, 36, 190 n. 8; Kron "Patriotic Heroes", 79, who remain neutral.

[22] Connelly, *The Parthenon Enigma*, 232-33.

[23] J. B. Connelly, J. B. 1996. "Parthenon and Parthenoi: a Mythological Interpretation of the Parthenon Frieze," *AJA* 100 (1996): 58-66; Connelly, *The Parthenon Enigma*, 165-66.

[24] Delphi 1584. Gloria Ferrari, *Alcman and the Cosmos of Sparta* (Chicago: The University of Chicago Press, 2008), 146-47.

importance of the daughters of Erechtheus in myth, cult and the iconography of the classical period.[25]

Fig. 4: The central scene of the east frieze of the Parthenon.

[25] In terms of the Parthenon frieze, the central scene of the east frieze is recognized by the majority of scholars as related to the Panathenaic festival, instead of the sacrifice of the daughters of Erechtheus. In terms of the "Akanthos column" in Delphi, besides the lack of any iconographic evidence of the daughters of Erechtheus, the identification of the Erechtheids/Hyakinthids with the dancing Hyades is problematic for a number of reasons. On this issue (and identification with Hyades), see Collard, Cropp and Lee, *Euripides: Selected Fragmentary Plays*, 194; Kearns, *The Heroes of Attica*, 61-62; Connelly, *The Parthenon Enigma*, 244. Contra Gantz, *Early Greek Myth: a Guide to Literary and Artistic Sources,* 128; Sourvinou-Inwood, *Athenian Myths and Festivals: Aglauros, Erechtheus, Plynteria, Panathenaia, Dionysia*, 123-34 who do not accept the identification of the Erechtheids/Hyakinthids with the Hyades.

Fig. 5: Delphi 1584. The Akanthus Column in Delphi.

Gender Aspects

Having examined the Athenian heroines who voluntarily sacrificed themselves for the salvation of the city during times of war, as well as the impact of this act in myth, cult and iconography, let us now turn on their relation to the Funerary Speeches (spoken for the war dead at the area of Kerameikos) and its gender aspects.

Several different ancient authors mention the brave act of self-sacrifice of Aglauros and the Erechtheids/Hyakinthids for the salvation of the city. Demosthenes (*Funeral Speech,* 27-29), in his speech for the Athenian war dead in the Battle of Chaeronea 338 BC against the Macedonians, uses the brave act and heroic self-sacrifice of the Erechtheids as an example for the communal good and the salvation of the city. Similarly, the orator Demades (*Hyper tes dodeketoias,* 37), talking on the self-sacrifice of the daughters of Erechtheus, attests that: "they triumphed over their feminine in their souls and that the weakness of their nature was made virile by devotion to the soil that reared them." In the epitaph for the war dead of the battle of Potidaea in 432 B.C. (*CEG, 10*), it is stated that: "the aether took the soul of the war dead, while the Earth took their bodies." It is very likely that this specific passage for the heroic death of the Athenian war dead at Potidaea was paralleled and recalled by the Athenians of the audience with the passage of Euripides' "*Erechtheus*" on the daughters of Erechtheus we examined above, where the goddess Athena translated the spirits of the Erechtheids to the aether.

Lykourgos (*Leoc.* 100), admitting the importance of the brave act of the daughters of Erechtheus, attests that the Athenians owe a debt to Euripides for passing the story of the self-sacrifice of the Erechtheids/Hyakinthids down to them, providing an example (*paradeigma*) and he further claims that the oath of Erechtheus' daughters was invoked alongside the oath of the ephebes and that of the Greeks at Plataea. The praise of the brave and heroic act of self-sacrifice of the Erechtheids/Hyakinthids continued undisrupted to the Roman period, where their acts were used as examples of bravery and valor. Characteristic cases are the testimonies of Cicero. In one passage (*Nat. D.* 3.19.5), Cicero attests that "most states have deified the brave for the purpose of promoting valor, giving as examples the daughters of Erechtheus," while in another passage (*Sest.* 21.48) he attests that "he will not be afraid of death, which even Athenian maidens, the daughters of King Erechtheus are said to have despised for the sake of their country." Taking into account the fact that the testimonies about the brave act of the Erechtheids/Hyakinthids continued undisrupted from the classical Greek period to the Roman times, we note the diachronic impact of their act and their importance as protectresses and city saviors during times of war. No matter the time period, classical Greek or Roman, the names of the daughters of Erechtheus (and their patriotic and brave self-sacrifice) were constantly recalled when the cities of Athens or Rome were threatened by war.

In terms of Aglauros (the daughter of Kekrops), we have already mentioned that several different sources: such as Demosthenes (19.303), Philochorus (*FGrH 328 F 105)* and Plutarch (*Alc.* 15.7-8), attest that the Athenian ephebes took their oath at the sanctuary of Aglauros, invoking her name as a witness to their oath. It has been suggested that the connection of Aglauros, spelled Agraulos by some sources: such as Philochorus (*FGrH*

328 F 105), Polyaenus (1.21.2), and Apollodorus (3.14.2), with the ephebes and the guarding of the city is further indicated by the etymology of her name. Aglauros' name has been both interpreted as "bright water," while Agraulos' as the one "living in the field" and Boedeker parallels the etymology of her name with a passage of Aeschylus (*Ag.* 559-62), where it is stated that "those who were guarding the city sleep in the open and are in danger of the dew/humidity of the day," bringing in mind and resembling the name of Agraulos.[26] On top of that, the entrance of the Athenian ephebes inside the cave of Aglauros, in order to take their ephebic oath, has been interpreted as an autochthonous act and has been paralleled with the birth of the Athenian ephebes on a secondary "autochthonous" level.[27]

Having examined the self-sacrifice of Aglauros and the Erechtheids/ Hyakinthids, we note that the orators repeatedly pointed out the examples of sacrificial heroines for the salvation of the city. Additionally, they were often paralleling their brave acts with the current soldiers, indicating that if women were capable of showing such bravery, at least the same courage is expected from men. In fact, in some cases, they clearly attest the gender aspects of the self-sacrifice of the virgins for the salvation of the city, emphasizing the impact of their brave act in relation to the males.

The role of women in antiquity is limited and they are mostly engaged with house activities, outside the public activities of the polis. Mythological examples and historical testimonies confirm this. Iphigeneia (Eur. *IA,* 1394) attests that "one man's life is worth more than that of thousands of women" and similarly, Aristotle (*Pol.* 1269b18; 1275b23; 1275a22-3) treats women and slaves, as in some ways, parallels. Women are excluded from citizenship and Loraux has highlighted that "there were no true feminine *Athenaioi* but merely *Attikai* gynaikes".[28] Given their exclusion from citizenship and limited role, regardless of being biologically essential for the existence of the city, women were expected to contribute less to the society.

In total contrast to this belief stand the female heroines who voluntarily sacrifice themselves for the salvation of the city. Their act of saving the city is the biggest paradox of all, given the very limited role of women in

---

[26] Deborah Boedeker, *Descend from Heaven: Images of Dew in Greek Poetry and Religion* (Chico, CA: Scholars Press, 1984), 104, 105-08, 108-09.
[27] Boedeker, *Descend from Heaven: Images of Dew in Greek Poetry and Religion*, 112.
[28] Nicole Loraux, *Les enfants d'Athéna: Idées athéniennes sur la citoyenneté et la division des sexes* (Paris: F. Maspero, 1981), 124-25.

antiquity. The heroic and brave act of the self-sacrifice of Aglauros and the Erechtheids/Hyakinthids, serves as a paradigm both for the males and the females, making them the ideal heroines and role-models for all the Athenian citizens. It has been shown that the masculinization of the heroines assimilated their actions to those of warriors and made their myth a more palatable example for men.[29] Kron highlights that the female heroic behavior is much more amazing compared to the male one, since women's nature is much weaker than men's.[30]

There is no doubt that the self-sacrifice of Aglauros and the Erechtheids/Hyakinthids made them the ideal patriotic heroines, situating their brave act in the very centre of the polis. We have already mentioned that their names were spoken at the Funerary Orations at Kerameikos, honoring the Athenian war dead. Besides Kerameikos, the brave self-sacrifice of the heroines, more specifically the one of the Erechtheids/Hyakinthids, was presented on the Acropolis, at the theater of Dionysus, where Euripides' play "*Erechtheus*" was performed and viewed by almost five or six thousand Athenians during the annual city Dionysia festival.[31] On top of that, we discussed that Aglauros was worshiped on the east slope of the Acropolis, where her shrine was placed and a Hyakintheion (shrine for the Hyakinthids) existed, although its exact location remains uncertain.

The presentation of the brave acts of the sacrificial virgins both at the Kerameikos (at the Funerary Orations) and the Acropolis (at the theater of Dionysos), as well as the cult/shrine of Aglauros on the Acropolis and the existence of the Hyakintheion, placed the Athenian heroines in both locations where the Athenian identity was shaped: the Acropolis and the Kerameikos.

The acts mentioned above situated the Athenian virgin women as the ideal examples and paradigms of bravery. The males, given their elevated and superior status compared to the females, were expected to act with the same (if not with more) courage and bravery during the battle. Given that, it seems that the self-sacrifice of the Athenian heroines for the salvation of the city, besides being a heroic act, also had gender aspects and offered a role model for women, since on a symbolic level women and men were equal. Viewed from an Anthropological and Gender aspect, females were situated in the

[29] Larson, *Greek Heroine Cults*, 104.
[30] Kron, "Patriotic Heroes", 83.
[31] Connelly, *The Parthenon Enigma*, 146.

very public sphere, central to the polis and the commonwealth, ideal examples of brave and patriotic citizens.

Abbreviations

BAPD: Beazley Archive Pottery Database.

FGrH = Jacoby, F. 1923. *Die Fragmente der griechischenHistoriker*. (FGrHist) Weidmann, Berlin.

Bibliography

Boedeker, Deborah. *Descend from Heaven: Images of Dew in Greek Poetry and Religion.* Chico, CA: Scholars Press, 1984.

Collard, Christopher, M. J. Cropp and K. H. Lee. *Euripides: Selected fragmentary plays*. Warminster: Aris & Phillips, 1993.

Connelly, J. B. 1996. "Parthenon and Parthenoi: a Mythological Interpretation of the Parthenon Frieze," *AJA* 100 (1996): 58-66.

Connelly, J. B. *The Parthenon Enigma*. New York: Alfred A. Knopf, 2014.

Dontas, George. "The True Aglaurion," *Hesperia* 52 (1983): 48-63.

Ferrari, Gloria. *Alcman and the Cosmos of Sparta*. Chicago: The University of Chicago Press, 2008.

Frame, Douglas. *Hippota Nestor*. Washington, DC/Cambridge: Center For Hellenic Studies, 2009.

Gantz, Timothy. *Early Greek Myth: a Guide to Literary and Artistic Sources*. Baltimore: Johns Hopkins University Press, 1993.

Kearns, Emily. *The Heroes of Attica*. London: Institute of Classical Studies, 1989.

Kearns, Emily. "Saving the City". In *The Greek City: From Homer to Alexander*, edited by Oswyn Murray and Simon Price, 323-44. Oxford: Oxford University Press, 1990.

Kron, Utah. "Patriotic Heroes". In *Ancient Greek Hero Cult*, edited by Robin Hägg, 61-83. Stockholm: Svenska Institutet i Athen, 1999.

Larson, Jennifer. *Greek Heroine Cults*. Madison: University of Wisconsin Press, 1995.

Loraux, Nicole. *Les enfants d'Athéna: Idées athéniennes sur la citoyenneté et la division des sexes*. Paris: F. Maspero, 1981.

Merkelbach, Reinhold. "Aglauros (Die Religion der Epheben)," *ZPE* 9 (1972): 277-83.

Mitsios, Ioannis. "The iconography and cult of heroes on the Acropolis of Athens from the late Archaic to the Classical period: contribution to the study of Athenian autochthony." Ph.D. diss., University of Athens, 2018.

Mitsios, Ioannis. "Ancient Pandemics in Mythical Athens: the Leokorai and the Hyakinthids," *Interface* 17 (2022): 85- 108.

Nagy, Gregory. *The Best of the Achaeans: Concepts of the Hero in Archaic Greek Poetry*. Baltimore: Johns Hopkins University Press, 1979.

Nagy, *Gregory. Greek Mythology and Poetics.* Ithaca: Cornell University Press, 1990.

Siewert, Peter. "The Ephebic Oath in Fifth-Century Athens," *JHS* 97 (1977): 102-11.

Sourvinou-Inwood, Christiane. *Athenian Myths and Festivals: Aglauros, Erechtheus, Plynteria, Panathenaia, Dionysia*. Oxford: Oxford University Press, 2011.

Wilkins, John. "The State and the Individual: Euripides' Plays of Voluntary Self- Sacrifice". In *Euripides, Women and Sexuality,* edited by Anton Powell, 177-94. London: Routledge, 1990.

THE APPROPRIATION OF WOMEN IN CLASSICAL LITERATURE DURING THE GREAT WAR (1914-18) BY THE *GIRLS' OWN PAPER*

PHYLLIS BRIGHOUSE

The Girl's Own Paper and Woman's Magazine, published by the Religious Tract Society between 1880 and 1956, was an up-market, popular monthly paper read by middle class, intelligent and often well-educated girls and women and its written narratives included stories and a broad range of articles.

Before examining the use of a group of articles employing Greek tragedy to provide role models which encouraged women to support Britain during the Great War, I begin with a brief snapshot of the British experience of this War. Military service in Britain was voluntary until conscription began in January 1916. Conscripted men fought in some of the War's fiercest battles and British casualties were horrendous. At the Battle of the Somme (July to November 1916) one hundred and twenty-five thousand British men were killed and two hundred and forty-five thousand injured. Between July and November 1917, in the Third Battle of Ypres at Passchendaele, a further seventy thousand British soldiers were killed and over two hundred thousand injured. By October 1917, therefore, enormous numbers of British women had suffered the loss or disablement of fathers, husbands, lovers, sons and brothers.

From October to December 1917, the paper published a series of three articles, "Women in Greek Literature," by the journalist Lily Watson.[1] These narratives co-opted the heroines depicted in ancient Greek literature

---

[1] Lily Watson, "Women in Greek Literature I," *Girls' Own Paper and Woman's Magazine*, October, 1917, 42-43; Lily Watson, "Women in Greek Literature II," *Girls' Own Paper and Woman's Magazine*, November, 1917, 67-8; Lily Watson, "Women in Greek Literature III," *Girls' Own Paper and Woman's Magazine*, December, 1917, 133-134.

in order to influence the attitudes and actions of British Women at a time of great suffering. The first article was preceded by an open letter from Handley Moule, Bishop of Durham, "To the Women of the Nation,"[2] and the last was accompanied by an Editorial.[3]

At the beginning of the twentieth century, such comparisons by Watson between the ancient and the modern worlds would not be considered unusual. Simon Goldhill, in *Victorian Culture and Classical Antiquity* (2011), comments that Classics had traditionally dominated secondary and higher education, and a classical education both formed and informed the mind of the (male) "figure of authority" and thus became "the furniture of the mind."[4] As the modern historian C.A. Hagerman comments, Classics contributed to "the official mind."[5] Goldhill also draws attention to the major revival of Classicizing art at the end of the Victorian era and the early twentieth century, including visual art such as architecture and paintings on public display.[6] British people at this time were surrounded by Classical models found in architecture, painting and public art.[7]

Watson's articles, together with the Bishop's letter and the Editorial, provide an insight into a significant period in British women's history concerning a paradigm shift on the place of women in society. An

---

[2] Handley C.G. Moule, "To the Women of the Nation," *Girls' Own Paper and Woman's Magazine*, October 1917, 41-42.

[3] Editorial, "The Fellowship of Service," *Girls' Own Paper and Woman's Magazine*, December 1917, 113-114.

[4] Simon Goldhill, *Victorian Culture and Classical Antiquity: Art, Opera, Fiction and the Proclamation of Modernity*. (Oxford: Oxford University Press, 2011), 2. Women had generally been excluded from such an education, but by 1913 this was challenged. In 1913, for example, Vera Brittain studied Greek for the test of competence in Greek (Responsions) demanded for entry to Somerville College, Oxford, and she sat her Responsions examination in March 1914. Vera Brittain, *Testament of Youth* (London: Virago, 1978), 67-68; 74.

[5] C.A. Hagerman, *Britain's Imperial Muse* (Basingstoke: Palgrave, Macmillan, 2013), 2.

[6] Goldhill especially identifies the paintings of J.W. Waterhouse. Goldhill, *Victorian Culture and Classical Antiquity*, 10, 23-64.

[7] For a range of Neoclassical examples of Classicizing art at this time see Phyllis Brighouse, "Classics, Empire and Didacticism, 1919-1939." (PhD diss., University of Liverpool 2022), 10-11. Two examples in printed matter in 1913 may be found in *Girls' Own Paper and Woman's Magazine*. These discuss Greek sculpture. See: Anonymous, "The Apollo Belvedere," *Girls' Own Paper and Woman's Magazine*, April, 1914, 368-370; and Anonymous, "The Diana of Versailles," *Girls' Own Paper and Woman's Magazine*, May 1914, 407.

examination of American author Kate Millet's *Sexual Politics* (1970), for example, reveals how European literature portrays men as having power over women.[8] Edward Said's *Orientalism* (1978) argues that western media use language to portray the cultures of Asia, the Middle East and North Africa in racist, imperialist terms.[9] Such examinations stimulated research which connected European male dominance over women through the power of language, such as Joanna de Groot's "'Sex' and 'Race' in the Construction of Language" (2000).[10] Such language had created a paradigm in Victorian Britain which confined women to domestic roles at home, the *oikos*. Beyond it, in the *polis,* men were occupied with careers and in the government and administration of the state. Just prior to the Great War, however, this paradigm was challenged, and articles in the Paper in late 1913 and early 1914 reflect the beginning of a paradigm shift.[11]

The Bishop's letter, Watson's articles and the Editorial provide insights on how, during the Great War, this paradigm shift became emphasised when large numbers of women were urged, for the sake of national survival, to make great sacrifices at home and also to leave their accepted "womanly" roles in the *oikos* and enter the male sphere of the *polis* beyond the home. In the Paper, Bishop Moule's letter appeals for women to make sacrifices and heroically work for the survival of the nation, and Watson's articles provide exemplars of women from Greek myth who sacrificed their futures and even their lives for the sake of family and state. An Editorial reinforces the Bishop's appeal. Together they indicate the acceleration of a major shift in attitudes to the place of women in society, with Watson's examples of women in Greek tragedy providing role models.

---

[8] Kate Millet, *Sexual Politics* (New York: Doubleday, 1970).

[9] Edward Said, *Orientalism (*London: Routledge & Kegan Paul, 1978).

[10] Joanna de Groot, "'Sex' and 'Race' in the Construction of Language" in *Cultures of Empire*, ed Catherine Hall (Manchester: Manchester University Press 2000), 37-60. Here, de Groot identifies draws attention to how a language of "truth" concerning gender was "verified" by Victorian academic researchers into anthropology and biology, and medical scholarship employed the same descriptions of physical attributes concerning both gender and race to "verify" the "scientific truth" of the superiority of western white men over women.

[11] For examples of articles in the paper on opportunities for women outside the home see Ida Smedley Maclean D.Sc., "Household Science in the Universities," *Girls' Own Paper and Woman's Magazine*, April 1914, 424-427; A Hospital Matron, "Wanted: Probationers," *Girls' Own Paper and Woman's Magazine*, January 1914, 244-245.

Bishop Moule's letter identifies and praises the qualities he finds in "Womanhood" in a domestic setting, but also assures women that by stepping outside their conventional roles in society in wartime they are serving both their God and their Nation. He expresses admiration for Womanhood's "claim on us men for chivalrous service," and asks that women also now serve their Nation chivalrously. God, he declares, wishes women to recognise the term "rights" as something **deserved by** others, and the word "duties" as something which women **owe to** others (my emphasis in both cases), and so "bear a fresh and living Witness" to Christ "in a new spirit."[12] "At this awful time in our nation's history," he says, a woman with domestic ties should "sacrifice yourself." Women should also support the actions of husbands and future husbands diligently, and patiently undertake domestic tasks which would perhaps be done by other women now serving their nation.[13] Any woman without such ties should "sacrifice yourself" in "teaching, nursing, headship of an institution, leadership of a good work or any other right walk of life." In this way they would "give something to God" and "something for others" so a woman could "rise in her own life and conduct towards … true Womanhood."[14] He concludes by asking for "*the daily and continuing sacrifice of lower aims in favour of the highest aims.*"[15] (Moule's italics).

This address was immediately followed by the first in Lily Watson's series "Women in Greek Literature." Here, Watson reinforces the Bishop's message by identifying the diligence and nobility involved in modern domestic work with the work and qualities of character in women depicted by Homer. First, Watson discusses Nausicaa's discovery of Odysseus after he is shipwrecked in the country of the Phaeacians. Although she is a princess, Watson points out that Nausicaa was not ashamed to join her companions in what Watson calls "laundry work," adding that Nausicaa, although a princess, "may, in her actions, serve as a model for present-day girls."[16] Homer, she comments, "is at pains to show that they really knew their business" and Watson describes this work in some detail,[17] implying that Nausicaa knew precisely how to launder clothing and implying this was

[12] Moule, "To the Women of the Nation", 40.
[13] Moule, "To the Women of the Nation", 40.
[14] Moule, "To the Women of the Nation," 41.
[15] Moule, "To the Women of the Nation," 41.
[16] Watson, "Women in Greek Literature I," 42.
[17] Watson, "Women in Greek Literature I," 42. *Cf.* Hom. *Od.*: 6.91-98.

a regular task.[18] Here, Watson is using an analogy from Greek myth to reinforce Bishop Moule's call for women to engage in domestic chores at a time of crisis without feeling such work is beneath them.

For Watson, Nausicaa's behaviour when she discovers Odysseus emphasises her compassion and courage, and again uses women in Greek myth to reinforce Bishop Moule's identification of courageous "womanliness" towards a man returning from a war. When her companions flee Nausicaa stands firm, listens to his story, scolds the girls for their cowardice, and provides Odysseus with clothes from their laundry pile.[19] She then urges him to seek further help from her mother, Queen Arete,[20] who then personally arranges for Odysseus' care and future comfort and listens to his story.[21] Watson then praises Odysseus' wife Penelope for her courageous loyalty to her husband, and her attention to domestic labor. Penelope, while patiently awaiting his return, undertakes the domestic task of weaving cloth in a practised manner before unravelling it again to keep her unwanted suitors at bay.[22] Similarly, Watson implies, modern women should work diligently and await the return of their own husbands.

Watson's emphasis throughout this first article concerns the courage of well-born women in ancient Greek literature and their willingness to undertake, in accordance with Bishop Moule's letter, what contemporary women might consider menial chores. Watson states:

> In today's world, it is a happy, modern development of thought which forbids the inclusion of any household toil under the contemptuous term 'menial' … in this we are actually coming nearer to the standard of these stately prehistoric times.[23]

Watson's aim is clearly to use the heroic qualities found in Greek heroines to inspire those same qualities in British women so they might not only endure loss and privation, but both make personal sacrifices by carrying out

---

[18] Athena chastises the girl in a dream for her dirty clothes, calling her lazy, and offering to help so that the work could be done quickly. *Cf.* Hom. *Od.:* 6.25-32. The task of washing clothes, therefore, might not be an isolated incident but equally it might be less than frequent.

[19] Watson, "Women in Greek Literature I," 42. *Cf.* Hom. *Od.:* 6.138-230.

[20] Watson, "Women in Greek Literature I," 42. *Cf.* Hom. *Od.*: 6.302-310.

[21] Watson, "Women in Greek Literature I," 43. *Cf.* Hom. *Od.*: 7.142-167.

[22] Watson, "Women in Greek Literature I," 43. *Cf.* Hom. *Od.*: 2.95-107, 19.139-150.

[23] Watson, "Women in Greek Literature I," 42.

duties, however menial, in support of the war effort., and support men of the family returning from the war.

Watson's second article published in November 1917 goes further and emphasises the endurance of the heroines of ancient Greece in the face of warfare and loss. This she associates directly with the losses faced by her readers in the darkest days of World War 1. She begins with the Trojan War as depicted in Homer's *Iliad*, and especially discusses *Iliad* book 6, lines 390 to 502, where Andromache bids farewell to her husband Hector before he goes into battle. As Watson emphasises, Andromache has already faced the loss of her mother, her father and all seven of her brothers,[24] and Watson quotes from the poem "The Monday Before Easter," by John Keble (1792-1866) a Church of England priest and poet:

> 'Father to me thou art, and mother dear
> And brother, too, kind husband of my heart'
> So speaks Andromache in boding fear
> Ere from her last embrace her hero art.[25]

In Watson's estimation, this scene in the *Iliad* sums up the experience of all wives when a husband leaves to fight in a war.[26] Those left behind, she says, cannot know if their loved ones will return. Hector is indeed killed by Achilles[27] and this, Watson comments, would find echoes in the hearts of many of the wives and mothers among her readers. However, while she commends Andromache's courage, Watson is also critical. Andromache

---

[24] Watson, "Women in Greek Literature II," 67. *Cf.* Hom. *Il.*: 6.414-28.

[25] Watson, "Women in Greek Literature II," 67. *Cf.* John Keble, "The Monday Before Easter," in *The Christian Year.* (Oxford: John Henry Parker, 1857. Fifty-third edition of 1827 first edition). The poet John Keble (1792-1866) was an English Anglican priest and one of the leading members of the Oxford Movement in the English High Church. Between 1831 and 1841 he was Chair of Poetry at Oxford, and *The Christian Year* comprises a collection of poems for Sundays and Feast days in the Christian year. By 1840 the text was in its nineteenth edition. That by 1858 this volume was in its fifty-third edition indicates that Keble's verses were extremely popular among High Church congregations during the first half of the nineteenth century.

[26] Watson, "Women in Greek Literature II," 67. As Hector, and Andromache, together with their baby son, watch the battlefield Andromache pleads with Hector to remain, and not make her a widow: she has already lost her mother, father and brothers. Hector says he understands but cannot, in honour, fail to join the battle. *Cf.* Hom. *Il.*: 6.509-529.

[27] *Cf.* Hom. *Il.*: 22.420-26; and 22.566-594.

would have kept Hector away from the battle, and safe; by contrast, the women of modern society hold to a higher law of duty.[28]

Watson then calls on Greek drama to identify role models for her readers from Greek myth, role models who have entered the *polis* from the *oikos* and made great sacrifices for their families and their nation. As W.K.C. Guthrie identifies in *A History of Greek Philosophy* (1962-81) ancient Greek dramatists show us "the limits and possibilities of language"[29] and in *Reading Greek Tragedy* (1986) Simon Goldhill emphasises "the role of language in the production of meaning," and comments on the way Greek tragedies "return again and again to the shifting, distorting qualities of language."[30]

Watson first turns to Euripides' *Andromache*, and as Justina Gregory identifies in "Euripidean Tragedy" (2008), while the Greek tragedians drew on myth and gnomic wisdom, Euripides' characters allude to the intellectual topics of the day.[31] It is in this sense that Watson appropriates the words and acts of heroines in Greek drama in times of warfare. By appropriating the language of heroines in Euripidean drama, Watson makes their meanings synonymous with the challenges faced by British women during the Great War. As Euripides had been born on the island of Salamis on the day of the Greek victory there during the Persian Wars,[32] she comments, the ideals found in tales of the war heard in his childhood would have inspired his dramas.

His audiences, Watson comments, could then, in turn, teach his readers, including modern female readers, about a "higher conception of women" than Homer's depictions of Nausicaa, Arete, Penelope and even Andromache. Homer's depictions, she comments, "belonged to a simpler age" than that of the Greek tragedians, while by contrast Euripides' female characters are more complex. This, Watson asserts, results in a new and higher conception of "woman,"[33] and Andromache in Euripides'

---

[28] Watson, "Women in Greek Literature II," 67.
[29] W.K.C. Guthrie, *A History of Greek Philosophy* (vol. 3. Cambridge: Cambridge University Press, 1969), 209.
[30] Simon Goldhill, *Reading Greek Tragedy* (Cambridge: Cambridge University Press, 1986), 2; 33-56.
[31] Justina Gregory, "Euripidean Tragedy," in *A Companion to Greek* Tragedy, ed. Justina Gregory (Oxford: Blackwell, 2008), 263.
[32] This is confirmed by Gregory. *Cf* Gregory, "Euripidean Tragedy," 252.
[33] Watson, "Women in Greek Literature II," 67.

Andromache has the courageous qualities needed save the life of her son.[34] Being an Edwardian and writing for the Religious Tract Society, however, Watson does not here reveal that this son was not Hector's child (who had been murdered by the Greeks) but the son fathered by her captor Neoptolemus.

Watson then turns to depictions of Agamemnon's daughter Iphigenia in Euripides' *Iphigenia at Aulis.* Assuming her audience does not know the plot, and recommending Arthur S. Way's translation, much of page 68 summarizes comments on the plot.[35] After outlining how Iphigenia is tricked by her father into travelling to Aulis where "the seer Calchas has declared that the sacrifice of Iphigenia was the only means of placating the goddess Artemis,"[36] Watson tells her readers,

> Iphigenia, at length, discovers the scheme and her appeal to Agamemnon is most pathetic…[37] and by tender memories of the past, she abjures him to spare her life. But braver thoughts come. Iphigenia perceives that her death is to be the salvation of her country and fearlessly resolves to die.[38]

Watson comments that "[t]he horror has only been for a moment. She [Iphigenia] springs to the height of divine self-devotion and steadfastly paces towards death."[39] At the very last moment, Watson confirms to her readers, the myth relates how Iphigenia is spirited away and replaced by a hind. This tragedy, Watson insists, reveals "filial piety, self-abnegation and especially the noble patriotism of Iphigenia have a lesson for the present day."[40] Watson here, is urging British women, like Iphigenia, to sacrifice their futures and even their lives for the sake of Britain during the Great War. Aristotle, in *Poetics*, criticizes the Iphigenia of *Iphigenia at Aulis* for her inconsistency in first rejecting, then accepting, her sacrifice,[41] but John Gibert concludes that that she changed her mind on the basis of insight and after reflection,[42] and Gibert's conclusion is consistent with both Bishop

---

[34] Watson, "Women in Greek Literature II," 67.
[35] Watson, "Women in Greek Literature II," 68.
[36] Watson, "Women in Greek Literature II," 68.
[37] Watson, "Women in Greek Literature II," 68. Watson here quotes Eur. *IA*, 1220-1225.
[38] Watson, "Women in Greek Literature II," 68. *Cf.* Eur. *IA*, 1468-1474, 1555-1556.
[39] Watson, "Women in Greek Literature II," 68.
[40] Watson, "Women in Greek Literature II," 68.
[41] Arist. *Poet.* 1454a31-33.
[42] John Gibert, *Change of Mind in Greek Tragedy* (Göttingen: Vandenhoeck & Ruprecht, 1995). *Cf.* Gregory, "Euripidean Tragedy," 260.

Moule's and Watson's wish that women should step outside the domestic sphere and actively contribute to the well-being of the nation, even at risk to their lives, rather than by merely supporting the courageous acts of then men in their families.

Significantly, however, although Watson acknowledges that Iphigenia is the heroine of two Greek tragedies, she does **not** discuss Aeschylus' *Agamemnon* where, in the view of Clytemnestra, Iphigenia is dragged unwillingly to the altar.[43] In *Agamemnon*, the Chorus recalls the sacrifice of Iphigenia with horror and as Simon Goldhill comments, Clytemnestra has the "ability to weave a net of words around a victim" and so lures Agamemnon to his death in vengeance for his sacrifice of an unwilling victim.[44] That tale does not, it seems, agree with the message of duty, patriotic loyalty and self-sacrifice that Watson and her publisher wish to transmit to twentieth century readers. That audience, during the previous month, had been shocked to read in the Paper about the execution by the German authorities of Nurse Edith Cavell on October 12, 1915. Cavell had risked and ultimately sacrificed her life by helping allied soldiers to escape from German-occupied Belgium. Frank Crane's article in the October issue of the paper, "The Awakening of the Ideal: A Message to You," acknowledges this sacrifice.[45] In the context of the Great War, Edith Cavell is depicted by Crane as a woman who had sacrificed not just her comfort but her life for the sake of her country, thus stepping out of the "womanly" *oikos* and into the *polis*. Reflecting the article's title, Crane urges British women to emulate Cavell, and "give expression to the ideal forces within ourselves and others" for the sake of their country.[46]

In the third article, published in December 1917, Watson considers Euripides' depictions of the sacrifices made by Polyxena, Makaria and Alcestis. Beginning with Euripides' *Hecuba* (Way employs the Roman nomenclature) Watson relates the story of how the Trojan queen Hecuba is told that her daughter Polyxena must be sacrificed at his tomb to placate the dead Achilles. Polyxena begs her mother not to resist, Watson points out, as she prefers death to the dishonorable life of a slave. The choice of death

---

[43] See the Chorus description of "a sacrifice strange, dread, lawless" in Aesch. *Ag*, 152-153, and Clytemnestra's speech at Aesch. *Ag*, 1411-1418 where she bitterly describes her "slaughtered" child rather than her "sacrifice."

[44] Goldhill, *Reading Greek Tragedy*, 2; 4.

[45] Frank Crane, "The Awakening of an Ideal: A Message for You," *Girls' Own Paper*, October 1917, 28.

[46] Crane, "The Awakening of an Ideal," 28.

before dishonor, Watson comments, would always have been Polyxena's choice, because Polyxena realizes that freedom, not life, is the greatest of all the blessings.[47] In her discussion of Euripides' *The Children of Heracles*, as with *Iphigenia at Aulis*, Watson does not assume her audience knows the plays, and she outlines the plot. Again, her summary emphasizes the example of sacrifice made by women, rather than by men. Although Watson does not quote specific lines, Makaria's example again reveals a woman who accepts danger to step outside the *oikos* and sacrifice herself in times of war for the good of the nation:[48]

> For a woman silence and discretion be fairest,
> And still tarrying in the home.[49]

The final example concerns Euripides' *Alcestis* and his depiction of Alcestis, the wife of Admetus, King of Pherae in Thessaly. Watson only briefly summarizes the plot, emphasizing the willingness of Alcestis to sacrifice herself for the sake of her husband.[50] For further study, Watson then recommends Robert Browning's accessible retelling of this myth in his poem, "Balaustion's Adventure," which would provide her readers with "the outline of [Alcestis'] story" together with "the transcript of the *Alcestis* ... with the poet's own comments."[51]

Watson has again clearly chosen plays consistent with Bishop Moule's letter. That Watson is using these narratives as role models for British women of the selfless sacrifices of heroines in Greek tragedy in times of conflict is indicated in the conclusion of her examination of Euripidean heroines:

---

[47] Watson, "Women in Greek Literature III," 134.

[48] Watson, "Women in Greek Literature III," 133. Watson relates how, after the death of Heracles, when King Eurystheus wished to kill his children, and how they fled to Athens (See Iolaus' explanatory opening speech in Eur. *Heraclid.*: 1-51), how Demophon, Theseus' son, refuses a judgement (*cf.* Eur. *Heraclid.*: 236-252), and how, then the gods demand the sacrifice of a royal woman as the price of victory for the fugitives in the resulting warfare, Heracles' daughter Makaria offers herself as a sacrifice so that her brothers might survive (*cf.* Eur. *Heracl.* 500-534).

[49] Eur. *Heracl.* 477-479. Way's translation. Makaria goes on to explain that although she is not the head of her household she has stepped outside this role for the sake of the people of her nation in a time of conflict (cf. Eur. *Heracl.* 480-483).

[50] Watson, "Women in Greek Literature III," 133.

[51] Watson, "Women in Greek Literature III," 133. *Cf.* Robert Browning, *Balaustion's Adventure: including a transcript from Euripides*. (London: Smith Elder, 1871).

> Surely it is worth while to study the literature which depicts such women as these! They are human in their clinging to early ties; but they realise that there is, for humanity, a "Higher than the love of happiness."[52]

Quoting Moule directly, she reminds her audience of the Bishop's call for the continuing sacrifice of lower aims for the highest ones noted above (see footnote 15),[53] adding:

> When we are studying these types of selflessness and devotion … we may be surely inspired with fresh eagerness to press forwards to a lofty ideal. Some may say, and with truth, … that it is easier to die than to live in sorrow, as is the present lot of countless women … but death – deprived of the Christian's bright hope – appeared in grim enough guise to the ancient world and "the transcendent Four" of Euripides are worthy of all honour.[54]

Watson, here, identifies the sacrifices for the family and for the nation made by the heroines of Greek drama as being models for her contemporaries in wartime. She then reaffirms the qualities of unselfishness and devotion to others by women in Greek literature and believes that by studying "these types of selfless devotion … we may surely be inspired with fresh eagerness to press forward to a lofty ideal."[55]

Watson then turns to Sophocles' depictions of Antigone.[56] She first considers *Oedipus at Colonus*. Here, she emphasizes Antigone's personal sacrifice and loyalty in remaining with her blinded father, likening this loyalty to that of Cordelia to her father in Shakespeare's *King Lear.*[57] Watson then enters an extensive discussion on Sophocles' *Antigone.* Here, in her plot summary, she emphasizes the courage and resolution as well as the loyalty of Antigone in her decision to risk her own death by returning to Thebes after the death of her father in order to bury her brother, in contravention of Creon's degree.[58] Watson especially comments on

---

52 Watson, "Women in Greek Literature III," 133.
53 Watson, "Women in Greek Literature III," 134. Watson quotes Moule directly but does not use his emphatic italics.
54 Watson, "Women in Greek Literature III," 134.
55 Watson, "Women in Greek Literature III," 134.
56 Watson, "Women in Greek Literature III," 134.
57 Watson, "Women in Greek Literature III," 134.
58 Watson, "Women in Greek Literature III," 134. Watson relates how Antigone risks her life to return to Thebes to ensure her brother does not remain unburied after his defeat in battle by their uncle Creon (lack of burial was shocking fate in ancient Greece, Watson here points out). This, Watson notes, was in defiance of Creon's declaration that her brother was a traitor, his order that the body should remain

Ismene's willingness to share her sister's fate when Creon condemns Antigone to death: Antigone's own loyalty to her sister which ensures Ismene does not also die. In the context of her message to her readers, Watson especially emphasises the nobility of Ismene's loyalty and courage as well as Antigone's. This double praise reinforces her message that by following these examples the sacrifices made by modern women in wartime modern women could also act nobly.

Sophocles' heroine Antigone together with Euripides' heroines Iphigenia, Makaria, Alcestis and Polyxena, Watson concludes, all display courage and a willingness to sacrifice their own comfort and even their lives for a higher cause. She then urges her readers to learn from the lives of the Greek heroines so they, too, might live selfless lives in times of warfare.[59] Taken together Watson's comments on these plays reinforce Bishop Moule's letter and his call for women to hold to their Christian faith and make great sacrifices during the war, both by supporting the sacrifices made by their husbands, fathers and brothers in the war effort and by entering the male sphere outside the home and contributing directly to the war effort.

That this was the policy of the paper is indicated by the December issue of the paper. Prior to Watson's final article, an editorial "The Fellowship of Service" was written by, or guided by, the paper's (female) editor Flora Klickmann. This directly addresses her female contemporaries. The Editorial asks:

> How little indeed would all the great ones in history have carried through, apart from fellow workers, who went before, or came after, or toiled loyally by their side. These are they who make great sacrifices and do noble deeds, but of whom the world seldom hears. They go undecorated and die unsung, in the ordinary services of life.[60]

The Editorial then comments how, in times of war:

> the same principle of co-operation holds good; in a home … an army, a business, or a Church, or the State, there must needs be a co-ordination of many services … [t]he efficiency of the whole depends on the efficiency of each part.[61]

---

unburied and his pronouncement that anyone defying this order would be buried alive. *Cf.* Soph. *Ant.* 162-210.

[59] Watson, "Women in Greek Literature III," 134.

[60] Editorial, "The Fellowship of Service," 113.

[61] Editorial, "The Fellowship of Service," 113.

The Editorial concludes that consequently "[e]ach may be a link in chain … and not mere isolated rings," so that "fellowship is a community of service."[62]

This, the Editorial asserts, leads to "a great fellowship" where "we can find the work of which we are capable," and that the war "has been a revelation in this respect," as "[n]ot in isolation but in association does life come to have meaning."[63]

Calling for the continuation of such co-operation after the war, the Editorial asks "Why should there not be the same zeal of service …on as vast and efficient a scale as the service of the wounded in time of war?"[64] The Editorial here projects the changes in the traditional roles of women into the future. It hopes that the changes forced on the nation's women would continue after the war, in line with the widening roles of British women indicated in the articles in the paper noted in footnote 11, above.

Watson's three articles emphasize that in wartime all women, ancient and modern, can find in themselves the qualities of courage, self-sacrifice, and a willingness to undertake what might be regarded as "menial tasks" in the home and outside it in support of the war effort. These articles are bracketed by similar messages from Bishop Moule and an Editorial. Alongside Watson's articles and these associated texts, the paper provides a wide range of contemporary models in which its readers might serve Britain. These include, for example, the work of Britain's Queen Mary which encouraged women to work in medicine, in nursing, in the home and in industry,[65] the work of Voluntary Aid Detachments,[66] professional nurses,[67] and the work of female telephonists.[68] Other women are shown as employed in "heavy

---

[62] Editorial, "The Fellowship of Service," 113.
[63] Editorial, "The Fellowship of Service," 113.
[64] Editorial, "The Fellowship of Service," 114.
[65] A Woman of the World, "The Queen and Women's Work," *Girls' Own Paper and Woman's Magazine*, June, 1918, 457-60.
[66] Mary Francis Billington, "Voluntary Aid Detachments," *Girls' Own Paper and Woman's Magazine*, November, 1917, 97-99.
[67] Mary Francis Billington, "The Nursing of Fighting Men," *Girls' Own Paper and Woman's Magazine*, December, 1917, 143.
[68] Mary E. Tongue, "The Girls Who Deserve a V.C: The Courage and Nerve shown by Telephone Operators," *Girls' Own Paper and Woman's Magazine*, January, 1918, 203-204.

work" and "dirty work," such as blacksmiths, chimney sweeps, and those working in farming, mining and fisheries.[69]

Watson's articles, when read alongside the Bishop's letter, the Editorial and articles on women's war work, provide an insight into a significant moment in women's history. At this time, the paradigm restricting women to the domestic role, the *oikos*, and restricting them in their interactions beyond it the *polis*, already challenged, broke down of necessity and these articles provide insights into this moment in history. During the Great War women were urged to make great personal sacrifices in the home and to work in the male sphere beyond it.

The British people were surrounded by Classical models, and Classics had a significant role in education. It was natural, therefore, that role models of great personal sacrifice by women in wartime should be drawn from the literature of the ancient world. The paper reveals how efforts by women took them out of the home to a greater extent than ever before and reflects forthcoming changes in the future of women's roles in society after the war ended.

Bibliography[70]

Ancient Texts

Aeschylus. *The Oresteia*, trans. Arthur S. Way: *Agamemnon*; *Choephoroe or The Mourners; Eumenides or The Reconciliation.* London, Macmillan & Co., 1908.

Aristotle. *Poetics.* London: Penguin, 1996

Euripides. *The Tragedies of Euripides in English Verse*, trans. Arthur S. Way, vol 1: *Alcestis*; *Medea*; *Hyppolytus*; *Hecuba*; *Ion; Suppliants.* London, Macmillan & Co., 1894.

Euripides. *The Tragedies of Euripides in English Verse*, trans Arthur S. Way, vol 2: *Andromache*; *The Children of Herakles*; *The Daughters of Troy*; *Electra; Helen; The Madness of Herakles.* London, Macmillan & Co., 1894.

[69] Anonymous, "Hard Jobs Being Filled by Women," *Girls' Own Paper and Woman's Magazine*, February, 1918, 242-244.

[70] The volumes of *Girl's Own Paper and Woman's Magazine* commence in October and are paginated from October to the following September. This allows for the binding of a volume's Papers in time for Christmas.

Euripides. *The Tragedies of Euripides in English Verse*, trans. Arthur S. Way, vol 3: *The Phoenician Maidens*; *Orestes*; *Iphigeneia in Taurica*; *Iphigeneia at Aulis*; *The Bacchanales*; *Rhesus*. London, Macmillan & Co., 1894.

Homer. *The Odyssey of Homer done into English Verse,* trans. Arthur S. Way, books 1-24. London, Macmillan & Co. Ltd., 1904.

Homer. *The Iliad of Homer in English Verse*, books 1-10 and books 11-24, trans. Arthur S. Way. London, Macmillan & Co. Ltd., 1910.

Sophocles. *Sophocles in English Verse*, trans. Arthur S. Way. *Oedipus the King; Oedipus at Kolonus; Antigone*. London, Macmillan & Co. Ltd., 1909.

Modern Texts

Anonymous. "Hard Jobs being Filled by Women." *Girls' Own Paper and Woman's Magazine* 38, no. 5 (February 1918): 242-244.

Billington, Mary Frances. "Voluntary Aid Detachments: What Good Organisation has Achieved in Directing Generous Service." *Girls' Own Paper and Woman's Magazine* 38, no. 2 (November 1917): 113-114.

Billington, Mary Frances. "The Nursing of Fighting Men: A Magnificent Record of Women's Efficiency." *Girls' Own Paper and Woman's Magazine* 38, no. 3 (December 1917): 143-147.

Brighouse, Phyllis. "Classics, Empire and Didacticism, 1919-1939." PhD diss., University of Liverpool, 2022.

Vera Brittain, *Testament of Youth*. London: Virago, 1978.

Browning, Robert. *Balaustion's Adventure: including a transcript from Euripides*. London: Smith Elder, 1871.

Crane, Frank. "The Awakening of an Ideal: A Message to You." *Girls' Own Paper and Woman's Magazine* 38, no. 1 (October 1917): 28.

De Groot, Joanna. "'Sex' and 'Race' in the Construction of Language." In *Cultures of* Empire, edited by Catherine Hall, 37-60. Manchester: Manchester University Press, 2000.

Editorial. "The Fellowship of Service." *Girls' Own Paper and Woman's Magazine* 38, no. 3. (December 1917): 113-114.

Gibert, John. *Change of Mind in Greek Tragedy*. Göttingen: Vandenhoeck & Ruprecht, 1995.

Goldhill, Simon. *Reading Greek Tragedy.* Cambridge: Cambridge University Press, 1986.

Goldhill, Simon. *Victorian Culture and Classical Antiquity: Art, Opera, Fiction and the Proclamation of Modernity*. Oxford: Oxford University Press, 2011.

Gregory, Justina. "Euripidean Tragedy." In *A Companion to Greek Tragedy*, edited by Justina Gregory, 251-270. Oxford: Blackwell, 2008.

Guthrie, W.K.C. *A History of Greek Philosophy.* Volume 3. Cambridge: Cambridge University Press, 1969.

Hagerman, C.A. *Britain's Imperial Muse.* Basingstoke: Palgrave, Macmillan, 2013.

Hospital Matron, A. "Wanted: Probationers," *Girls' Own Paper and Woman's Magazine* 35, (January 1914): 244-245.

Keble, John. "The Monday Before Easter." In *The Christian Year.* Oxford: John Henry Parker, 1857. Fifty-third edition of 1827 first edition.

Maclean, Ida Smedley, D.Sc. "Household Science in the Universities." *Girls' Own Paper and Woman's Magazine*, 35, no. 7 (April 1914): 424-427.

Millet, Kate. *Sexual Politics.* New York: Doubleday, 1970.

Moule, Hanley, G.C. (Bishop of Durham). "To the Women of the Nation." *Girls' Own Paper and Woman's Magazine* 38, no. 1 (October 1917): 40-41.

Said, Edward. *Orientalism*. London: Routledge & Kegan Paul, 1978.

Tongue, Mary E. "Girls Who Deserve the VC: The Courage and Nerve shown by Telephone Operators." *Girls' Own Paper and Woman's Magazine* 38, no. 4 (January 1918): 203-204.

Watson, Lily. "Women in Greek Literature: Part I." *Girls' Own Paper and Woman's Magazine* 38, no. 1 (October 1917): 42-43.

Watson, Lily. "Women in Greek Literature: Part II." *Girls' Own Paper and Woman's Magazine* 38, no. 2 (November 1917): 67-68.

Watson, Lily. "Women in Greek Literature: Part III." *Girls' Own Paper and Woman's Magazine* 38, no. 3 (December 1917): 133-134.

Woman of the World, A. "The Queen and Women's Work." *Girls' Own Paper and Woman's Magazine* 38, no. 9 (November 1917): 457-460.

'NOT A FEW HOPLITES': THE EVOLUTION OF ILLYRIAN INFANTRY, 5TH TO 3RD CENTURY BC.[1]

FLORIAN D. FEIL

Introduction

Illyria, situated along the south-east Adriatic coast between the Akrokeraunian mountains and the promontory of Diomedes, lay not only on the periphery of the Greco-Roman world, but also remains among the more marginal subjects in modern scholarship. This is especially true for Illyrian military history, a topic which has been treated almost exclusively in comments on particular passages in the literary sources.

Because of this, Illyrian infantry has been characterised in diametrically opposed ways: older scholarship, mostly relying on Thucydides, nearly unanimously assumed that they were light-armoured and undisciplined throughout the Classical-Hellenistic period.[2] A notable exception is Alberto Gitti's perceptive analysis of Illyrian infantry in Polybius' account of the Battle of Medion in the late third century, but it seems to have remained

[1] I would like to thank Graham Wrightson for organising this stimulating conference, Daniel Yamanian for technical support in pre-recording the paper, the participants for the fruitful discussion of the ideas presented there and Frank Daubner, Sean Manning and Graham Wrightson for commenting on the written version.

[2] Helmut Berve, *Das Alexanderreich auf prosopographischer Grundlage: Erster Band: Darstellung* (Munich: Beck, 1926), 139; Timothy Griffith, *Mercenaries in the Hellenistic World* (Cambridge: University Press, 1935), 68; Marcel Launey, *Recherches sur les armées hellénistiques* (Paris: Boccard, 1949), 417. Max Fluß, "Illyrioi," in *Paulys Realencyclopädie der classischen Altertumswissenschaft: Supplementband 5*, ed. Wilhelm Kroll (Stuttgart: Metzler, 1931), 338-39 tries to harmonise Polybius' third-century Illyrians with the fifth-century Illyrians of Thucydides.

unknown to Anglophone scholarship.[3] His interpretation, however, was shared and extended by Frank Walbank, who concluded from all Polybian descriptions of third-century Illyrian warriors that "the Illyrian armour was heavy" and emphasised their discipline and close formation,[4] which was accepted by Sylvie Le Bohec.[5] Going even further, and based upon Diodorus and Arrian, Nicholas Hammond argued that the Illyrians adopted phalanx tactics and the Greek panoply already in the early fourth century; Pierre Cabanes shared his interpretation, but both their analyses were not extended into the third century.[6] Stemming from these antecedents, there is a strand of scholarship which assigns phalanx tactics to 'the Illyrians', although such statements are often not embedded in a chronological context.[7] More recent works, however, do not reflect previous scholarly debate, but instead revert to the idea of fourth-century Illyrian infantry being light and flexible.[8] These differing assessments seem to be largely due to

---

[3] Alberto Gitti, *La politica dei re Illiri e la Grecia fino all'intervento Romano* (Milan: Popolo d'Italia, 1936), 15-20, 32-36, 64-67.

[4] Frank Walbank, *A Historical Commentary on Polybius: Vol. 1: Commentary on Books I-VI* (Oxford: Clarendon, 1957), *ad* Polyb. 2.3.5.

[5] Sylvie Le Bohec, "Les soldats illyriens au service des rois de Macédoine," in *L'Illyrie méridionale et l'Épire dans l'Antiquité – II*, ed. Pierre Cabanes (Paris: Boccard, 1993), 228.

[6] Nicholas Hammond, "Alexander's campaign in Illyria," *JHS* 94 (1974): 81, 86; *id.*, "The Battle between Philip and Bardylis," *Antichthon* 23 (1989): 1 n. 2; *id.*, *The Macedonian State: Origins, Institutions, and History* (Oxford: Clarendon, 1989), 100; most extensively in *id.*, "What may Philip Have Learnt as a Hostage in Thebes?," *Greek, Roman and Byzantine Studies* 38, no. 4 (1997): 362-65. Pierre Cabanes, *Les Illyriens de Bardylis à Genthios: IVe-IIe siècles avant J.-C.* (Paris: SEDES, 1988), 134-35, 189-90.

[7] Aleksandar Stipčević, *The Illyrians: History and Culture*, trans. Stojana Čulić Burton (New Jersey: Noyes, 1977), 170, still based entirely on Gitti, *La politica*, 15-20; John Wilkes, *The Illyrians* (Oxford: Blackwell, 1992), 126. Unfortunately, both Stipčević and Wilkes, arguably the most commonly known books about ancient Illyria, offer only one passing comment each. Richard Gabriel, *Philip II of Macedonia: Greater than Alexander* (Washington, D.C.: Potomac, 2010), 106-09 is more extensive.

[8] Edward Anson, "The Introduction of the *Sarisa* in Macedonian Warfare," *Anc. Soc.* 40 (2010): 60; Timothy Howe, "Plain Tales from the Hills: Illyrian Influences on Argead Military Development," in *The History of the Argeads: New Perspectives*, eds. Sabine Müller et al. (Wiesbaden: Harrassowitz, 2017), 100, where only Illyrian light infantry is mentioned in the fourth century (*cf.* also the focus on Illyrian mobility on p. 103-05). He seems to assume (p. 103) that the only heavy infantry in Illyria was the Greeks sent by Dionysius and the 500 Illyrians wearing Greek panoplies, which makes heavy infantry in Illyria seem like a transient, foreign

generalisations being made from incomplete evidence, as well as an implicit view of Illyrian military institutions as largely static.

Therefore, to contribute to a clearer understanding of these aspects of Illyrian history, the following pages aim to provide a chronological and comprehensive analysis of the literary evidence on Illyrian infantry from the fifth to third century BC. Of course, it needs to be emphasised that written sources, even if they remain fundamental for our broader picture of Illyrian warfare, cannot answer all our questions and may even raise additional ones, such that extensive study would have to be devoted to the archaeological evidence or the socio-economic and political background and consequences of the military developments discussed here. These important but far-reaching questions lie beyond the scope of the present paper and can at best be hinted at in the following, but it is nevertheless hoped that an examination of the literary sources for Illyrian infantry will stimulate scholarly discussion about the history and archaeology of Illyria, as well as open up possibilities for further research on Balkan military history.

With regards to the wider Balkan perspective, it is of course necessary to keep in mind that the term 'Illyrians' came to be used precisely because it simplified the complex local and regional identities in the south-western Balkans.[9] As both ethnic configurations and military institutions are subject to constant change, assigning fighting-styles to ethnic umbrella terms is inherently problematic. But it is also unavoidable, for whenever ancient historiographers care to mention military matters in 'Illyria', they simply speak of 'Illyrians', and given the patchy evidence, we are at their mercy. Therefore, it seems better to presently use this problematic way of speaking, also in hope of possible future refinements, rather than not investigate the military developments in this region at all.

---

phenomenon, since Arrian's Illyrian hoplites are not mentioned in the discussion of Alexander's campaign in Illyria (p. 106-7). Most recently, Graham Wrightson, *The Battles of Antiochus the Great* (Barnsley: Pen and Sword, 2022), 81 has characterised third-century Illyrians as light infantry. – The most recent book on Illyria, Andreas Lippert and Joachim Matzinger, *Die Illyrer: Geschichte, Archäologie und Sprache* (Stuttgart: Kohlhammer, 2022), does not discuss Illyrian infantry.

[9] *Cf.* Wilkes, *Illyrians*, 86-87.

The Fifth Century: Illyrian Peltasts

The earliest and probably most famous mention of Illyrian warriors dates to the later fifth century and is to be found in Thucydides' story of the Spartan general Brasidas.[10] In 423, Brasidas and his hoplite army joined Perdikkas, the ruler of Lower Macedonia, on an expedition against Arrhabaios of Lynkestis, a kingdom in Upper Macedonia. Perdikkas had also hired a large band of Illyrian mercenaries, who had a reputation of being ferocious fighters.[11] But while Brasidas and Perdikkas were already in Lynkestis and awaited the Illyrians' arrival, Arrhabaios won over the Illyrians, and upon hearing this, Perdikkas and Brasidas resolved to retreat the next day. In the night, however, Perdikkas' men panicked and fled, leaving Brasidas' force unsupported. At dawn, with the enemy attack impending, Brasidas formed the hoplites into a hollow square, with his light troops in the centre and a picked force of hoplites in the rear to ward off attacks, and planned to march off.[12]

In this situation, Thucydides ascribes an exhortation speech to Brasidas which describes the actual weakness of the Illyrians: "They have no order which would make them ashamed to leave a place when pressed, and as flight and charge are equally honourable among them, their bravery is untested."[13] Of course, this somewhat topical statement, from a fictional motivational speech and the mouth of somebody whose "enticing lies" were notorious,[14] need not be taken at face value. But both the preceding formation, designed not to fight one decisive battle but to ward off constant harassment while on the march, and the following narrative clearly show that Brasidas was fighting against flexible, light-armoured warriors who did not use a phalanx formation and were unable to sustain close combat against hoplites.[15] Especially the detail that the hoplites, fulfilling Brasidas' orders, sallied out from the square, "received them when they attacked and repelled them but, when they became quiet, retired" is in fact the usual way for

---

[10] Thuc. 4.124-128.

[11] Thuc. 4.125.1.

[12] Thuc. 4.125.2-3.

[13] Thuc. 4.126.5: οὔτε γὰρ τάξιν ἔχοντες αἰσχυνθεῖεν ἂν λιπεῖν τινὰ χώραν βιαζόμενοι ἥ τε φυγὴ καὶ ἡ ἔφοδος αὐτῶν ἴσην ἔχουσα δόξαν τοῦ καλοῦ ἀνεξέλεγκτον καὶ τὸ ἀνδρεῖον ἔχει. – As standard translations often render military-related details somewhat loosely, translations are by the author.

[14] Thuc. 4.108.5. On this speech, see Simon Hornblower, *A Commentary on Thucydides: Vol. 2: Books IV-V.24* (Oxford: Clarendon, 1996), *ad* Thuc. 4.126.

[15] Thuc. 4.127-128.2. *Cf.* the literature in n. 2 above, and Anson, "*Sarisa*," 59-60.

hoplites to fight off peltasts.[16] The Illyrians he encountered are best described as such, thus conforming with the common way of warfare on the northern periphery of the Greek world.[17]

The Fourth Century: Illyrian Hoplites

We should not assume, however, that the ways of war in these 'barbarian' regions were static. In fact, the same passage of Thucydides provides us with a glimpse of what was to come, for Arrhabaios had a sizeable force of "Lynkestian hoplites",[18] clearly distinct for Thucydides from the ethnic Greeks who made up the hoplite force in Perdikkas' army.[19] These Lynkestian hoplites must have been what the later Argead rulers Archelaos probably,[20] and Philip surely, went on to create in their realms, a phalanx force of non-Greek hoplites.[21] In the case of the late fifth-century hoplites in Lynkestis, their existence is all the more remarkable since the landlocked Lynkestians apparently lived in villages, not in Greek-style *poleis*.[22]

---

[16] See below.

[17] *Cf.*, e.g., Thuc. 2.79, 4.124.1, 5.10.; Xen. *Hell.* 5.2.14, 6.1.19. See Anson, "*Sarisa*," 60-61; Sławomir Sprawski, "Peltasts in Thessaly," in *Iphicrates, Peltasts and Lechaion*, eds. Nicholas Sekunda and Bogdan Burliga (Gdańsk: Foundation for the Development of Gdańsk University, 2014); Jan Best, *Thracian Peltasts and their Influence on Greek Warfare* (Groningen: Wolters-Noordhoff, 1969).

[18] Thuc. 4.124.2, on which see, e.g., Peter Brunt, "Anaximenes and Alexander I of Macedon," *JHS* 96 (1976): 151-52; Nicholas Hammond and Timothy Griffith, *A History of Macedonia: Vol. 2: 550-336 B.C.* (Oxford: Clarendon, 1979), 147-48.

[19] Thuc. 4.124.1.

[20] Thuc. 2.100.2, on which see, e.g., Hammond and Griffith, *History of Macedonia 2*, 138-40; Eugene Borza, *In the Shadow of Olympus: The Emergence of Macedon* (Princeton: University Press, 1990), 162-66; Sabine Müller, *Die Argeaden: Geschichte Makedoniens bis zum Zeitalter Alexanders des Großen* (Paderborn: Ferdinand Schöningh, 2016), 164-72.

[21] Another early non-Greek hoplite force could be inferred if it were assumed that Anaximenes in his notorious fragment on the military reforms of an "Alexander" of Macedon (*BNJ* 72 F 4, and see the commentary there for an introduction to the problem and a selection of the vast bibliography) meant Alexander I or II. This complicated issue cannot be discussed here, but I subscribe to the view that this Alexander is III.

[22] Thuc. 4.124.4. Urbanisation may have been underway in Lynkestis, as it was in neighbouring regions of Upper Macedonia (*cf.* Diod. 16.4.4, 16.4.7; Hammond and Griffith, *History of Macedonia 2*, 93-95; Miltiades Hatzopoulos, "The Cities," in *Brill's Companion to Ancient Macedon*, ed. Robin Lane Fox (Leiden and Boston: Brill, 2011); Georgia Karamitrou-Mentessidi, "Aiani – Historical and Geographical

Thus, if military institutions did not necessarily correspond to the ethnic and sociocultural dichotomies often over-emphasized by Greek writers, we should not be surprised to find hoplites in Illyria as well. The first passage pointing in this direction comes from Diodorus, who writes in his description of an alliance between Dionysius I of Syracuse and the Illyrians in the early fourth century BC:[23]

> Since the Illyrians were at war, he dispatched to them an allied force of two thousand soldiers (στρατιώτας) and five hundred Greek suits of armour (πανοπλίας Ἑλληνικάς). The Illyrians distributed the suits of armour among their best warriors and mixed the soldiers with their own (τοὺς δὲ στρατιώτας κατέμιξαν τοῖς ἰδίοις στρατιώταις).

Two points are relevant for the issue of Illyrian infantry, the panoplies and the allied soldiers. Panoplies, of course, consisted of a cuirass, helmet, and the large, heavy hoplite shield usually called *aspis*. This heavy and somewhat cumbersome gear would have been ill-suited to the mobility-centered, individualistic fighting style described by Thucydides. Of course, hoplite armour did not automatically necessitate phalanx fighting tactics,[24] but it would surely have suggested as much, especially since the connection between hoplite and phalanx was firmly established in the early fourth century.[25] The Archaic Greek concept of a heavy-armoured warrior with a large *aspis* providing cover to an unarmoured skirmisher[26] was apparently foreign to fourth-century Illyria, as in this case, the best Illyrian warriors should already have had suitable armour. We may therefore assume that the Illyrians followed the only existing model of what to do with hoplite armour, that is, fighting like Greek hoplites. Thus, it seems scholars have been right to take Illyrians wearing hoplite armour as an indication of them fighting in phalanx formation.[27]

Context," in *Brill's Companion to Ancient Macedon*, ed. Robin Lane Fox (Leiden and Boston: Brill, 2011).

[23] Diod. 15.13.2.

[24] See, e.g., Donald Kagan and Gregory Viggiano, "The Hoplite Debate," in *Men of Bronze: Hoplite Warfare in Ancient Greece*, eds. Donald Kagan and Gregory Viggiano (Princeton: University Press, 2013); Cezary Kucewicz, Matthew Lloyd and Roel Konijnendijk, "'Not Many Bows'? Light-armed Fighters of the Tenth through Fourth Centuries," in *Brill's Companion to Greek Land Warfare Beyond the Phalanx*, eds. *id*. (Leiden and Boston: Brill, 2021), 213-15.

[25] Fernando Echeverria Rey, "Hoplite and Phalanx in Archaic and Classical Greece," *CPhil* 107, no. 4 (2012): 303 and *passim*.

[26] Kucewicz, Lloyd and Konijnendijk, "'Not Many Bows'?", 213-15.

[27] See n. 6 above.

Another hint is provided by the detail that the Illyrians "mixed together" the auxiliary forces sent by Dionysius with their own.[28] For clearly, Dionysius would not send even more peltasts to Illyria, and the number of his men is too high for a valuable special corps like Gallic warriors or Cretan archers; so the soldiers he sent were probably hoplites, usually fighting in a phalanx. Mixing these soldiers with men who did not use the same tactic would have meant purposely relinquishing their unique strength, which seems difficult to imagine. On the other hand, if the Illyrians already had a phalanx, mixing the Greek hoplites with their own men would have strengthened the whole formation without wasting the advantage offered by their allies' skills. This would also explain why Dionysius took the unusual step of sending hoplite armour to barbarian allies: apparently, he expected them to be able to make good use of it.

At this point, it is worth emphasizing that, whether or not this was stimulated by Dionysius' support, the Illyrians embarked on an impressive series of raids and conquests in the following decades. Without recounting the details of their wars against Epirus and Macedonia,[29] the early fourth century BC undoubtedly marked a heyday of Illyrian power. The charismatic leadership of king Bardylis, who succeeded in uniting many Illyrian tribes, is surely part of the explanation, but the adoption of a more efficient fighting tactic like, e.g., the phalanx may well have been another factor contributing to Illyrian power (not unlike future developments in Macedon). This seems even more likely as not only the beginning of fourth-century Illyrian ascendancy is marked by the arrival of Greek hoplites and hoplite armour, but also the end of Illyrian supremacy coincides precisely with the appearance of Philip's superior Macedonian phalanx.

In fact, the momentous battle between Bardylis and Philip of Macedonia in 359 provides further indication that the Illyrians had abandoned peltast-style warfare and adopted phalanx tactics.[30] The Macedonian cavalry, apparently having defeated their Illyrian counterpart, rode around the Illyrian flank, while the Macedonian phalanx advanced against their front; "but the

---

[28] *Cf. LSJ s.v.* καταμείγνυμι.

[29] See William Greenwalt, "Macedonia, Illyria and Epirus," in *A Companion to Ancient Macedonia*, eds. Joseph Roisman and Ian Worthington (Malden, M.A.: Wiley-Blackwell, 2010).

[30] Diod. 16.4.3-7; Front. *Strat.* 2.3.2; Hammond, "The Battle between Philip and Bardylis"; Graham Wrightson, "The Macedonian Heavy Infantry of Philip II and Alexander the Great: Professionals in a World of Amateurs," in *Warfare in the Ancient Mediterranean World*, ed. Edward Anson (Leiden: Brill, 2024, in press).

Illyrians, forming themselves into a square (πλινθίον), bravely joined battle."[31] In this formation, the Illyrians firmly held their ground until one of their flanks, consisting of weaker warriors, eventually gave way under the combined attacks of Philip's cavalry and phalanx.[32] But although the Illyrians lost the battle, their formation is noteworthy: forming a square was the natural reaction of a surrounded phalanx,[33] but implementing this tactic with several thousand warriors[34] and during battle, right before the infantry forces clashed, required a discipline and training that seems alien to the Illyrians of Brasidas' day. Moreover, Bardylis' Illyrians must have been equipped and trained for melee combat to sustain the Macedonian onslaught for an extended period of time, whereas the undisciplined light enemies of Brasidas would probably have tried to extricate themselves by simply running away from what would have been an unwinnable battle for them.

These hints at Illyrian phalanx infantry are explicitly confirmed by none other than Arrian who, in his account of Alexander the Great's campaign in Illyria, writes that the Illyrians "held the high ground with many horsemen, many javelin-men and slingers, and not a few hoplites (ὁπλίταις δὲ οὐκ ὀλίγοις)."[35] Faced with an uphill battle against this dangerous combination of troops, Alexander quickly manoeuvred his phalanx around in many difficult formations as in a parade and then suddenly led it to attack the Illyrians, who were terrified by their opponents' military skills and withdrew. Unfortunately, this means we do not hear any further details about the military of Alexander's Illyrian foes, but the expression "ὁπλῖται" chosen by Arrian to describe the Illyrian close-combat infantry is revealing. Of course, Arrian calls Greek heavy infantry by the same name, but he also

---

[31] Diod. 16.4.6.

[32] Front. *Strat*. 2.3.2.

[33] For phalanxes forming squares, see, besides the example of Brasidas above, Xen. *An*. 1.36, 4.19-20, 28, 43; Gabriel, *Philip II*, 106-107. Anson, "*Sarisa*," 61 rightly emphasises that the square was necessitated here by the flanking manoeuvre of the Macedonian cavalry, but aditionally, discipline and training were needed for re-arranging such a large army.

[34] Diod. 16.4.4 gives the figure of 10,000 picked Illyrian infantry.

[35] Arr. *Anab*. 1.5.12. Arrian's account of Alexander's campaign in Illyria is based on eye-witness testimony, probably Ptolemaios Lagu: Hammond, "Alexander's Campaign in Illyria," 77; Albert Brian Bosworth, *A Historical Commentary of Arrian's History of Alexander: Vol. 1: Commentary on Books I-III* (Oxford: University Press, 1980), 51. Hammond, "Alexander's Campaign in Illyria," 81, 86 notes the hoplites, but Nicholas Hammond and Frank Walbank, *A History of Macedonia: Vol. 3: 336-167 B.C.* (Oxford: Clarendon, 1988), 44 write "heavy-armed infantry". Howe, "Plain Tales" does not discuss the passage.

applies it to Macedonian phalangites, Persian *kardakes* and Indian infantry.[36] Greek, Macedonian, Persian and Indian infantry are all credited with fighting in a phalanx, meaning a compact formation for close combat,[37] so the same should apply to Illyrian hoplites, too. This reinforces the impression given by the two previous passages from Diodorus and, taken together, permits the conclusion that in the fourth century, the Illyrians fought as hoplites in a phalanx formation.

Illyrian Infantry in the Third Century

In the century after Alexander, we are unfortunately confronted with what seems to be a total lack of information on Illyrian infantry; for the late third century, on the other hand, Polybius provides us with a surprising abundance of details. Describing the Battle of Medion in 231 BC, in which 5,000 Illyrian infantry attacked the Aetolian army, which was at that time besieging the Akarnanian town of Medion, Polybius writes:[38]

> [The Illyrians,] using the order customary among them, advanced in companies[39] against the Aetolian camp. But the Aetolians [...] drew up the greater part of their hoplites and cavalry right in front of their camp on the level ground, but with a part of their cavalry and the light-armed occupied the high ground on the right before the palisades, which was strategically well-placed. But the Illyrians, falling upon the light-armed troops with a charge, pushed them away with their mass and the weight of their formation and forced the horsemen, who fought together with the light-armed, to retreat to the heavy arms.[40] Then, making a charge from the high ground against the enemy deployed in the plain, they quickly routed them, since the Medionians from the town also threw themselves upon the Aetolians at the

---

[36] Macedonian phalangites called hoplites: Timothy Griffith, "ΜΑΚΕΔΟΝΙΚΑ: Notes on the Macedonians of Philip and Alexander," *PCPS* 184 (1956), 4 n. 2. At *Anab*. 2.8.6, Arrian subdivides the Persian ὁπλιτικὸν at Issus into Greeks and *kardakes*, and concerning the latter specifies: ὁπλῖται δὲ ἦσαν καὶ οὗτοι. Additionally, an unspecified πλῆθος ψιλῶν τε καὶ ὁπλιτῶν is placed behind the Greeks and *kardakes*. – Indian infantry is called hoplites at Arr. *Anab*. 5.15.6.

[37] At Issus, the *kardakes* are the "barbarians in phalanx formation" in line with the Greek mercenary hoplites in Arr. *Anab*. 2.8.8, thus part of the "phalanx" of Darius at 2.8.10. The Indian hoplites form a πυκνὴ φάλαγξ at Arr. *Anab*. 5.16.2.

[38] Polyb. 2.3.2-7. Besides Walbank, *Commentary 1*, *ad loc*., see the excellent analysis of Gitti, *La politica*, 32-36, 64-67, with a map.

[39] Polyb. 2.3.2: χρησάμενοι τῇ παρ᾽ αὐτοῖς εἰθισμένῃ τάξει προῆγον κατὰ σπείρας.

[40] Polyb. 2.3.5: οἱ δ᾽ Ἰλλυριοὶ τοὺς μὲν ἐλαφροὺς ἐξ ἐφόδου προσπεσόντες τῷ τε πλήθει καὶ τῷ βάρει τῆς συντάξεως ἐξέωσαν, τοὺς δὲ μετὰ τούτων ἱππεῖς συγκινδυνεύοντας ἠνάγκασαν ἀποχωρῆσαι πρὸς τὰ βαρέα τῶν ὅπλων.

same time. And they killed many of them, captured even more, and got possession of their arms and all their baggage.

Although we are not informed about the Aetolians' numerical strength, it is clear that they, besieging Medion with their full force,[41] will have outnumbered the Illyrians substantially. An additional risk for the Illyrians was that they, having just arrived in their small *lemboi* boats the night before, surely had no cavalry and apparently no relevant number of skirmishers. Their force seems to have consisted of melee infantry alone, which made battle against an opponent with a more well-rounded army dangerous. Not without reason, therefore, the Aetolians were "astonished by the daring of the Illyrians",[42] and apparently expected to be able to severely hinder their advance with just a part of their famous cavalry[43] and the light infantry if they occupied the high ground. This turned out to be a misjudgement, as these forces could not withstand the Illyrians' charge, which they effected in spite of the disadvantageous terrain. The speedy ascent will have been facilitated by the Illyrians' traditional battle-order of distinct companies, κατὰ σπείρας,[44] which, as Xenophon knew, allowed smaller units to push through passageways which would have been too narrow for a larger phalanx formation.[45]

The *speira*-formation may also have offered the Illyrians an advantage over enemy hoplites, since it would have allowed them to concentrate the force of their attack on specific points in the phalanx, thereby creating gaps which could throw the whole formation into confusion. Overall, of course, the Illyrians' second charge is somewhat less impressive, as it was aided by both terrain and the Medionians' rear-charge against the Aetolians, and as the Aetolian heavy infantry was generally not held in high regard;[46] but routing a ready phalanx supported by cavalry and skirmishers is a remarkable feat in any case.

---

[41] Polyb. 2.2.7: στρατεύσαντες πανδημεί.

[42] Polyb. 2.3.3.

[43] Polyb. 18.5.4-5.

[44] We do not know what size Illyrian *speirai* were or if they were based on descent, as Walbank, *Commentary 1*, *ad* Polyb. 2.3.2, 2.66.5 supposed. For Polybius' description (2.66.5) of 1,600 Illyrians alternating with 5,000 Makedonian phalangites at Sellasia, however, the Illyrian units must have been shallower than the Macedonian ones at least there and then, otherwise their presence on the front would have been negligible.

[45] *Cf.* Xen. *An.* 4.8.10-19, and the discussion of Sellasia below.

[46] Polyb. 18.5.5.

Following up their victory over the Aetolians, 5,000 Illyrian infantry, having taken the Epirote capital of Phoinike by treachery, defeated the larger part of the Epirote army in open battle in 230, again taking many prisoners. Later that year, 10,000 Illyrians were prepared to face the combined relief armies of the Aetolians and Achaeans at Helikranon near Phoinike, although they eventually retreated because of revolts in Illyria. Although Polybius gives no details, these events illustrate the military self-consciousness of the Illyrians, which appears well-founded as the Epirots in the battle of Phoinike surely outnumbered them and, crucially, must have been greatly superior in cavalry, which the Illyrians probably lacked.[47]

A few years later, the Illyrians make another prominent appearance in Polybius' account of the Battle of Sellasia in 222 BC, where the Macedonian king Antigonus Doson fought against Cleomenes III of Sparta.[48] The Spartan army, roughly 20,000 strong, occupied a fortified defensive position on two neighbouring hills, Euas and Olympos, with Cleomenes holding the right hill Olympos with one half of the army, consisting of 6,000 Spartan hoplites[49] and light mercenaries. His brother Eukleidas defended the left hill Euas with the other half, made up of 4,000 Macedonian-style phalangites[50] and unspecified allied troops, probably both light and heavy. The cavalry and some skirmishers were posted in the valley between the hills.

Antigonus, who commanded a much larger force of nearly 30,000 men, of which 10,000 were Macedonian phalangites and 1,600 Illyrians,[51] resolved to attack both hills simultaneously. Arraying his troops, Antigonus himself took command over half of the Macedonian phalanx on the left wing

---

[47] Phoinike: Polyb. 2.5.8, 6.9, and *cf.* 2.4.9, 3.1. Helikranon: Polyb. 2.6.3-5. Arriving again in their small *lemboi* boats, the Illyrians surely carried no horses with them, although they may have captured a few at Phoinike. This exemplary situation may provide another hint at why Illyrian infantry was so good as a stand-alone force, even against armies with skirmishers and cavalry: If an Illyrian raiding party arrived on a hostile shore, they often will have faced not only numerically superior, but also more well-rounded armies, against which they had to fight with bare infantry, as horses and numerous light-armed will often have often exceeded the carrying capacities of their vessels. *Cf.* also Florian Feil, "A Note on the *sibyna*, an Illyrian Weapon Similar to the Boar Spear," *Iliria* 46 (2022): 78-79.

[48] Polyb. 2.65-69; Plut. *Cleom.* 28. See Walbank, *Commentary 1*, *ad* Polyb. 2.65.7 for a map (the Spartan army faced north, the Macedonians south), and Wrightson, *Antiochus*, 80-82.

[49] Plut. *Cleom.* 28.8.

[50] Plut. *Cleom.* 11.3.

[51] Polyb. 2.65.4.

opposite Cleomenes and also stationed his cavalry between the hills; but on his right wing, opposing Eukleidas, Antigonus deployed the Illyrians in the first line together with his 5,000 Macedonian bronze shield phalangites, their units alternating with one another. Behind this first line, Antigonus stationed two more lines of allied troops, one being composed of Akarnanians and Cretans, the last of Achaean infantry.[52] Their place in the first line indicates that the Illyrians could to fight in extended melee against Macedonian-style phalangites and other heavy infantry, because had they receded from the enemy frontline, they would have opened gaps in the Macedonian line more than ever, which cannot have been desired; thus, the Illyrians were intended to function as heavy infantry once the opposing frontlines met.[53]

It is clear that the Illyrians were assigned a crucial role in the battle plan, for in the night before battle, they are singled out by Polybius as bivouacking in full order along the front line and, at a flag signal given from Antigonus' wing, beginning the attack against the Euas.[54] They did not, however, charge headlong into the enemy, but moved orderly and in concert with all the troops on their wing, which is demonstrated by the fact that when enemy skirmishers attacked the Achaeans on the now exposed flank and rear of the Macedonian right, the advance of the whole Macedonian right was temporarily slowed down.[55]

At this juncture, Polybius – perhaps unjustifiably[56] – reproaches Eukleidas for not having moved downhill to counter-attack the advancing opponent, as in his opinion, this would have thrown the enemy formation into confusion and "deprived them of the peculiar advantage of their weaponry and formation".[57] In his commentary on the passage, Frank Walbank takes

[52] Polyb. 2.66.5.
[53] In Polyb. 18.28.10, where such an arrangement of alternating units of *sarisophoroi* and other infantry is attributed to Pyrrhus, the non-phalanx units are σημαῖαι, that is, *manipuli* of Italian infantry, in all likelihood also intended to fight against the Roman-Italian heavy infantry in close combat and thus themselves heavy. For the Illyrians as heavy infantry, see also n. 68 below.
[54] Polyb. 2.66.10-67.1.
[55] Polyb. 2.67.
[56] *Cf.* Polyb. 2.69.6, where Cleomenes tears down the palisade before advancing his phalanx. Since Antigonus took pains to ensure that the Illyrian ascent would be surprising and quick, he may actually have intended their presence to prevent Eukleidas from undisturbedly moving out his phalanx and attacking the ascending troops.
[57] Polyb. 2.68.5: τὸ τοῦ καθοπλισμοῦ καὶ τῆς συντάξεως ἰδίωμα.

this as a reference to the Illyrians specifically, but it seems more likely that Polybius means the whole line of alternating Macedonian and Illyrian units.[58] For due to their different armament and peculiar arrangement, the units would have been able to support each other in the ascent, with the more flexible Illyrians protecting the *sarisophoroi* when the latter's formation was disturbed, and falling back on them when pressed too hard. Moreover, the smaller units themselves would have been especially useful in quickly climbing the hill,[59] and an attack during the ascent could – at least theoretically – have caused the *speirai* to lose contact and thus the ability to support one another, as they did to great effect in the ensuing battle on the hilltop.

For when the Macedonian right reached the plateau of the Euas, Polybius does speak of Eukleidas' men being pushed back "by the weight of the armament and formation"; but *pace* Walbank, who took this to refer to the Illyrians in particular, Polybius does not specify whose armament and formation he means.[60] The Illyrians are only singled out in the following clause as quickly occupying the ground yielded by the receding enemy, while soldiers being pushed out of their position by the weight of their opponents' armament and formation most naturally evokes the fearsome offensive power of the Antigonid *sarisa* phalanx, which was conceived as especially 'heavy'.[61] This is not to exclude that the Illyrians may also have aided the Macedonians in pushing back the enemy phalanx, but their specific strength seems to have lain, according to Polybius' wording, in moving up more quickly than the Macedonians and exerting constant pressure on the enemy line while it was slowly giving ground. The Macedonian phalanx, on the other hand, had to keep its more fragile formation intact and may thus have advanced a bit less aggressively, but more steadily, thereby preventing the enemy from flanking the Illyrian *speirai* as they pushed forward. It may have been for such or similar reasons

[58] Of course, if any troop type depended on an undisturbed formation to make good use of its weaponry, it was the *sarisa* phalanx: *Cf.* Plut. *Aem.* 20.4.

[59] *Cf.* n. 45 above.

[60] Polyb. 2.68.9: λοιπὸν ὅσον ἐκ ποδὸς ἐπιέσθησαν τῷ βάρει τοῦ καθοπλισμοῦ καὶ τῆς συντάξεως, εὐθέως οἱ μὲν Ἰλλυριοὶ τὴν κατάστασιν ἐλάμβανον, οἱ δὲ περὶ τὸν Εὐκλείδαν τὴν ὑπὸ πόδα διὰ τὸ μὴ καταλείπεσθαι τόπον εἰς ἀναχώρησιν καὶ μετάστασιν ἑαυτοῖς. Walbank, *Commentary 1*, *ad loc.*

[61] Polyb. 18.29-30, Plut. *Aem.* 19.1. Specifically at Sellasia, Polybius speaks of τὸ βάρος τῆς τῶν Μακεδόνων τάξεως (2.69.8) and the Macedonians using τὸ τῆς ἐπαλλήλου φάλαγγος ἰδίωμα (2.69.9), which verbally echoes the sentences relating Eukleidas' supposed mistake and defeat.

that Antigonus resorted to the unusual deployment at Sellasia, and the outcome justified the special role he assigned the Illyrians: the Spartan forces on Euas were soon pushed down the far side of the hill and fled, which forced Cleomenes, threatened with being surrounded, to abandon his defensive position and engage Antigonus' phalanx head-on, in turn leading to a crushing defeat for the Spartan army.[62] Antigonus' victory, therefore, was not in the smallest part due to the effectiveness of his Illyrian allies.

The last noteworthy descriptions of Illyrian infantry in action are to be found in Polybius' fifth book, where he describes the lightning campaigns of the Macedonian king Philip V in 218 against the Aetolians and Spartans. In the army with which Philip advanced into the undefended Aetolian heartland in forced marches and destroyed the Aetolian central sanctuary at Thermos, he had approximately 750 Illyrians, although he seems to have asked for more.[63] Despite their small number, however, the Illyrians are regularly mentioned as aiding in or performing crucial tasks, especially on the retreat, when they defended the lighter infantry in the Macedonian rear against encroaching Aetolian skirmishers. Retreating from Thermos, Philip's rear-guard, consisting of some light Akarnanians and 1,200 mercenaries, was attacked by 3,000 Aetolian skirmishers. But behind a hill along his marching route, Philip had prepared an ambush of "the Illyrians and the most suitable peltasts" who, as the Aetolians passed them while harassing the Macedonian rear, charged and routed the enemy.[64] Similarly, when Philip bypassed the Aetolian border *polis* of Stratos, the mercenaries in the Macedonian rear were attacked and pinned down by Aetolian infantry, cavalry and Cretan mercenary archers, but when the Illyrians came to their support, the Aetolians fled and were chased back to Stratos.[65] Just like at Medion, we note that the Illyrians were perfectly able to fight all kinds of enemy troops and pursue a fleeing foe.

After the successful Aetolian campaign, Philip took his army, including the Illyrians, to Lakonia, bypassed Sparta and again plundered the country. On his return, however, he was opposed at the narrows between Sparta and the

---

[62] Plut. *Cleom*. 28.8.
[63] Polyb. 5.4.3: The Illyrians arrived in 15 *lemboi*, one *lembos* normally carrying 50 warriors (Polyb. 2.3.1). Internal troubles in Illyria meant fewer warriors could be sent abroad, which bespeaks the political fragility and comparatively low manpower of the Illyrian kingdom.
[64] Polyb. 5.13.5-6.
[65] Polyb. 5.14.3-7.

Menelaion east of the city.[66] The Spartan king Lykurgos had himself occupied the steep hills at the Menelaion with 2,000 men, while the rest of the Spartan army was ready to sally out from the town when Philip passed by. The Macedonian king therefore needed to clear the enemy position at the Menelaion and advanced against the hills with his mercenaries, peltasts and Illyrians.[67] First, the he ordered the mercenaries to attack alone, who fared badly because of the enemy's superior armament and position, "but when Philip drew up the peltasts behind the fighting men as a reserve and, having outflanked [Lykurgos' men] with the Illyrians, made an attack from the flanks, […] those about Lykurgos, terrified by the attack of the heavy arms, turned round and fled."[68] Philip secured the hills with the Illyrians[69] and, as the Spartan main army remained quiet, led off his army unopposed.

These passages show that the Illyrians were a valuable component of the Macedonian army also outside major battles: whenever fighting took place on uneven terrain, the Illyrians were employed, together with peltasts and/or mercenaries, and provided crucial support whenever the lighter mercenaries came in distress. Their service was especially valuable against the numerous and proficient Aetolian skirmishers whom the Illyrians, once together with some peltasts and once alone, succeeded in chasing away. This was surely due to the Illyrian custom of aggressively charging the enemy in separate *speirai*, which gave them a local advantage over scattered and unarmoured skirmishers[70] and allowed individual units to support each other; additionally, the Illyrian flanks and rear would have been protected by friendly light-armed against enemy encirclement. The Illyrians' flexibility

---

[66] Polyb. 5.20-22.7.
[67] Polyb. 5.22.9.
[68] Polyb. 5.23.3-4: ἐπεὶ δὲ τοὺς μὲν πελταστὰς ὁ Φίλιππος ὑπέβαλε τοῖς ἀγωνιζομένοις, ἐφεδρείας ἔχοντας τάξιν, τοῖς δ' Ἰλλυριοῖς ὑπεράρας ἐκ πλαγίων ἐποιεῖτο τὴν ἔφοδον, τότε συνέβη τοὺς μὲν παρὰ τοῦ Φιλίππου μισθοφόρους ἐπαρθέντας τῇ τῶν Ἰλλυριῶν καὶ πελταστῶν ἐφεδρείᾳ πολλαπλασίως ἐπιρρωσθῆναι πρὸς τὸν κίνδυνον, τοὺς δὲ παρὰ τοῦ Λυκούργου, καταπλαγέντας τὴν τῶν βαρέων ὅπλων ἔφοδον, ἐγκλίναντας φυγεῖν. – Walbank, *Commentary 1*, *ad loc.* thinks that the "attack of the heavy arms" refers to the peltasts, but it is clear (and emphasized by μὲν – δέ) that they were positioned as a reserve (ἐφεδρεία, *cf.* Walbank, *Commentary 1*, *ad* Polyb. 2.66.7), while those who made the attack (ἔφοδος) were the Illyrians. This conforms with other allusions to their 'heaviness' at Medion and Sellasia (see above), but more cannot be inferred from the term, as Polybius uses βαρέα ὅπλα or τὰ βαρέα τῶν ὅπλων to mean 'heavy infantry' in general, that is, all troop types except skirmishers and cavalry (*cf.*, e.g., Polyb. 3.73, 3.102.2).
[69] Polyb. 5.23.6.
[70] *Cf.* Polyb. 2.3.5.

and movement speed, therefore, were probably roughly equal to those of light-armed mercenaries and peltasts, as the different units would otherwise have lost contact easily during prolonged and fluctuating combat on broken terrain.

Illyrian Arms and Armour

Given the frequent emphasis on the Illyrians' flexibility, we may ask just how heavy their armament actually was. Generally speaking, it seems natural that the Illyrian peltasts of the fifth-century were very lightly equipped and that the melee infantry of the fourth and third century was better protected. A close reading of the sources gives some additional hints: it has already been suggested that Dionysius sending hoplite armour to the Illyrians indicates that he perceived them as lacking the necessary armour for phalanx combat.[71] Moreover, if Dionysius' panoplies were given to the best Illyrian warriors, we get the impression that this degree of protection was normally never reached in early fourth-century Illyria. It does not, however, automatically follow that the Illyrians, once having received the panoplies, generally adopted this kind of armour[72] – Dionysius may have perceived the lack of armour as more problematic than the Illyrians themselves, who could have adapted to all hoplites using lighter armour. For living near the notable *poleis* of Apollonia and Dyrrhachion, the Illyrians must have been aware of the existence of panoplies already before.[73] But many Greek hoplites, and Macedonian phalangites normally, fought with light or no armour,[74] so the same should have worked for the Illyrians, too. Thus, if the minimal equipment for fourth-century Greek

[71] Fluß, "Illyrioi," 338.
[72] Thus the literature in n. 6 above.
[73] Cabanes, *Les Illyriens*, 189: "Les Illyriens de Bardylis étaient encore émerveillés lorsque Denys de Syracuse leur fournissait cinq cents panoplies d'hoplites […]" may be misleading.
[74] Quantifying the use of different types of body armour by Greek hoplites is never easy (*cf.* Adam Schwartz, *Reinstating the Hoplite. Arms, Armour and Phalanx Fighting in Archaic and Classical Greece* (Stuttgart: Franz Steiner, 2009), 96-97). The *communis opinio*, based on Anthony Snodgrass, *Arms and Armour of the Greeks* (Ithaca: Cornell University Press, 1967), 90-93, 109 and John Anderson, *Military Theory and Practice in the Age of Xenophon* (Berkeley: University of California Press, 1970), 26-28, that fourth-century hoplites generally wore no body armour anymore is certainly wrong (*cf.* Griffith, "The Macedonians"), but the works cited do show that many hoplites were (nearly) unarmoured. – For the armour of Philip's phalangites, see Polyaenus, *Strat.* 4.2.10; Griffith, "The Macedonians"; Hammond and Griffith, *History of Macedonia 2*, 421-24.

hoplites was just a large, heavy shield and a spear, we unfortunately cannot, from the adoption of phalanx tactics by contemporary Illyrians and Arrian's use of the term hoplites for them, infer anything specific about Illyrian armament.

Obviously, these inferences cannot, on their own, support much argumentative weight, but they seem to receive some confirmation from Polybius. Firstly, Polybius could imply a similar absence of metal body armour in the Illyrian armament when he distinguishes between Illyrians and *thorakitai*, lit. "men wearing breastplates".[75] Even so, as third-century Illyrians were not classed with the unarmoured skirmishers (εὔζωνοι), but with the heavy infantry (βαρέα ὅπλα),[76] they may have worn some kind of armour, such as helmets, greaves and, for protecting the body, chainmail, leather or linen armour, but no *thorakes*.

Secondly, and more importantly, comparatively light armour could explain why the Illyrians were such an efficient force on broken terrain and in cooperation with light mercenaries and/or peltasts. In fact, the Illyrians seem to have been heavier than the mercenaries, but lighter than the peltasts; for on his advance into Aetolia, Philip put the mercenaries in the vanguard, followed by the Illyrians, the peltasts, and finally the phalangites.[77] Also in the skirmish at the Menelaion, the Illyrians were sent to support the mercenaries, while the peltasts were held back as the last, and therefore probably heaviest, reserve.[78] Since Hellenistic peltasts wore *thorakes* or lighter armour,[79] this again suggests that the Illyrians wore only lighter armour and no *thorakes*.

That much, however, seems to be all that could be teased out of the literary evidence. An examination of archaeological sources and especially actual depictions of Illyrian warriors would of course be needed to supplement this, but unfortunately, on the only depiction of Illyrian warriors known so far, the so-called Gradishtë belt plate from a third to second-century tomb in Selca e Poshtme, southeastern Albania, the warriors' upper bodies are

---

75 Polyb. 11.11.5, 14.1, 15.5.
76 See n. 68 above.
77 Polyb. 5.7.11.
78 Polyb. 5.23.3-4.
79 Ael. *Tact.* 12; 48; Nep. *Iphicr.* 1.

completely covered by their shields.[80] When, on the other hand, archaeological sources are concerned, the picture is clear: weapon burials were traditional throughout Classical-Hellenistic Illyria, but while spears and, to a lesser extent, helmets, swords and axes are common,[81] metal body armour is virtually unknown. In fact, the only relevant piece seems to be a coat of mail found in the same elite tomb as the Gradishtë belt plate, thus also dated to the second half of the third century.[82]

Helmets, on the other hand, are much more frequently attested, most notably on the Gradishtë belt plate, where they are worn by both the foot-soldier and the horsemen. They are depicted as having a round top, a neck-guard, large cheek-pieces and a flashy crest extending backwards down to the shoulders. These helmets are surely to be identified with a type of helmet that is in fact called 'Illyrian' since, although originating in the Archaic Peloponnese, it is extremely common all over Illyria, where it continued to be used into the Hellenistic period, long after it fell out of favour elsewhere.[83] Thus, the combined archaeological evidence points to helmets being used frequently by Illyrians, including Illyrian hoplites. Furthermore, greaves are attested once in Hellenistic Illyria,[84] and just like the helmets, they would have given the Illyrians additional protection without encumbering them. In all these considerations, of course, we have to keep in mind that grave inventories do not precisely represent the world of the living, and metal body armour could just have been too precious for use as a grave good. But although the archaeological data cannot prove a near-total absence of metal body armour in Illyria, it does point in the same direction as the literary sources.

---

[80] See Stipčević, *The Illyrians*, 203 or Wilkes, *The Illyrians*, 236 for photos. – The Vace belt plate, cited by Anson, "*Sarisa*," 60 (see Stipčević, *ibid.* for a photo) is irrelevant, as it comes from Slovenia.

[81] Stipčević, *The Illyrians*, 169-78; Wilkes, *The Illyrians*, 238-41; Lippert and Matzinger, *Die Illyrer*, passim.

[82] Neritan Ceka, *Qyteti Ilir pranë Selcës së Poshtme* (Tirana: Akademia e Shkencave e RPS të Shiqipërisë, 1985), 170, who notably judges it to be an eastern import. Three archaic *thorakes* from modern day Slovenia (Stipčević, *The Illyrians*, 174; Wilkes, *The Illyrians*, 239-40) and a putative padded vest (of which only metal ornaments are actually preserved) from the Archaic tomb 2/1 in Ilijak, Bosnia (Lippert and Matzinger, *Die Illyrer*, 47-48) cannot be associated with Illyria in the common, pre-Caesarian sense of the term.

[83] Stipčević, *The Illyrians*, 173-74; Wilkes, *The Illyrians*, 108; Lippert and Matzinger, *Die Illyrer*, 64-66 with a map.

[84] Lippert and Matzinger, *Die Illyer*, 102.

Moreover, depending on the respective military environment, the absence of armour need not be a disadvantage. For just like in Thessaly, Thrace and pre-Philippic Macedonia,[85] skirmishers will have made up a vast majority in local Illyrian levies,[86] and strange as it may sound, this would have given Illyrian hoplites a good reason to fight with little to no armour. For Brasidas' tactic in Lynkos, the Battle of Lechaion[87] and countless other instances have amply demonstrated that a force of hoplites, if attacked by skirmishers, needed some men to sally out and chase the enemy away to deny them the possibility of shooting at the dense body of men from close range. The hoplites did this out of self-perceived necessity,[88] and it was not an idle pursuit: Xenophon himself says that hoplites could actually catch skirmishers if they kept chasing them long enough or if the peltasts let them come too close – which, we must imagine, they often did.[89]

Thus, while heavy armour could undoubtedly offer hoplites an advantage in close combat against other hoplites,[90] it would have brought them a disadvantage if the enemy had many skirmishers, since heavy equipment would have slowed them down in pursuit, retreat and flight. We can thus imagine a military environment in which hoplites, since their main threat was not armoured hoplites, but light skirmishers, did not normally seek to be as heavily armoured as they could, but on the contrary as lightly as possible. Naturally, such light armour would also have been advantageous on difficult terrain like the hills of Illyria, where the enemy would take refuge.[91] The constant fighting against skirmishers and on broken terrain could also have caused third-century Illyrians to eventually abandon the unified phalanx formation and fight in smaller units, retaining the benefits of a close formation but gaining crucial mobility and thus tactical flexibility.

This would explain why the Illyrian warriors of the late third century were such a remarkably efficient infantry force. Their impressive performances in Greek battles, however, make it all the more disappointing that, as far as

---

[85] See n. 17 above.

[86] At Arr. *Anab.* 1.5.12, the javelin-men and slingers are many, the hoplites not few.

[87] Xen. *Hell.* 4.5.11-18, with Sekunda and Burliga (eds.), *Iphicrates, Peltasts and Lechaion.*

[88] Xen. *An.* 3.3.8, 12; 4.1.26.

[89] Xen. *An.* 3.3.9, 15; *Hell.* 4.3.22. It was noteworthy generalship on Iphikrates' part to order his peltasts to flee when the sallying hoplites were still far away (Xen. *Hell.* 4.5.15).

[90] *Cf.* Plut. *Tim.* 28.1.

[91] For heavy armour turning into a liability on difficult terrain, *cf.* e.g. Plut. *Tim.* 28.5-8.

I can see, we have no detailed description of any of the Illyrians' many confrontations with the Romans. In general, Rome's Illyrian wars are treated briefly in the surviving sources, where the Illyrians' speedy defeats and subjection under the authority of Rome are emphasised.[92] But even given the fragmentary state of early Latin literature, as well as the near total loss of Livy's descriptions of the Illyrian wars, there are indications that Illyrian warriors could in fact put up a fight against the Roman legions.

In the First Illyrian War, Polybius tells us that the Romans lost "not only many soldiers, but also some tribunes and a quaestor" in the siege the (unidentified) Illyrian town of Nutria,[93] and also the early second-century poet Ennius mentioned Illyrian warriors in a now isolated verse: "The Illyrians resist, stabbing with *sicae* (curved short swords) and *sibynae* (an Illyrian weapon similar to the boar spear)".[94] The line preserves unique information about Illyrian weaponry and thus should have been embedded in a somewhat detailed battle narrative. This can be taken to mean that third-century Illyrian warriors did make an impression upon contemporary Romans, although later Roman authors almost universally chose not to preserve the record.

There is, however, one notable exception: In an excerpt from Cassius Dio's Roman history, we read that during the First Illyrian War, the Romans, apparently in late autumn of 229,[95] "went up to a place above the sea and came off badly at a hill called Atyrion".[96] In the given historical context, this can only mean that the Romans launched a substantial operation with their land forces, probably with the greater part of their 22,000-strong consular army, and were defeated by the Illyrians.[97] Unsurprisingly, Atyrion hill is not located, but since we know from Polybius that the Illyrian queen Teuta had retreated to Rhizon (modern-day Risan in the easily defensible bay of Kotor, Montenegro), and since the most logical step for the Romans

---

[92] Polybius spends just one paragraph, 2.11, on the whole First Illyrian War, and two on the second, emphasizing that the Romans conquered the purportedly impregnable stronghold of Dimale in just one week (3.18.5), while Livy, after relating that the Third Illyrian War (44.30.1-32.5) was finished in thirty days, comments that "this was the only war of which the end became known in Rome before the beginning".

[93] Polyb. 2.11.13.

[94] Enn. *ap*. Paul. 453 Lindsay (= *ann*. 526 Skutsch, with commentary): *Illyrii restant sicis sibynisque fodentes*. See Feil, "*sibyna*".

[95] *Cf*. Polyb. 2.11.12-16: The siege of Issa was already lifted, and the following makes clear that winter was coming.

[96] Cass. Dio frg. 49.7.

[97] The strength of the land forces: Polyb. 2.11.7.

would have been to go after her, it can be suspected to lie on the bay of Kotor, which happens to be dominated by remarkable hills above the sea.

According to Cassius Dio, this victory and the approaching winter induced Teuta to hope for a complete Roman withdrawal, and only when she saw that the Romans were wintering in Illyria, winning over her subjects and collecting local troops, she sued for peace in early spring, before the next campaigning season.[98] This continuation of the war into 228 is reported, but not explained by Polybius; of course, it is easy to see why the Achaean historian glossed over the inglorious episode. But the defeat of the legions at Atyrion hill also explains why only one of the consuls, Fulvius Centumalus, who had commanded the fleet, celebrated a naval triumph over the Illyrians; the initial successes of the land forces, achieved in cooperation with the fleet, were obviously overshadowed by their more recent setbacks.[99]

The battle of Atyrion and the siege of Nutria thus illustrate that the Illyrians were in fact capable of defeating the Romans, and on this reading, it seems significant that Rome's victory in the First Illyrian War was predominantly due to other factors than the fighting power of the legionnaires: Ruthless Roman diplomacy, which included encouraging defections from Teuta, was crucial, along with an overwhelming naval superiority and the organisational capacity to winter in the newly conquered territory.[100] The soldiers of Italy, more heavily armoured than Illyrian infantry and nearly as flexible, both individually and in their *manipulae*, were probably better suited for fighting against the Illyrians than a phalanx of hoplites, but that was not enough to guarantee military or political victory.

This is also suggested by the events of the second Illyrian war in 219:[101] The Roman consular army, surely numerically superior, stormed the southern Illyrian stronghold of Dimale suspiciously quickly,[102] thus causing the neighbouring tribes, who already had previous Roman affiliations, to surrender. In the following battle for the island of Pharos, the Illyrian

---

[98] Cass. Dio frg. 49.7; Polyb. 2.12.3.

[99] *Fasti triumphales*, p. 101 Degrassi.

[100] I hope to discuss the political events of the First Illyrian War elsewhere.

[101] Polyb. 3.18-19.

[102] Polyb. 3.18.5 gives seven days; of course, there would have been ample opportunity for local friends of Rome (for these, *cf.* Polyb. 2.11.11 about the very tribe around Dimale) to further the operation, as well as for Polybius to re-glorify it (*cf.* the following note).

garrison consisted of 6,000 chosen warriors, inferior to the consular army at least in quantity. Unwilling to engage in a lengthy siege of the homonymous and heavily fortified *polis* of Pharos, the Romans employed a stratagem (which must also have involved collaborators with local knowledge): They landed the greater part of their troops in remote bays of the island at night and, in the morning, drew out the Illyrian garrison by landing a small force near the *polis*. The Roman main army, having united and marched across the island, then showed up to occupy a hill behind the Illyrian army, cutting off their retreat to Pharos. With their usual discipline, the Illyrians turned around and attacked the enemy in good order, but when the Roman detachment from the coast fell upon their rear, they were thrown into confusion and eventually fled. The Romans proceeded to conquer Pharos by treachery and razed it to the ground, thus ending the Second Illyrian War.[103]

Conclusion

After Rome's victory in its second Illyrian war, nothing is heard anymore about the fierce Illyrian warriors, which is why this event also marks the end of this study. Even if it has become clear that the evidence is not at all abundant, it is now possible to reconstruct, at least along general lines, the evolution of Illyrian infantry from the fifth to the third century BC. Thucydides describes late fifth-century Illyrians as flexible, individualistic fighters, essentially peltasts who are unable to fight a hoplite phalanx in close combat. In the fourth century, however, a change has clearly occurred, since the Illyrians can use hoplite armour, mix Greek hoplites with their soldiers, form a large defensive square when surrounded, and are explicitly said to have hoplites by the time of Alexander the Great, all of which suggests that they fought in a hoplite phalanx. In the third century, a further tactical development is observable, since the Illyrians' traditional battle formation now consists of distinct companies which are quick and flexible even on broken terrain. These characteristics, as well as their tactical roles in Antigonid armies, suggest that third-century Illyrians were only lightly armoured, but nonetheless very capable in close combat: With their charges, they defeat skirmishers, cavalry and hoplites, and they cooperate effectively with skirmishers, peltasts and Macedonian phalangists alike.

[103] With the supposed storming of Pharos (Polyb. 3.19.12), Polybius may well flatter the consul Aemilius Paullus, ancestor of his patron Scipio Aemilianus, so the less glorious version of Cassius Dio (in Zon. 8.20) is more credible.

The Illyrians' amazing flexibility, of course, raises the question, lingering since the introduction, of how to properly designate Illyrian infantry in both ancient Greek (peltasts/hoplites) and modern (light/heavy infantry) terms. For the fifth and fourth century, of course, this seems fairly unproblematic; Illyrian warriors of the fifth and fourth centuries can be called peltasts and hoplites, or light and heavy infantry, respectively with no subtleties being lost in translation. A characterisation of third-century Illyrian infantry along these lines, however, is bound to be tricky, for it has to take into account that the designations 'heavy' and 'light' can not only refer to differing quality of armour, but also to different tactical roles.[104] 'Tactically heavy' infantry has the main function of engaging in close combat, while 'tactically light' infantry is primarily used for protecting comrades, harassing enemies, fighting on rough terrain, and other tasks which require flexibility and/or speed. Armament and tactical role obviously tend to correlate – 'tactically heavy' infantry often being heavily armoured for increased protection, and 'tactically light' infantry often wearing little to no armour for increased flexibility – and are therefore frequently conflated. In the present case, however, it seems more profitable to maintain the distinction between armament and tactical role to work out how exceptional third-century Illyrian infantry was.

According to their armament, they are readily classified as medium infantry, since they carried hoplite shields and often wore helmets, but no metal body armour; thus, they were a bit more lightly armoured than peltasts. This, however, and their native training allowed them to fulfil the tactical roles of both light and heavy infantry, as they were able to sustain extended melee against other heavy infantry, but no less apt at pursuing fleeing foes and fighting on broken terrain. To be sure, this makes them similar to contemporary peltasts, but while it would be tempting to simply call them 'light-armoured peltasts', such an *interpretatio* (*pseudo-*)*Graeca* suggests the opposite of Polybius' frequent allusions to the Illyrians' "heaviness", and in the one instance where Polybius speaks periphrastically about the Illyrians, he actually calls them "heavy infantry".[105] Moreover Polybius, other than Arrian, never speaks of Illyrian hoplites (or peltasts, for that matter), which does seem to be telling. If Polybius refrained from such an interpretation and maintained a distinction between peltasts and Illyrians, it seems prudent to follow him. Thus, it appears as no coincidence that third-century Illyrians were not normally classed together with other infantry or

---

[104] Graham Wrightson, *Combined Arms Warfare in Ancient Greece* (London/New York: Routledge, 2019), 16-18.

[105] See n. 68 above.

characterised by a generic troop type-designation, but are usually identified by their ethnic denomination alone. By this time, the military developments in Illyria had brought forth a distinct kind of warrior whose signature trait was his unmatched versatility.[106] This explains why Illyrian mercenaries were so frequently and diversely employed by the Antigonid kings, and it makes the swiftness of Rome's victories over Illyria even more impressive.

Bibliography

Anderson, John. *Military Theory and Practice in the Age of Xenophon.* Berkeley: University of California Press, 1970.

Anson, Edward. "The Introduction of the *Sarisa* in Macedonian Warfare," *Ancient Society* 40 (2010): 51—68.

Berve, Helmut. *Das Alexanderreich auf prosopographischer Grundlage: Erster Band: Darstellung*. Munich: Beck, 1926.

Bosworth, Albert Brian. *A Historical Commentary of Arrian's History of Alexander: Vol. 1: Commentary on Books I—III*. Oxford: University Press, 1980.

Cabanes, Pierre. *Les Illyriens de Bardylis à Genthios: IVe—IIe siècles avant J.-C.* Paris: SEDES, 1988.

Ceka, Neritan. *Qyteti Ilir pranë Selcës së Poshtme.* Tirana: Akademia e Shkencave e RPS të Shiqipërisë, 1985.

Echeverria Rey, Fernando. "Hoplite and Phalanx in Archaic and Classical Greece." Classical Philology 107, no. 4 (October 2012): 291—318.

Feil, Florian. "A Note on the *sibyna*, an Illyrian Weapon Similar to the Boar Spear." *Iliria* 46 (2022): 67—83.

Fluß, Max. "Illyrioi." In *Paulys Realencyclopädie der classischen Altertumswissenschaft: Supplementband 5*, edited by Wilhelm Kroll, 311—45. Stuttgart: Metzler, 1931.

Gabriel, Richard. *Philip II of Macedonia: Greater than Alexander.* Washington, D.C.: Potomac, 2010.

Gitti, Alberto. *La politica dei re Illiri e la Grecia fino all'intervento Romano*, Milan: Popolo d'Italia, 1936.

Greenwalt, William. "Macedonia, Illyria and Epirus." In *A Companion to Ancient Macedonia*, edited by Joseph Roisman and Ian Worthington, 279—305. Malden, M.A.: Wiley-Blackwell, 2010.

Griffith, Timothy. *Mercenaries in the Hellenistic World.* Cambridge: University Press, 1935.

[106] *Cf.* also Feil, "*sibyna*," 78-79.

—. "MAKEΔONIKA: Notes on the Macedonians of Philip and Alexander." *Proceedings of the Cambridge Philological Society* 184 (1956): 3—10.

Hammond, Nicholas. "Alexander's Campaign in Illyria." *Journal of Hellenic Studies* 94 (1974): 66—87.

—. "The Battle between Philip and Bardylis." *Antichthon* 23 (1989): 1—9.

—. *The Macedonian State: Origins, Institutions, and History*. Oxford: Clarendon, 1989.

—. "What May Philip Have Learnt as a Hostage in Thebes?" *Greek, Roman and Byzantine Studies* 38, no. 4 (October 1997): 355—72.

Hammond, Nicholas and Timothy Griffith. *A History of Macedonia: Vol. 2: 550—336 B.C.* Oxford: Clarendon, 1979.

Hammond, Nicholas and Frank Walbank. *A History of Macedonia: Vol. 3: 336—167 B.C.* Oxford: Clarendon, 1988.

Hornblower, Simon. *A Commentary on Thucydides: Vol. 2: Books IV—V.24*. Oxford: Clarendon, 1996.

Howe, Timothy. "Plain Tales from the Hills: Illyrian Influences on Argead Military Development." In *The History of the Argeads: New Perspectives*, edited by Sabine Müller, Tim Howe, Hugh Bowden and Robert Rollinger, 99—111. Wiesbaden: Harassowitz, 2017.

Kagan, Donald and Gregory Viggiano. "The Hoplite Debate." In *Men of Bronze: Hoplite Warfare in Ancient Greece*, edited by Donald Kagan and Gregory Viggiano, 1—56. Princeton: University Press, 2013.

Kucewicz, Cezary, Matthew Lloyd and Roel Konijnendijk. "'Not Many Bows'? Light-armed Fighters of the Tenth through Fourth Centuries." In *Brill's Companion to Greek Land Warfare Beyond the Phalanx*, edited by Roel Konijnendijk, Cezary Kucewicz and Matthew Lloyd, 205—35. Leiden and Boston: Brill, 2021.

Launey, Marcel. *Recherches sur les armées hellénistiques*. Paris: Boccard, 1949.

Le Bohec, Sylvie. "Les soldats illyriens au service des rois de Macédoine." In *L'Illyrie méridionale et l'Épire dans l'Antiquité: Vol. 2*, edited by Pierre Cabanes, 225—30. Paris: Boccard, 1993.

Lippert, Andreas and Joachim Matzinger, *Die Illyrer: Geschichte, Archäologie und Sprache*. Stuttgart: Kohlhammer, 2022.

Schwartz, Adam. *Reinstating the Hoplite: Arms, Armour and Phalanx Fighting in Archaic and Classical Greece*. Stuttgart: Franz Steiner, 2009.

Snodgrass, Anthony. *Arms and Armour of the Greeks*. Ithaca: Cornell University Press, 1967.

Stipčević, Aleksandar. *The Illyrians: History and Culture*. Translated by Stojana Čulić Burton. New Jersey: Noyes, 1977.

Walbank, Frank. *A Historical Commentary on Polybius: Vol. 1: Commentary on Books I—VI*. Oxford: Clarendon, 1957.

Wilkes, John. *The Illyrians*. Oxford: Blackwell, 1992.

Wrightson, Graham. *Combined Arms Warfare in Ancient Greece: From Homer to Alexander the Great and his Successors*. London and New York: Routledge, 2019.

—. *The Battles of Antiochus the Great: The Failure of Combined Arms at Magnesia that Handed the World to Rome.* Barnsley: Pen and Sword, 2022.

—. "The Macedonian Heavy Infantry of Philip II and Alexander the Great: Professionals in a World of Amateurs." In *Warfare in the Ancient Mediterranean World*, edited by Edward Anson. Leiden: Brill, 2024 (in press).

A New Interpretation of the Battle of Panion

Graham Wrightson

Polybius' description of the Battle of Panium is mainly a critique of the description of another historian, Zeno's account of the same battle. In modern analyses of the battle scholars follow Polybius' description and continue to discount Zeno's version of events.[1] Yet in the context of the

[1] Belazel Bar Kochva, *The Seleucid Army* (Cambridge: CUP, 1976),146-57 has the fullest discussion in sticking with Polybius' criticisms and is the one on which all other scholars still rely. Paul Johstono, "No Strength to Stand": Defeat at Panium, the Macedonian Class, and Ptolemaic Decline," in *Brill's Companion to Military Defeat in Ancient Mediterranean Society*, edited by Jessica H. Clark, and Brian Turner, (Leiden: Brill, 2017), 162-187 continues to follow Bar Kochva in arguing for the battle taking place either side of the river (on which more later) without ever explaining why. No other detailed reconstructions take place within depth analysis. Mike Roberts, and Bob Bennett, *Twilight of the Hellenistic World* (Barnsley: Pen and Sword, 2012), 152-4 discuss Bar Kochva's analysis without ever addressing his problematic assumptions in detail. They do, however, include a different location for the battle, though in doing so ignore Zeno that the Egyptian right wing was on a hill, the very information that influenced Bar Kochva's description. Other scholars do not really attempt to reconstruct the battle: Werner Huss, *Agypten in hellenistischer Zeit 332–30 v. Chr.* (Munchen: C.H.Beck, 2001), 490–91; Michael Taylor, *Antiochus the Great* (Barnsley, UK, 2013), 89–92; John Grainger, *The Syrian Wars*. Mnemosyne suppl. 320. (Leiden: Brill, 2010) 257–60. Rather they focus on the significance of the defeat for the Egyptians: Maurice Holleaux, *Etudes d'epigraphie et d'histoire grecques, III: Lagides et Seleucides.* (Paris: Adrien-Maisonneuve, 1942), 317–35; Edouard Will, *Histoire politique du monde hellenistique*. 2 vols. (Nancy: Points, 1967), II.108–12, 118–20; Huss, 489–501; Günther Hölbl, *History of the Ptolemaic Empire.* Trans. T. Saavedra. (New York: Routledge, 2001), 136–37; Arthur Eckstein, *Mediterranean Anarchy and the Rise of Rome* (Berkeley: UCP, 2006), 292–94; Grainger, 245–71; Dov. Gera, "Ptolemy Son of Thraseas and the Fifth Syrian War," *AncSoc* 18 (1987): 63; Pierre Jouguet, *Macedonian Imperialism, and the Hellenization of the East*, Trans. M. R. Dobie. Reprint. (New York: Routledge, 1996), 186; R. Malcolm Errington, *A History of the Hellenistic World, 323–30 BC* (Malden, MA: Blackwell, 2008), 196.

many other Hellenistic battles that saw armies deploy elephants, cavalry and a sarissa phalanx together, Panium appears to be no different. Despite Zeno's confusing narrative, and despite Polybius' claims to the contrary, we can piece together the armies and the events of the battle as described by Zeno. Through a comparative analysis of the theory of combined arms in Hellenistic armies, we can rehabilitate Zeno and deliver him safely from Polybius' withering critique.

Polybius uses Zeno's description of the battle of Panium as an example of the mistakes other historians make. He has four main criticisms:

1. Zeno incorrectly adds into the battle two sons of Antiochus, both also named Antiochus.
2. Zeno is incorrect to state that the cavalry and elephants were stationed directly in front of the phalanx and yet that the elephants received the phalanx as it was pushed backwards.
3. Zeno states that Scopas fled when he saw Seleucid cavalry returning in the rear but that he was also the last to leave the field.
4. Zeno states that the Aetolian cavalry on the Egyptian left wing was put out of action when attacked by elephants because they were unused to the sight of them but had remained unbroken since the beginning.

All of these criticisms have a reasonable explanation when placed in the context of the standard deployment of Macedonian style armies based on the sarissa phalanx fighting in a system of combined arms alongside elephants, light infantry, elite heavy infantry and heavy and light cavalry.[2]

[2] On how Hellenistic armies fought and functioned see especially Christopher Matthew, *An Invincible Beast: Understanding the Hellenistic Pike Phalanx in Action* (Barnsley: Pen and Sword, 2015); Graham Wrightson, "Macedonian Armies and the Perfection of Combined Arms," in T. Howe, E. Garvin, and G. Wrightson eds. *Greece, Macedon and Persia: Studies in the Social, Political and Military Consequences of Conquest Societies in Honour of Waldemar Heckel* (Oxford: Oxbow, 2015), 59-68. On how this applies to Seleucid armies see Graham Wrightson, *The Battles of Antiochus the Great: The Failure of Combined Arms at Magnesia that Handed the World to Rome* (Barnsley: Pen and Sword, 2022), xi-11. See also Graham Wrightson, *Combined Arms Warfare in Ancient Greece from Homer to Alexander the Great and his Successors* (Oxford and New York: Routledge, 2019) for the development of combined arms in Greek warfare culminating in its perfection under Alexander and his Successors.

Let us examine the criticisms by following the events of the battle as described by Zeno.

Polybius begins his description (16.18-19) of Zeno's account:

> He has, however, paid so little attention to facts that his recklessness and lack of experience are again unsurpassed. 4 Undertaking in the first place to describe Scopas's order of battle he tells us that the phalanx with a few horsemen rested its right wing on the hills, while the left wing and all the cavalry set apart for this purpose stood on the level ground.

This is a normal deployment for Hellenistic armies with phalanx in the center flanked by cavalry. Scopas choosing to maximize the topography by securing his right flank on some hills is a time-honored military stratagem. So far so good, nothing unusual here from Zeno, in fact it is as expected.

Polybius continues:

> 5 Antiochus, he says, had at early dawn sent off his elder son, Antiochus, with a portion of his forces to occupy the parts of the hill which commanded the enemy, and when it was daylight he took the rest of his army across the river which separated the two camps and drew it up on the plain, placing the phalanx in one line opposite the enemy's center and stationing some of his cavalry to the left of the phalanx and some to the right, among the latter being the troop of mailed horsemen which was all under the command of his younger son, Antiochus.

Again, the Seleucid deployment is exactly what you would expect. Antiochus III placed his phalanx opposite the enemy phalanx, no surprises there, and stationed cavalry on either side of the phalanx to protect its flanks just as generals did in every single Hellenistic battle. The only unusual deployment is Antiochus sending his son at dawn hours before the battle to occupy commanding hills. This is again a strategically sensible decision to make use of the topography and race to take the high ground before the enemy. There are hundreds of battles globally throughout history where generals sought to gain the same advantage and made the exact same decision.[3]

---

[3] There are many battles that feature a hill in their name, perhaps most famously to Americans Bunker Hill or Hamburger Hill, and of those battles without hill in their name Gettysburg. For Canadians it is the Battle for Hill 70 in World War 1. In England perhaps Edgehill is the most significant battle named after a hill, and of course at the Battle of Hastings a hill was of crucial tactical importance. In Greek

Let us examine that topography. This is how Zeno through Polybius describes the two armies, accepting that the elephants and associated troops were on the flank of the phalanx not in front.

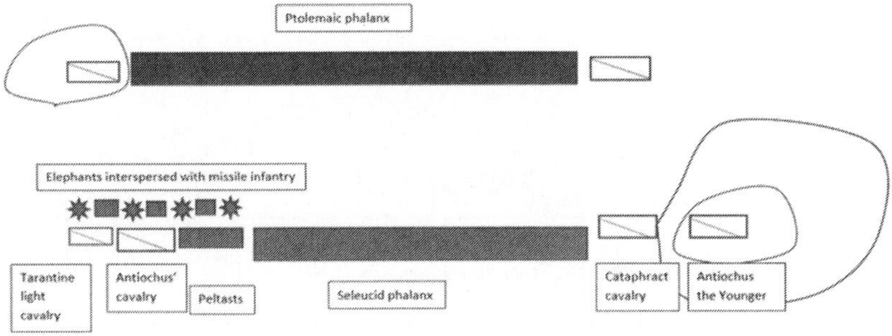

Fig. 1: Army dispositions

Antiochus, coming from the northeast, crossed the river and occupied the foothills with his right wing. Getting there first, or surprising Scopas by sending his son to cross the river at night, Antiochus gained control of the high ground at the foot of Mount Hermon sloping down to the Egyptian position in the plain below. Scopas also secured his right flank on some high ground.

Looking at the topography of Panium and the Banias river, Scopas likely took position resting his flank on the site of the modern village of Snir. This is on a hill, as Zeno through Polybius describes the Egyptian right wing. It prevented Antiochus advancing further south into the plain, secured the army's right flank on the precipitous riverbank, and preserved an avenue of retreat southwards (see the map in figure 3, below).

warfare, at the Battle of Cynoscephalae Philip V's enthusiasm to beat the Romans to the high ground caused a disastrous split in his phalanx and cost him the battle.

Fig. 2: The precipitous banks of the river facing north from Snir.

Moreover, Zeno's description implies that Scopas was surprised by Antiochus sending a force to occupy the heights. If this was the case Scopas perhaps planned to use them himself to hem Antiochus in by the river and thus maximize the advantage of his smaller army by minimizing the advantage of Antiochus' cavalry and elephants. That possible plan is represented by the green line on the map below with both flanks placed on elevated ground.

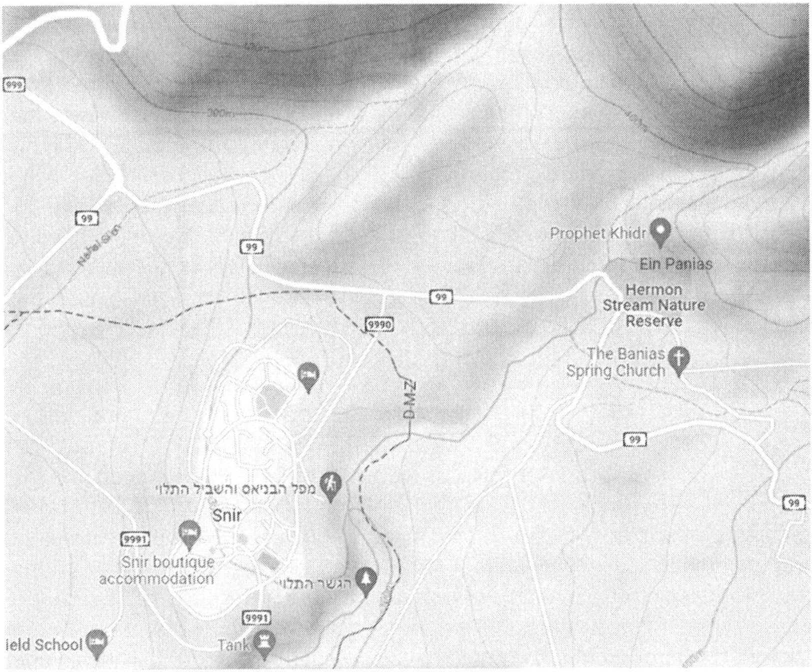

Fig. 3: Map of the topography of the battle site with the line representing where Scopas could have drawn up his line to block Antiochus and rest both his flanks on hills.

This position was raised and protected from outflanking by the extremely difficult terrain of the river valley (see Figure 1 above). Unfortunately for Scopas, Antiochus moved his cavalry to the hills at night and thus forced Scopas to set up his left-wing position back down in the plain. This ceded the right flank advantage to Antiochus and is what eventually decided the battle.

Polybius (12.19.7) states that ten stades (roughly 6000 feet) holds a phalanx of 16,000. But this is in its most open order as described in the tactical manuals. Asclepiotous (4.3) is clear that when marching on the enemy the phalanx was forced into a gap of 3 feet for each rank, but took up an even tighter 1.5 feet in a stationary formation when the enemy is attacking. Having marched a unit of volunteers armed with sarissas up hills and over ditches, I can confirm that the manuals are right that marching with 3 feet gaps is achievable when the sarissa is lowered in attacking formation.[4] The body naturally turns sideways when holding the sarissa with two hands across the body with feet facing 90 degrees from the front of the unit, and so takes up less space. When marching it is natural to turn your hips and walk normally with the sarissa resting on your right hip. The open order of 6 feet as Asclepiodotus records (4.3), is the more natural order when marching since it prevents soldiers and sarissas becoming entangled and this must be the organization to which Polybius refers here. But when closing on the enemy, marching in open order with spaces of six feet per man is extremely open and would create gaps for the enemy to run into – something that Polybius (18.31.5-6) himself famously states was disastrous for a phalanx.

As with everything, scholars usually take Polybius at his word, yet he betrays constant confusion about exactly how a sarissa phalanx fought in battle.[5] Moreover his description of the width of the Macedonian battle line is part of his criticism of Callisthenes' account and Polybius completely misunderstands how a phalanx marched to engage the enemy. He has no concept that 6 feet per man is far too open a formation for a battle and that the description only covers a march in open order not the later fighting in close order. Even when Polybius argues for the phalanx adopting a close order (12.21.3) he calculates the phalanx with a breadth of 20 stades, yet in close order a phalanx of 16,000 would only occupy 5 stades. Polybius is certainly misled by Callisthenes' exaggerated numbers, but it is also possible that he simply does not understand how a phalanx marches and deploys to battle otherwise he could easily correct Callisthenes as modern historians do.[6] Callisthenes does not mention a change to close order for battle presumably because his audience of soldiers who actually fought in a phalanx regularly already knew that would happen.[7]

---

[4] For more on these experiments and for photographs see Wrightson 2022.

[5] See recently for example Paul Johstono, *The Army of Ptolemaic Egypt 323 to 204 BC: An Institutional and Operational History*. (Barnsley: Pen and Sword, 2020), 226. He includes Polybius' statement without even any discussion to the contrary. See Wrightson, *The Battles of Antiochus* for preliminary criticisms of this famous passage in Polybius.

[6] Walbank (Frank Walbank, *A Historical Mommentary on Polybius. Vol. 2.* (Oxford and New York: OUP, 1967), 2.373-4) has an excellent discussion of Polybius' misunderstandings, the terrain, and Callisthenes' exaggerated numbers.

[7] Graham Wrightson, "The Nature of Command in the Macedonian Sarissa Phalanx," *AHB* 24 (2010): 80-83 gives a detailed analysis of Polybius' misunderstanding.

The map in Figure 4 (below) illustrates where the two battle lines drew up based on Zeno's description. I have argued elsewhere that a sarissa phalanx of 16,000 took up just over 3000 feet (roughly five stades) in close order (1000 ranks of 16 with 3 feet space for each rank).[8] That is just over half a mile in width. Johstono argues that the two phalanxes numbered closer to 25,000 at Panium.[9] If that is the case, then the phalanxes took up 4,687 feet in close order (as they would be when fighting each other). Johstono states that the phalanx took up 1600 meters, or 5,250 feet.[10] That gives space for a phalanx of 28,000 fighting in 1,750 ranks of 16 with 3 feet space for each rank, not the 25,000 Johstono describes.

Polybius (12.18.3-6) describes a cavalry squadron drawn up 8 riders deep and that an equal gap must exist between squadrons for wheeling. He states that one stade (roughly 600 feet) is enough for only 800 horsemen. Yet as Walbank rightly points out, there is no other evidence for squadrons leaving such large gaps between them, and Polybius is likely simply incorrectly copying a Roman manipular formation.[11] Thus, without the gap between each squadron that Polybius assumes, 800 cavalry would occupy 300 feet width per squadron – half of Polybius' width because there is no equal gap. Thus, cavalry in 100 ranks of 8 takes up about 3 feet per rank. If we take Antiochus' largest extant troop totals at Magnesia of around 12,000 cavalry,[12] they would need 4,500 feet total. It is more likely his cavalry was somewhere in between this and the total at Raphia a few years before of 6,000. After this battle he added a force of cataphracts clearly described as a decisive force at Panium, which numbered 6,000 at Magnesia and 1,500 at Daphnae. If we posit 10,000 cavalry of Antiochus at Panium (since his left wing force was only a small squadron of Tarantines and his 2,000 strong Royal cavalry), these would need 3,750 feet. In my view this is an overestimate of the size of the cavalry at Panium, but it demonstrates that the two armies could fit entirely on the plain below the foothills of Mount Hermon.

---

[8] Wrightson "Nature of Command," 80.

[9] Johstono "Defeat at Panium," 167-8.

[10] Johstono "Defeat at Panium," 168.

[11] Walbank 2.369.

[12] See Johstono "Defeat at Panium," 167 provides a useful table listing all the cavalry in Antiochus' army.

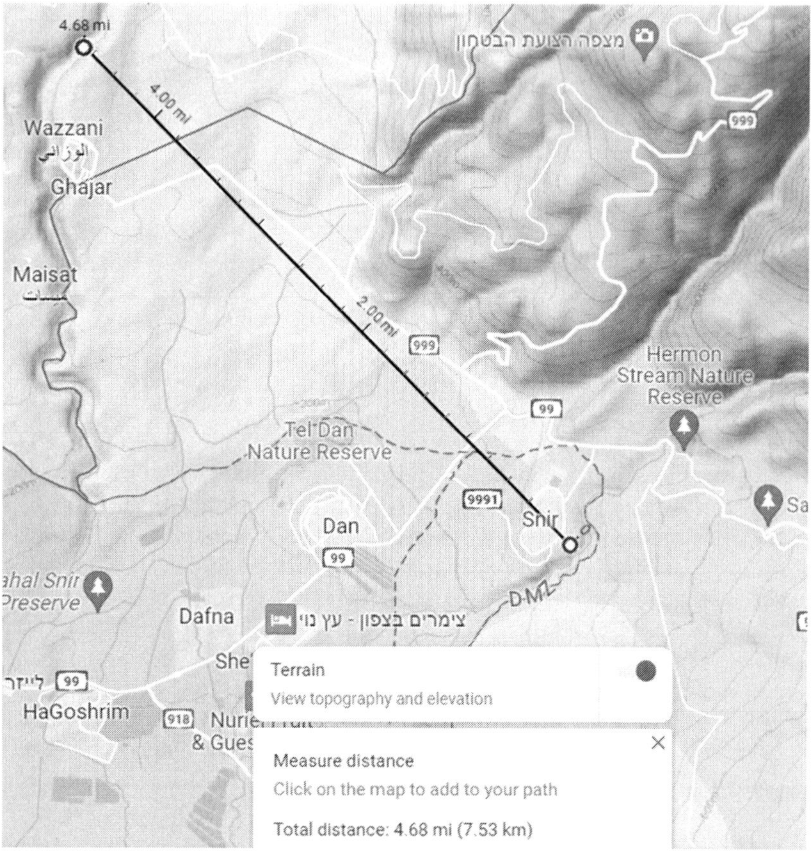

Fig. 4: Map of the plain showing the line from the steep slopes beside Snir to the large rise at the other end of the plain near Ghajar. This shows that the entire plain has a width of 4.68 miles and can accommodate a large battle line.

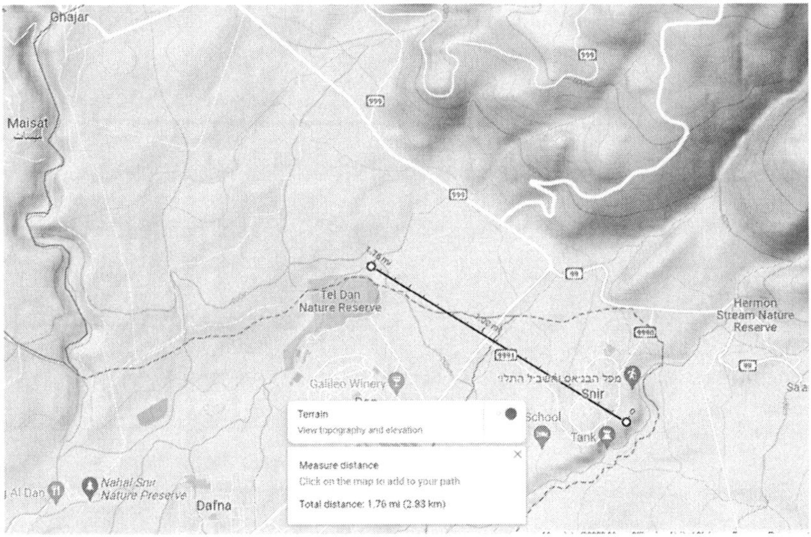

Fig. 5: Map of the plain from the steep slopes beside Snir to the slight hill at Tel Dan that encompasses the eastern half of the plain. This shows that half the plain has a width of 1.76 miles or 2.83 km and can accommodate a large battle line.

The battle lines as drawn on Figure 6 (above) take up around 8,250 feet or 2.5km, just over one mile and three quarters, which would be more than enough space to deploy a phalanx with cavalry and elephants on each wing. A phalanx of 25,000 taking up 4,687 feet in close order would give 3,562 feet for the cavalry on each wing of the Egyptian battleline. Roughly 1,781 feet or 545 meters on each wing.

The Seleucids approached from the northeast and camped by the river. Antiochus' son then took the heavy cavalry to cross the river at night and claim possession of the high ground on the slopes of Mount Hermon. The rest of the Seleucid army drew up to the east with the left wing bounded by the steep banks of the river.

The Egyptian position allowed for the right wing to enjoy the protection of the steep riverbank. Scopas could command the plain and deny Antiochus open passage into the rest of Syria. The problem is that this position was lower than Antiochus', thus ceding him the advantage in the battle of fighting down a gentle slope. Scopas likely decided that he did not have enough cavalry to face Antiochus equally on both flanks. So, he determined that using the precipitous riverbank to protect his right flank, he would be able to station more cavalry on his left wing in the open plain.

As is evident on the map, this position also gave the advantage of a bend in the river. This river bend would force Antiochus to shorten his line only to have to extend it once he got past the bend. This would give Scopas the opportunity to send cavalry to outflank the Seleucid line or at least force Antiochus into a last-minute positional adjustment before the armies engaged.

With the left flank of the Egyptian line ending at Tel Dan, the small river makes a natural boundary. The rise at Tel Dan provides a useful elevated position on the flank and at this point the plain curves to the west with the elevation decreasing to the south behind the Egyptian lines providing a clear avenue of retreat.

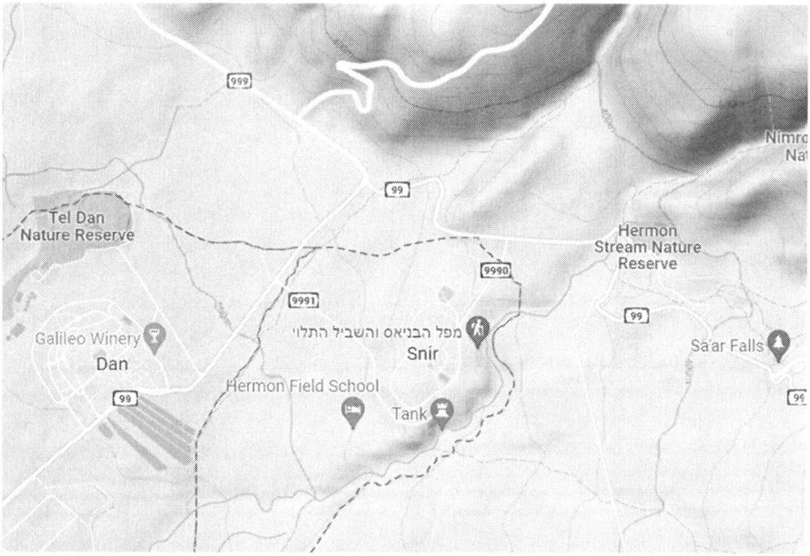

Fig. 6: The two battle lines with the Egyptian line in red and the Seleucid in blue. The Egyptian right-wing rests on the precipitous slope around Snir and the left at the river near Tel Dan where the plain curves to the west. The Seleucid right rests on the slopes of Mount Hermon. The Seleucids approached from the northeast and camped on the other side of the river, probably between Sa'ar Falls and the Hermon stream Nature reserve.

Bar Kochva, in his seminal work on the Seleucid Army, argues based on one word of Polybius' account of Zeno, ὑπωρείας - foothills, that the Egyptian army drew up on either side of the river with their right flank resting on foothills.[13] Most scholars follow Bar Kochva's argument in reconstructing the battle taking place on both sides of the river, including Johstono's recent article on the battle.[14]

I agree that the only substantial foothills in the area are on the eastern side of the river, but there is nothing in Zeno's account as described by Polybius that suggests there was any separation between the armies let alone that they fought on two sides of a river. Moreover, other than the special case at the Battle of Sellasia, there is no other instance of a Macedonian style army fielding a sarissa phalanx choosing to deploy on both sides of a river.[15] It would serve no tactical or strategic purpose to do so and would hamstring the coordination of the battle as it did at Sellasia.[16] Why would Antiochus choose to send cavalry over the river when he could just face and defeat each half of the Egyptian army in turn on either side of the river?

There is no strategic reason for Scopas to set up the army on either side of the river. His entire goal was to face Antiochus in battle and prevent him passing deeper into Egyptian controlled Syria. He could choose the battlefield and the positioning. There is no way Antiochus would have marched past Scopas' position without facing him since if he did Scopas could then cut his communication, supply lines and line of retreat north. Thus, Antiochus arrived at the river and learned of Scopas' position on the other side. He correctly assessed that the high ground over the river was advantageous, that there were two areas of rising ground, and that if he ceded command of both to Scopas, victory would be more difficult to obtain. To even the odds Antiochus determined to seize the heights in the north, the direction from which he came. By doing so both armies ended up with their right flank resting on higher ground, though the hills on

---

[13] Bar Kochva, 146-157. River discussion: 150-155.

[14] Johstono, "Defeat at Panium." Walbank's commentary does not discuss the problem (2.523). Johstono's impressive work on the Egyptian army unfortunately does not go as far as Panium and ends with the Battle of Raphia in 217 despite the chapter title listing 204 as the end point, Johstono, *Army*, 227-262.

[15] Cleomenes did so at Sellasia to separate the Macedonian army and as the weaker army defend a natural chokepoint. See Joseph Pietrykowski, *Great Battles of the Hellenistic World* (Barnsley: Pen and Sword, 2009), 170-180.

[16] As one might expect, at Sellasia the weaker half of the Spartan army collapsed in isolation even before the stronger part led by Cleomenes had joined battle with the Antigonid phalanx: Polyb. 2.65-9.

Antiochus' right were steeper and more significant. This led to the standard Hellenistic deployment of each side relying on their right-wing cavalry for victory in the battle charging down from an elevated position.

Johstono follows Bar Kochva almost directly in placing the lines north-south across the river and having a 4km long battle line, following to the letter Polybius' description of a phalanx in open order. Yet the Google satellite image he provides of the battlefield (Figure 7 below) shows that if the lines deployed northwest by southeast they could easily fit on the plain. The lines are my own mirroring those on Figure 6 above.

Fig. 7: Johstono's map of the battlefield from Mount Hermon looking South.

From this image it is clear to see that Zeno's description of the Egyptian right wing resting on hills is valid. You can clearly see the raised ground on the right end of the red line, at the modern village of Snir. I firmly believe there is no reason at all to suggest the battle took place on either side of the river. This makes the engagement fit the established pattern of all other Hellenistic battles involving sarissa phalanxes.

Roberts and Bennett argue that the two battle lines drew up on a north-south axis after the Seleucid army crossed the river but they fail to account in this position for the hills on the Egyptian right wing that Zeno clearly describes.[17] They do, however, as I do, provide an argument for discounting

[17] Roberts and Bennett, 152-4.

Bar Kochva's theory of the river in between "based on the fact that it would be very unconventional for a battle in this era to be fought in two parts separated by a river."[18]

Polybius continues with the Seleucid army deployment:

> 7 Next he tells us that the king posted the elephants at some distance in advance of the phalanx together with Antipater's Tarantines, the spaces between the elephants being filled with bowmen and slingers, while he himself with his horse and foot guards took up a position behind the elephants.

This Seleucid deployment again follows exactly standard practice. The elephants are stationed as a flank guard of the phalanx with light cavalry to their left. Missile infantry stand between the elephants as extra defense, just as they did in the enemy line at Alexander the Great's first encounter with elephants at the Battle of the Hydaspes in India.[19] The king, Antiochus, with his two elite units of cavalry and infantry stood behind the elephant screen ready to attack the enemy line wherever and whenever an opportunity arose. This is the standard deployment of almost every Hellenistic army.

In fact, Antiochus' deployment on the left flank actually mirrors his deployment at Raphia, with his personally led horse guards immediately behind a flank guard of elephants.[20] The difference this time is that he also stationed missile infantry between the elephants for extra flexibility and had light cavalry, the Tarentines, to shield the flank of the elephants and heavy cavalry when they charged. Such a deployment of missile infantry, light cavalry, elephants, and heavy cavalry had become standard in Hellenistic armies of the late third century.

Antiochus reinforced his left flank personally with his royal guard units and elephants knowing that the main Egyptian thrust would come against that position. Furthermore, he reduced the total frontage of his left-wing by

---

[18] Roberts and Bennett, 155.

[19] Arr. *Anab.* 5.11-18; Plutarch *Alex.* 60; Diod. Sic. 17.87-88; Curt. 8.13-14; Polyaenus, *Strat.* 4.3.22. For the best discussion of elephants in the battle see Jenna Rice, "Animals in Ancient Greek Warfare: A Study of the Elephant, Camel, and Dog" (PhD. diss., University of Missouri – Columbia, 2020), 44-8.

[20] Polyb. 5.82-6. For descriptions of the battle see Pietrykowski, 181-194; Roberts and Bennett, 80-100. See also Johstono's exhaustive work on the Egyptian army at Raphia, Johstono, *Army*, 227-262; and on the Seleucid side recently Wrightson, *The Battles of Antiochus*, 12-31.

placing his cavalry and elite heavy infantry behind the elephants. If Scopas did take up position on the riverbank, Antiochus' flank would have abutted the river and reduced the width of the battle line. It also prevented Antiochus from being able to send cavalry to outflank Scopas' position. This is likely why Scopas would have chosen this location.

This topographical location also explains perhaps why Antiochus needed to tempt the Egyptian phalanx out of position and make a gap for his elephants and cavalry to charge into. Unlike in battles on flat plains, the river prevented space to send the cavalry on a charge on the flank of the enemy and thus Antiochus had to come up with another way to try to expose the flank of the enemy phalanx.

The only unusual thing, and something that seems to confuse Polybius, is that Zeno describes the elephants as standing some distance 'in front of the phalanx.' Polybius takes this to mean that Zeno is completely wrong since it makes no sense for elephants and cavalry to fight directly in front of the phalanx. But Zeno may not have said that exactly. The Greek Polybius uses is προτάξαι.[21]

In no battles for which we have a detailed description do historians state what happens with elephants stationed in front of the phalanx. The battles of Paraetacene, Gabiene, Ipsus and Magnesia describe elephants drawn up in front of the phalanx,[22] just as Zeno's description at Panium. Yet none of these battles, which are otherwise (excepting Ipsus) described clearly in our sources, provide a description of how these two elephant contingents fought

---

[21] Walbank does not include a discussion of the term, but he does rightly caution against evaluating Polybius' "criticism of Zeno's account of Panium simply on the basis of what P. says; he may have read Zeno carelessly (e.g. on the two Antiochi (18. 8, 19. 9) or what happened to Antiochus' light-armed, cavalry, and elephants before the phalanxes met (18. 10))" Walbank, 2.523.

[22] The accounts of the two battles of Paraetacene and Gabiene describes the elephants in front of the phalanx, but does not say what happens to them in the battle before, during, or after the two phalanxes engaged. Paraetacene (Diod. Sic. 19.26-31) and Gabiene (Diod. Sic. 19.39-43; Plut. *Eum.* 16). For the fullest secondary discussion of these battles see A. M. Devine, "Diodorus' Account of the Battle of Paraitacene (317 B.C.)," *Ancient World* 12 (1985): 75–86; and A. M. Devine, "Diodorus' Account of the Battle of Gabiene," *Ancient World* 12 (1985): 87–96. See also Pietrykowski, 91-112. Ipsus (Plut. *Demetr.* 28-9; App. *Syr.* 55.). Magnesia (Livy 37.40; App. *Syr.* 7.32).

and where the large beasts went when the two phalanxes engaged directly in the center.

I have argued elsewhere concerning Ipsus that elephants drawn up in front of Sarissa phalanxes makes absolutely no tactical or strategic sense.[23] Champion agrees in his analysis of Paraetacene and doubts the presence of elephants there at all.[24] I have also argued elsewhere that the Roman descriptions of elephants in front of the phalanx at Magnesia refer to the elephants shielding the phalanx as it formed a defensive square to retreat to the camp rather than an initial deployment of phalanx units on either side of the elephants.[25]

Other historians, when attempting to reconstruct the events of this battle, take Polybius at his word for his criticisms of poor old Zeno. They too then try to explain like Polybius how the phalanxes of each side can fight against each other when the Seleucid cavalry, elite infantry, missile infantry, Tarantines and elephants fought in front of them. Usually, the explanation is the creation of gaps in the Seleucid phalanx to allow the elephants, cavalry and light troops through before they engaged with the Egyptian phalanx, or as they retreated. Yet just as discussed by Polybius,[26] such a maneuver would be foolhardy to say the least.

The strength of the phalanx was in presenting an unbroken line of sarissas to the enemy. Any gaps in the formation of the phalanx could prove fatal, as so famously happened at Pydna.[27] Only on one occasion, at Gaugamela, does an ancient historian describe the phalanx making lanes in its formation, but this is almost certainly a mistake.[28] Moreover, any gaps big enough to let through a hundred elephants as well as accompanying soldiers would be too large to close again before the enemy phalanx or missile troops covered

---

[23] Wrightson, "Macedonian Armies," 66.

[24] Jeff Champion, *Antigonus the One-Eyed: Greatest of the Successors* (Barnsley: Pen & Sword, 2014), 57.

[25] Wrightson, *The Battles of Antiochus*, 126-9.

[26] Polyb. 18.31.

[27] See Paul Johstono and Michael Taylor, "Reconstructing the Battle of Pydna," *Greek, Roman, and Byzantine Studies* 62 (2022): 44–76; Graham Wrightson, *The Third Macedonian War and the Battle of Pydna: Perseus' Neglect of Combined-arms Tactics and the Real Reasons for the Roman Victory* (Barnsley: Pen and Sword, 2023).

[28] See Waldemar Heckel, Carolyn Willekes and Graham Wrightson, "Scythed Chariots at Gaugamela," in E. Carney and D. Ogden, eds. *Philip II and Alexander the Great: Father and Son, Lives and Afterlives.* (Oxford: OUP, 2010), 103-113.

the distance between the lines. And what would be the point of missile troops stationed with the elephants if the elephants were to march through gaps in the phalanx? Did the light infantry follow through the same gaps after the elephants or go through before the elephants? Polybius makes no mention of this, but it is important to consider in order to get a complete picture of events.

We also have no statement that the Seleucid phalanx fled in disorder, which would have allowed elephants to come through their line. Zeno's description simply states that "the elephants received the Seleucid phalanx and did great service."[29] In the context of a Hellenistic battle, what greater service is there to a phalanx than performing a devastating flanking attack on the enemy phalanx just as it bears down? This assistance must have taken place on the flank of the phalanx not in its front. A frontal assault to the phalanx would not have been disastrous since the dense hedge of sarissas facing the elephants would disconcert the animals just as it did in preventing horses from charging a phalanx head on. Even the first time a phalanx faced elephants, at Alexander's Battle of the Hydaspes, the length of the sarissas disconcerted the elephants and the phalanx remained unbroken.[30] Jenna Rice argues convincingly that elephants as herd animals were a danger to friends more than foe when faced with an obvious threat that would prompt the lead animal to hesitate.[31]

Nor is there any reason evident in the rest of Zeno's account as preserved by Polybius as to why Antiochus would use such an unusual organization. He was not facing any enemy elephants directly in opposition, there were no cavalry or skirmishers in front of the Egyptian phalanx that the elephants could disperse, and the elephants would not charge the enemy phalanx directly in order to create confusion unless Antiochus did not expect his own phalanx to be able to survive against the Egyptian.

It is impossible to find any comparison in Hellenistic warfare that provides a reason as to why Antiochus would put elephants directly in front of the phalanx. The closest is the Indian deployment against Alexander the Great at the Hydaspes where Porus stationed his infantry in between the elephants all along his line.[32] But he did this because elephants were the main method

---

[29] Polyb. 16.19.1

[30] See Rice, 44-8.

[31] Rice, 74-81.

[32] Arr. *Anab*. 5.11-18; Plut. *Alex*. 60; Diod. Sic. 17.87-88; Curt. 8.13-14; Polyaenus, Strat. 4.3.22.

of warfare in India and he perhaps knew that his infantry would not stand up to facing the sarissa phalanx alone. Even at that battle the elephants were not able to break up the phalanx with their attacks despite the advantage of the Macedonian phalanx never having faced elephants before. At the Battle of the Hydaspes the elephants did charge Alexander's phalanx, but despite creating gaps and causing damage it did not take long for the sarissas and missile troops to put a large number of elephants out of action and turn them disastrously against their own side's troops in mad panic.[33] Jenna Rice argues convincingly that Alexander chose not to use his elephants in battle because the danger they posed to the order of his army in a charge was almost as great as any damage they might cause to the enemy.[34]

Polybius is not alone in becoming confused by Zeno's account. Modern historians believe Polybius' criticisms of Zeno entirely and reconstruct the battle with elephants in front of the Seleucid phalanx. Historians try and justify this weird positioning without finding a satisfactory answer. Just as Ockham's razor suggests, the simplest and best solution to this problem is that Zeno never meant that the elephants were directly in front of the phalanx. He likely said they were ahead of it, but meant off to the side in their usual position.

In my view, all the descriptions of elephants stationed in front of the phalanx merely depict a screen while the phalanxes deployed on the battlefield.[35] This would prevent the enemy skirmishers or cavalry from harassing the important and cumbersome phalanx when it was vulnerable in an open order while marching. In the Hellenistic battles where both armies fielded almost identical troops, each side sent the elephants ahead to screen the battlefield deployment of the phalanx. Before the battle began these elephants then marched to their positions on the army's flanks to act as a defensive shield for the cavalry and light troops and the phalanx in the center. All our surviving accounts describe the ensuing elephant-on-elephant combat on the

[33] See Rice, 46-8.
[34] Rice, 48-9.
[35] Most armies used a screen to shield their deployment and prevent the enemy attacking when their units were not ready. Elephants would make a perfect army screen as their size prevented any enemy penetration to attack the units while deploying. Once the army was ready the elephants would move to their positions on the wings. The later sources misunderstood what their sources described and erroneously stationed elephants in front of the phalanx for the whole battle instead of just at the beginning during deployment. Alternatively, it is possible that Roman sources familiar with the Carthaginian uses of elephants in the centre of their whole line in front of their infantry described Hellenistic armies in the same way.

wings (such as at Paraetacene and Gabiene) but never in the center in my view because there never were elephants in the center in front of each phalanx.

Thus, perhaps Zeno's much criticized description actually preserves the reason for this confusion. The elephants drew up before the phalanx while it was deploying, and then moved to the flank. Perhaps Zeno's use of this word follows standard descriptions of the other battles that seem to suggest elephants stood in front of phalanxes. They drew up there initially but then moved aside later. It is also possible that elephants described as stationed in front of infantry actually stood only in front of non-sarissa-armed infantry. This would make sense as it added extra frontal defense for those infantry and extra flank defense for the phalanx in the center, but did not impede the phalanx itself.

Even if elephants did stand in front of the phalanx at the start of the battle, on no other occasion other than here at Panium does a historian describe elite heavy infantry and cavalry also standing between the elephants and the phalanx. This is what makes Zeno's description entirely nonsensical. Yet, at the same time, it is this information that suggests he is describing the standard flank deployment of these units behind the elephants but beside the phalanx.

Polybius, and perhaps Zeno also, misunderstood how all Hellenistic battles were fought, Polybius misread Zeno's account or him his own source, and then Polybius criticized Zeno's ensuing description. There is nothing that proves Zeno stated that the elephants, light infantry, elite heavy infantry, and cavalry were directly in front of the phalanx, and there is much to doubt this positioning. In fact, most evidence suggests that they were ahead and off to the side just as they were in my view in all other battles of this period.[36]

---

[36] On occasion elephants deployed behind the main battle line, but that was only for specific purposes. At Ipsus the allies used a late elephant blockade to prevent Demetrius' victorious cavalry from returning to the battle (Plut. *Demetr*. 28-9; App. *Syr*. 55.). Still the best analysis of the battle is Bar Kochva, 105-110. At Gaza Ptolemy knew the Egyptian elephants would not stand up to the larger Antigonid ones (Diod. Sic. 19.80-84; Plut. *Demetr*. 5) but used caltrops to neutralize Demetrius' elephants. Kistler (John Kistler, *War Elephants* (Westport, CT: Praeger, 2006), 61) summarizes the effectiveness of this device well: "Because elephants are heavy, they cannot simply jump up to free themselves from traps holding two feet: normally they only lift one foot at a time. When two feet are impaled, an elephant is

Following Hellenistic standard deployments, Zeno likely suggested that Antiochus' left wing, which he is describing here in its usual deployment, stood a little in front of the position of the phalanx. These units did not stand directly in front of the phalanx but off to the side, though they were further forwards and nearer the enemy line. This is not unusual. In fact, it is the normal deployment if Antiochus utilized an oblique formation. As discussed in detail elsewhere, the oblique battle line was the standard tactical deployment of almost all Hellenistic battles.[37]

Usually, the oblique formation involved advancing the attacking cavalry on the right wing and holding back the defensive left wing with the phalanx drawn up at an angle in between. Antiochus' phalanx did not draw up at an angle to the Egyptian line. Thus, the elephant positioning here implies that Antiochus advanced his left wing only. As we will see below, Zeno's description later in Polybius' account suggests that Antiochus had a specific reason for advancing his elephants on his left flank only.

Polybius' description demonstrates that Zeno described a normal Hellenistic battle:

> 8 Such being their positions as laid down by him, he tells us that the younger Antiochus, whom he stationed in command of the mailed cavalry on the plain opposite the enemy's left, charged from the hill, routed and pursued the

---

trapped in place. The screaming of the wounded elephants caused the unharmed pachyderms to flee, disrupting Demetrius' own cavalry horses. Ptolemy sent his archers and javelin throwers forward to kill the mahouts, clearly intending to capture, and not kill, the elephants. Seeing this major setback, and fearing that their horses might step on more of these traps, Demetrius' cavalry fled the field, and his infantry joined them in full flight." At Magnesia the Romans did not have enough elephants to warrant their deployment in the main battle line. On all other occasions elephants always began the battle and were best served at protecting the flanks. See Wrightson, "Macedonian Armies," and see especially Rice's excellent analysis of elephants in Hellenistic battles: Rice, 51-210.

[37] Perhaps the most decisive use of the oblique formation was Philip's defeat of the Greeks at Chaeronea where the Thebans succumbed to their own tactic (Diod. Sic. 16.85–86; Polyaenus, *Strat.* 4.2.2; Plut. *Alex.* 12 & *Dem.* 20). See Wrightson, "Macedonian Armies"; Wrightson, *Combined Arms Warfare*; Wrightson, *The Battles of Antiochus*, 25-6. At Paraetacene Antigonus planned the usual oblique formation but his left-wing cavalry commander insubordinately charged ahead first and lost the battle, though Antigonus rescued a draw by his actions on the other wing (Diod. Sic. 19.26-31). The most concise secondary discussion of the battle is Devine 1985a.

> cavalry under Ptolemy, son of Aeropus, who commanded the Aetolians[38] in the plain and on the left,

To commence the battle Antiochus, the son of the king, led the Seleucid right wing heavy cavalry, particularly the cataphracts, in a devastating charge against the Egyptian left-wing cavalry on the plain. That this Seleucid right-wing cavalry stood initially on a hill meant that their charge against the enemy in the plain was downhill. This added to the impetus of the mailed cataphracts to make their charge even more fearsome and irresistible. The forces of the Egyptian left wing commander Ptolemy, son of Aeropus, fled in a complete rout. Again, this was the standard outcome of the initial right wing heavy cavalry charge in most Hellenistic battles. The exact same thing happened when Antiochus himself led the cavalry charge against Ptolemy at the Battle of Raphia.[39]

Zeno's description continues with the standard Hellenistic battle:

> 9 and that the two phalanxes met and fought stubbornly,

In every Hellenistic battle, following the cavalry charge on one or both wings, the sarissa phalanx of both sides fought each other in the center. This clash was usually the main one that decided the outcome of the battle. Whichever side's phalanx came out on top usually won. Most times it was enough for one side to survive long enough for their own cavalry support to attack the enemy phalanx in the rear and cause their retreat.[40] There are only a handful of notable exceptions where a phalanx could lose its cavalry support yet still win the battle by defeating the opposing phalanx. And this happened only if the victorious enemy cavalry did not press its advantage

---

[38] The Egyptian army certainly did not contain only Aetolians. Polybius (or Zeno) listing Ptolemy commanding the Aetolians is probably an error. He was the Egyptian cavalry general commanding the left wing and was an important Macedonian-descended personality in the Ptolemaic kingdom. The smaller contingent of Aetolian troops likely fought on the other wing as a phalanx regiment and a cavalry squadron. It was this unit of cavalry protecting the Egyptian right flank that Antiochus' elephants scared away. See Johstono 166-8.

[39] Polyb. 5.82-6. Ptolemy's cavalry fled when attacked by Antiochus. On the other wing the Egyptian cavalry moved around the elephant screen and routed the defensive Seleucid cavalry.

[40] At Ipsus the Antigonid phalanx held out for a long time waiting for Demetrius' cavalry to return but it never did. Plut. *Demetr.* 28-9; App. *Syr.* 55.

to attack the phalanx in the flank or rear.[41] Again, there is nothing unusual about Zeno's description.

Polybius then demonstrates his confusion about Zeno's description and perhaps his confusion about how all Hellenistic battles were fought:

> 10 forgetting that it was impossible for them to meet as the elephants, cavalry, and light-armed troops were stationed in front of them.

Polybius believes that Zeno described the elephants, light infantry and cavalry standing directly in front of the sarissa phalanx. Yet as discussed above Zeno's account may not in fact have said that, or if it did Polybius should have been able to infer what Zeno meant from knowledge of similar Hellenistic battles.[42]

Polybius continues his confused rendition of the battle stemming from his misunderstanding of the original placement of the elephants:

> 19 1 Next he states that the phalanx, proving inferior in fighting power and pressed hard by the Aetolians, retreated slowly, but that the elephants were of great service in receiving them in their retreat and engaging the enemy. 2 It is not easy to see how this could happen in the rear of the phalanx, or how if it did happen great service was rendered.

If the elephants were to the side of the phalanx the two phalanxes could, and did, clash in the center just as Zeno said. The Seleucid phalanx, when hard-pressed retreated slowly back up the slope drawing the Aetolians forwards. The elephants, stationed to the side could then be of great support to the Seleucid phalanx by attacking the flank of the Egyptian phalanx as it passed them.

---

[41] The one most relevant to the Battle of Panium is Raphia where Antiochus forced the Egyptian cavalry from the battlefield but pursued them too far and never used his advantage to outflank the Egyptian phalanx. By the time he returned to the battle the Egyptian phalanx had forced the Seleucid phalanx into retreat.

[42] Historians generally view Antiochus' deployment at Magnesia against the Romans as also having elephants deployed in front of the phalanx (Livy 37.40; App. *Syr*. 7.32). However, I believe that these Roman accounts describe only what the Seleucid line looked like through the mist after the phalanx formed a defensive square punctuated by the elephants. This is what would give the impression of elephants as towers in a castle wall, as the Roman sources describe, and would be a sensible deployment when the phalanx was attempting to withdraw to the camp when shorn of cavalry flank support. See Wrightson, *The Battles of Antiochus*, 115-131.

In the battle Zeno states that the Seleucid phalanx fell back drawing the Ptolemaic phalanx forwards. If this was indeed a planned maneuver, as I argue, it would have had an even greater effect since the Seleucids would have withdrawn uphill conveying a more conclusive impression that they were being pushed back and in retreat. It is similar to Philip II's maneuver at the Battle of Chaeronea where the Macedonian phalanx decided the engagement by feigning a withdrawal up a hill only to stop at a given signal and charge downwards onto the confused enemy Greek phalanx.[43] The tactic of a feigned withdrawal was a standard part of Macedonian army practice.[44] The surprise tactic seems to have been especially favored by the Seleucids, as argued by Bar Kochva,[45] and was first on display at the Battle of Ipsus when Seleucus' elephants blocked Demetrius' victorious cavalry from returning.[46]

It is also clear from Polybius' description that Antiochus planned for his right-wing cavalry to be the focal point of his attack and offensive battle

---

[43] Diod. Sic. 16.85-86; Polyaenus, *Strat.* 4.2.2; Plut. *Alex.* 12 & *Dem.* 20. There is only one detailed account of the battle from Polyaenus, who is the only source to mention the feigned withdrawal. For secondary discussions of the battle see Nick Hammond, "The Two Battles of Chaeronea: 338 BC and 86 BC," *Klio* 31 (1938): 186–218; Frank Adcock, *The Greek and Macedonian Art of War* (Berkeley and Los Angeles: California University Press, 1957); Paul Rahe, "The Annihilation of the Sacred Band at Chaeronea," *AJA* 85 (1981): 84–7; James Ashley, *The Macedonian Empire: The Era of Warfare Under Philip II and Alexander the Great, 359–323 BC* (Jefferson, NC: McFarland, 1998); John Buckler and Hans Beck, *Central Greece and the Politics of Power in the Fourth Century BC* (Cambridge: Cambridge University Press, 2008); Wrightson, *Combined Arms Warfare*, 171-3. For the battle in the context of the other battles of the Macedonians see Pietrykowski.

[44] See Borja Antela-Bernardez, "Furious Wrath: Alexander's Siege of Thebes and Perdiccas' False Retreat," in *Ancient Warfare: Introducing Current Research* (*IAWC* vol. 1), eds. Geoff Lee, Helene Whittaker & Graham Wrightson (Cambridge Scholars Press, 2015), 94-106; Wrightson, *The Battles of Antiochus*, 27-8.

[45] Bar Kochva (204) argues that the feigned withdrawal and surprise ambuscades were common tactics of the Seleucids, but the sources for the battles he lists as proof of this do not state that such surprises occurred. These are Cyrrhestica (286 under Seleucus I), against the Galatians (275 under Antiochus I), against Molon (220 under Antiochus III), and Elasa (160 under Demetrius I). These are all major battles not very well depicted in the sources. There are certainly other minor examples that Bar Kochva ignores just as Antiochus at Atabyrium.

[46] Plut. *Demetr.* 28-29; App. *Syr.* 55. W.W. Tarn was the first to argue this as Seleucus' intention: W. W. Tarn, "Two Notes on Seleucid History: 1. Seleucus' 500 Elephants," *JHS* 60 (1940): 87 n. 1; Tarn, *Hellenistic Military and Naval Developments* (Chicago, Ares, 1975), 68-9. Bar Kochva gives a more nuanced analysis including the terrain and Demetrius' intentions: Bar Kochva, 109-110.

plan, just as in most Hellenistic battles.[47] The unit of mailed cavalry, better known as cataphracts, excelled at charging the enemy cavalry and forcing them to retreat by sheer weight of armor and horses. That Antiochus gave command of this force to his son demonstrates that he was preparing his eldest son for his future role as overall commander, when he became king, since Seleucid kings almost always led the principal heavy cavalry charge in person, just as Antiochus had done himself at Raphia.[48]

Polybius also misunderstands how easy it was to tell friend from foe in a confrontation between two phalanxes:

> 3 For once the two phalanxes had met it was not possible for the elephants to distinguish friend from foe among those they encountered.

Since the length of the sarissas of the two phalanxes would maintain a three- or four-foot gap between the front ranks of each unit distinguishing each unit would be easy, one facing one way one the other. Polybius seems to view a battle in terms of swordsmen where the battle is a chaotic melee. With phalanx confrontations it is always clear which way each phalanx faces. Our reconstruction of elephants to the side also explains why the elephants did not confuse friend from foe and cause casualties in the Seleucid phalanx by mistake. The elephants would only attack the flanks of the Egyptian phalanx while the Seleucid one continued backing away.

This reconstruction is the only way that the following part of Polybius' description makes sense:

> 4 In addition to this he says the Aetolian cavalry were put out of action in the battle because they were unaccustomed to the sight of the elephants.

The Aetolian cavalry was on the right of the Egyptian phalanx shielding its vulnerable flank. Thus, the only way that the elephants could attack the

---

[47] The few occasions where a general chose to make the main heavy cavalry charge on the left wing are noteworthy in their small number. At Gabiene Eumenes attempted to neutralize Antigonus' elite heavy cavalry by facing him directly with his own best troops on his left wing. Unfortunately, it backfired for him (Diod. Sic. 19.39-43; Plut. *Eum.* 16). The most concise secondary discussion of the battle is Devine 1985b. On this battle and its connection with the earlier Battle of Paraetacene see Wrightson, *Combined Arms Warfare*, 204-10. At the Battle of the Hellespont Eumenes was able to kill the enemy general when the elite cavalry forces directly opposed each other (Diod. Sic. 18.29-32; Plut. *Eum.* 7). On this battle see Wrightson, *Combined Arms Warfare*, 203-4.

[48] Polyb. 5.82-6.

Aetolian cavalry was if the elephants were opposite them on the Seleucid left wing. As the Egyptian phalanx advanced the Aetolian cavalry galloped alongside possibly into a gap between the Seleucid phalanx and left-wing cavalry. As the Aetolians came forward Antiochus launched the elephants against them. Antiochus likely knew before the battle that being unaccustomed to elephants the Aetolian cavalry would flee and expose the flank and rear of the Egyptian phalanx and so he adjusted his plan accordingly. He advanced his elephants only on the wing where the Aetolians stood so as to screen his flank better by disconcerting the opposing cavalry. In my view it is likely that Antiochus always intended for his phalanx to pull back up the slope so as to tempt the Aetolian cavalry forward in range of the elephants and thus, in defeating that cavalry, expose the Egyptian phalanx to a flank attack. So far, all of Zeno's description makes sense in comparison to other Hellenistic battles.

Polybius' next detail, which actually led to confusion on his part, adds support to this hypothesis:

> 5 But the cavalry posted on the right remained unbroken from the beginning as he says himself, while the rest of the cavalry, which had been assigned to the left wing, had been vanquished and put to flight by Antiochus.

The cavalry on the right was the Aetolian cavalry. It had remained unbroken since the start of the battle since Antiochus, the king, had not sent his left-wing cavalry to attack it yet. It was perhaps protected from a flanking attack by the steep ravine. All the other cavalry had indeed fled the field earlier after the attack of the Seleucid right-wing cavalry or had begun to do so. The Antiochus Polybius mentions here is clearly the son of Antiochus who led the cataphract charge on the right wing.

Polybius then reveals his confusion about Zeno's description when he misunderstands which cavalry fled the elephants and where they did so:

> 6 What part of the cavalry, then, was it that was terrified by the elephants in the center of the phalanx,

The cavalry terrified by the elephants was the Aetolian unit on the Egyptian right wing. They were definitely not in the center of the line in front of the phalanx, and they are not to be confused with the rest of the Egyptian cavalry on the left wing that had fled. As discussed above, that confusion stemmed from believing the elephants stood directly in front of the phalanx. Polybius also seems to forget that in all Hellenistic battles, generals placed cavalry on both wings of the army to shield the phalanx's vulnerable flanks.[49] He seems to think that Scopas had stationed all his cavalry on his left wing, which had fled when attacked by the cataphracts. Like every other Hellenistic general, Scopas had cavalry on both his wings, and it was his right-wing cavalry that remained unbroken because it had not yet been attacked.

The one difficulty with this interpretation is that not having Zeno's original account we do not know exactly when he described the right-wing cavalry as unbroken. In my reconstruction they were unbroken until they encountered the elephants. If he said this after his description of the Aetolian cavalry fleeing the elephants, then the Aetolians must have been in a different part of the battle. However, without having access to the original we must attempt to reconstruct the battle from the evidence of Polybius' account and comparison to other battles. In this sense I believe that the most sense comes from the Aetolian cavalry on the right remaining unbroken until it faced the elephants when the entire unit fled en masse.

Polybius continues his confusion from misunderstanding the elephant position when he fails to grasp what Antiochus was doing in the battle:

> and where was the king all this time and what service did he render in the action with the horse and foot he had about him, the finest in the army? We are not told a single word about this.

This is the easiest confusion to rectify. Zeno's description is clear that Antiochus stood with his best guard units behind the elephant screen. Again, this is the standard deployment and mirrors exactly Antiochus' position at Raphia only on the other wing.[50] Since Polybius believes the elephants stood directly in front of the phalanx, then he also believes Antiochus and his best troops stood in between those two units. He is right that it would be strange for Antiochus and his best men to stand directly in front of the phalanx. There is no parallel for that in any other Hellenistic battle. This is more proof, if needed, that Zeno's account actually described the standard

[49] The exceptions to this are after the late third century when the Kingdom of Macedon did not have enough cavalry to station some on each wing and so kept them only on one side such as at Sellasia (Polyb. 2.65-9; Plut. *Cleom.* 27-8; Plut. *Phil.* 6), Cynoscephalae (Polybi. 18.21.1-3; Livy 33.6), and Pydna (Livy 44.40-42; Plut. *Aem.* 18-22). At this latter battle, the king Perseus retreated with his cavalry apparently having never joined the attack. The defeated Macedonian infantry were so annoyed with the inaction of the cavalry that they even physically assaulted them (Plut. *Aem.* 23.1).

[50] Polyb. 5.82-6.

deployment of elephants in front of the king and the left-wing heavy cavalry and elite guard troops off to the side of the phalanx.

The point remains that Zeno does not describe what Antiochus did personally in the battle. Polybius perhaps does not grasp how total the Seleucid victory was on this flank. The simplest solution is that Antiochus did not do anything in the battle because he did not need to do so. Antiochus waited with his best troops until the elephant engagement raged before intending to lead his elite forces on an assault on the enemy flank around the side of the elephants. This exact maneuver had caused the Ptolemaic left-wing cavalry to disintegrate completely at Raphia and there is no reason why Antiochus would doubt the same thing would occur again. At Raphia, the Egyptians defended their flank with their own elephants precipitating an elephant-on-elephant battle.[51] At Panium, the elephant charge was so successful in breaking the Egyptian wing because there were no opposing elephants and so Antiochus did not need to lead his men around an elephant confrontation. The retreat of the Aetolian cavalry at the onset of the elephants had already won Antiochus the initiative in the battle on that flank. He then enjoyed the luxury of keeping his best troops in reserve until the opportune moment. Why commit expensive elite veteran troops to battle if it already goes your way?

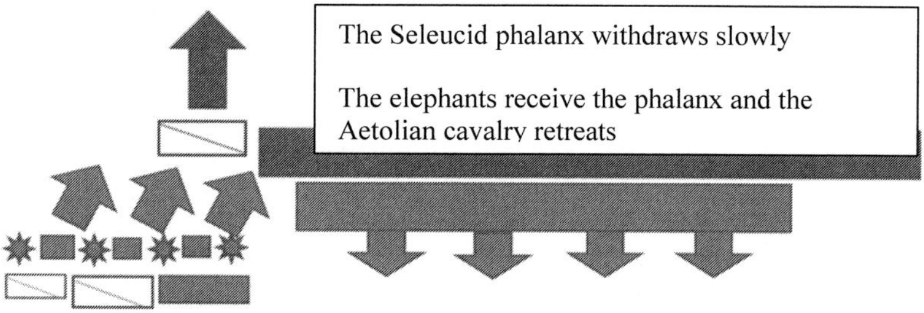

Fig. 8: The actions of the left side of the battle.

Nonetheless, Zeno's later statement that the Egyptian phalanx was surrounded by cavalry, infantry, and elephants perhaps suggests that Antiochus simply led his troops to attack Scopas' phalanx from the rear while the elephants attacked the flank and the Seleucid phalanx attacked from the front. This is exactly what you would expect a general to do in such a situation. However, this statement follows the appearance of Antiochus the Younger's cavalry in the rear and the likeliest, and simplest, explanation

[51] Polyb. 5.82-6.

is that it was this cavalry force that surrounded the Egyptian phalanx and attacked from the rear.

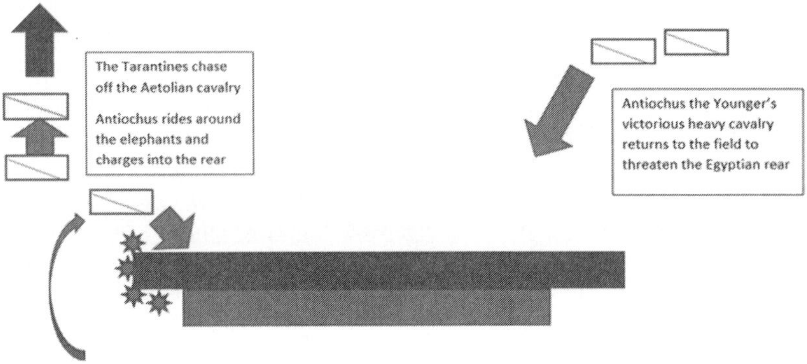

Fig. 9: The end stages of the battle where the Ptolemaic phalanx is surrounded and threatened by the returning cavalry of Antiochus the Younger

After his critique of the elephant position, we get to Polybius' main problem with Zeno's account, the two sons of Antiochus:

> 8 Where again was the king's elder son, Antiochus, who had occupied positions overlooking the enemy with a part of the army? 9 Why! according to Zeno this young man did not even take part in the return to the camp after the battle; naturally not, for he supposes there were two Antiochi there, sons of the king, whereas there was only one with him in this campaign.

As Polybius understood Zeno, Antiochus III's elder son, named Antiochus, led the advance cavalry controlling the hills while his younger son, also named Antiochus, commanded the right wing in the battle later that day. Polybius correctly censures Zeno for this error since only Antiochus' elder son named Antiochus was at the battle probably aged 21. Antiochus' second son was named Seleucus and was 18 at the time of Panium, though we do not know if he was present. His youngest son, who received the name Antiochus only when he became king later, was only 15 at the time, perhaps younger, and was likely too young to serve in the army and certainly too young to command the entire right wing in place of his elder brother(s). Moreover, the youngest son's name was initially Mithradates and he was only renamed Antiochus when he became king in 175. Thus, Zeno should only be describing one son named Antiochus, hence Polybius' censure of his description. Zeno, or his source, perhaps did not use the phrase 'the Younger' to distinguish between two sons of Antiochus but between two people at the battle called Antiochus, one the king the other younger one his son. Zeno perhaps did get confused and thus Polybius rightly censures him.

Polybius seems to completely misunderstand Zeno, or Zeno misunderstood his own source. Just as I do above, when describing the actions of Antiochus III, the king, and of his son Antiochus, it is easier to utilize the phrase 'the younger Antiochus' in reference to the son. Perhaps when describing the initial occupation of the high ground by the cavalry Zeno referred to this son as Antiochus III's eldest son, also called Antiochus. This was likely the first time he appeared in Zeno's account and so he was explaining to his audience who this different Antiochus was. Later, when Zeno describes the actions of the right-wing cavalry, he states that 'the younger Antiochus' commanded there. He does not mean the younger son of the king named Antiochus but the younger Antiochus fighting at the battle. This seems to me to be a natural way of describing the son of Antiochus in the battle. Polybius believed Zeno made an error in introducing the second younger son, when in reality the younger comparison adjective refers to the son being younger than his father.

The younger Antiochus first led the nighttime advance force to occupy the hills. The next day when his father brought the whole army across the river, Antiochus the Younger then took his natural place commanding the right-wing cavalry in the army, which rested on the same hills that he had occupied the night before presumably with some of the very same cavalry force. It was standard in Seleucid armies for the heir to the throne to command the cavalry wing just as Antiochus I did at the Battle of Ipsus in 301.[52] Antiochus the Younger then led the usual right wing cavalry charge with the cataphracts and later returned from his pursuit of the routing Egyptian cavalry to threaten the rear of the Egyptian left wing.

Rabbi Benjamin Scolnic raises the intriguing possibility that Zeno relied on a Seleucid source from the reign of Antiochus IV, the very son who was too young to be in the battle.[53] Scolnic argues convincingly that Antiochus IV engaged in propagandistic accounts that retroactively inserted him as the hero into the narrative of the battle of Panium at the expense of his now dead older brother Antiochus, and perhaps even at the expense of his father

[52] Plut. *Demetr.* 28-9; App. *Syr.* 55.
[53] Benjamin Scolnic, "Second-Hand Propaganda: Polybios and Zeno on the Role of Antiochos IV at the Battle of Panion," in *Seleukid Ideology: Creation, Reception and Response. Seleukid Perspectives*, volume 1, edited by Altay Coskun, and Richard Wenghofer (Stuttgart: Franz Steiner Verlag, 2023), 177-200.

Antiochus III. If Zeno's account did indeed not provide any details on the actions of Antiochus III in the battle it is perhaps because the propagandistic account removed his actions in favor of those of Antiochus IV. Scolnic also suggests elsewhere that the placement of Antiochus III on the left wing for the only time in any of his battles cleared the way for Antiochus IV to insert himself as the hero on the right wing, the traditional place of honor in Seleucid battles.[54] It is conjectural, but it does present an interesting assessment of the nature of Zeno's sources and the problems with examining this battle.

In my view, Antiochus III commanded the left wing for the only time in his career because that was the most important tactical location on the field. He let his eldest son and heir gain experience in leading a cavalry charge on the right wing. The battle went so well on the left wing after the elephant charge that Antiochus did not need to lead his elite troops into the battle and thus risk their deaths or injuries, especially after his son returned with the cavalry to surround the Egyptian phalanx. Despite not knowing precisely what Zeno said in his account, or what sources he used, we can relatively confidently state that there was only one son of Antiochus at the battle. This eldest son of the king was also named Antiochus and was perhaps referred to as Antiochus the Younger.

Polybius continues his confusion into the conduct of the Egyptian general Scopas:

> 10 And can he explain how Scopas was both the first and the last to leave the field? For he tells us that when he saw the younger Antiochus returning from the pursuit and threatening the phalanx from the rear he despaired of victory and retreated; 11 but after this the hottest part of the battle began, upon the phalanx being surrounded by the elephants and cavalry, and now Scopas was the last to leave the field.

We can make sense of Zeno's description of Scopas' actions if we consider what Antiochus himself did with his best cavalry after the elephant charge was so successful. Though Antiochus did not need to support the elephants, which presumably remained entangled with the Egyptian phalanx, he likely would not remain completely inactive. Antiochus loved the thrill of the cavalry charge and would probably have wanted to get into the thick of the action.[55] As Polybius states, it would be surprising if Antiochus did not

[54] Personal discussion with the author.

[55] Antiochus' thrill of the charge actually cost him in his two most important battles of Magnesia and Raphia. See Wrightson 2022.

make use of his best troops somehow. The easiest and most tactically astute maneuver was to ride around the elephants and attack the rear of the Egyptian phalanx now shorn of cavalry support. Once he was sure the Aetolian cavalry had fled completely, perhaps chased by the light Tarantine cavalry, Antiochus could launch all his elite forces into the rear of the Egyptian phalanx.

Polybius states that in the battle, following the elephant attack and the arrival of Antiochus the Younger to the rear, that "after this the hottest part of the battle began, upon the phalanx being surrounded by the elephants and cavalry." It was at this point that Scopas left the battlefield. The elephants are clearly those from the Seleucid left wing that plunged into the phalanx. The cavalry is perhaps the ones that Antiochus the Younger brought back in the rear having routed the Egyptian left wing, thus surrounding the Egyptian phalanx in the center as Polybius states. Perhaps Antiochus III also led his cavalry and elite infantry around the elephants to join in the encirclement of the phalanx. Polybius' statement cannot confirm one way or the other which cavalry forces surrounded the Egyptian phalanx.

Moreover, that Polybius reports Zeno describing the elephants surrounding the Egyptian phalanx confirms that the elephants were not directly in front of the phalanx. They could not have moved from in front of the Seleucid phalanx to suddenly be surrounding the rear of the Egyptian phalanx. Clearly the elephants began the battle off to the side and moved to attack the flank and rear of the Egyptians when the Seleucid phalanx gradually fell back.

When Polybius reports Zeno's statement concerning Scopas that "when he saw the younger Antiochus returning from the pursuit and threatening the phalanx from the rear he despaired of victory and retreated," he perhaps means that Scopas gave the order for the whole army to retreat. It is common for historians to use the name of the general to refer to the whole army. Caesar is a famous master of that in his *Gallic Wars*.[56] With one wing of the

[56] Pick any section of the *Gallic Wars*, but for example, 3.14 "Caesar had taken several towns by assault, when he perceived that all his labor availed nothing, since the flight of the enemy could not be checked by the capture of towns, nor damage done to them; accordingly he determined to await the fleet." He did not take the towns by himself but his army did so. And 3.28: "In the days then remaining Caesar set to work to cut down the forests, and, to prevent any flank attack on troops unarmed and unprepared, he placed all the timber felled on the side toward the enemy, and also piled it as a rampart on both flanks." Caesar did not cut down the

phalanx surrounded already by elephants and perhaps also cavalry from Antiochus' victorious left wing, as soon as he saw the younger Antiochus' cavalry reappearing on the battlefield behind his left-wing threatening to completely cut-off his avenue of retreat, Scopas ordered his army to withdraw. Then we can add in the statement, "Scopas was the last to leave the field." He gave the command to retreat when he saw the younger Antiochus' cavalry threatening his army's withdrawal to the camp and then retreated himself only once the rest of his army had successfully disengaged and made it out to safety.

Unless the Egyptians failed to watch their rear and did not see the return of the Seleucid cavalry, Scopas would try to extract what was left of his army before his rear was completely surrounded by that victorious force. To do any less would be military suicide. Once surrounded there would be no way to win and no way to survive. That Scopas escaped with a large chunk of his army suggests that an avenue of retreat was available. Scopas escaped with 10,000 soldiers to the coastal city of Sidon to the northwest of the battle site, suggesting that his retreating army escaped on its left wing that faced that direction leaving the surrounded right wing to its fate.[57] This suggests that Scopas withdrew the army before Antiochus the Younger's cavalry could charge into the fray as his forces would have completely surrounded the Egyptian phalanx.

Moreover, since Scopas retreated northwest towards Sidon, rather than directly southwards towards Egypt back the way he came, that must have been the only avenue open to him. This makes sense if we envision the defeated Egyptian left wing cavalry taking the easiest route to flee directly south. Antiochus the Younger's cavalry chased them far enough to make sure they could not return. Antiochus then rode back towards the main battle approaching from the south up the valley. This force would block Scopas' retreat south and mean that he had no choice but to flee northwest between the hills covering the ground where his left wing had originally stood.

Livy states that Ptolemaic units still fought Antiochus in the summer of 198. Some sought shelter in the citadel of Jerusalem; others in Phoenicia, Samaria, and the Decapolis.[58] Jerusalem and Samaria are directly to the

---

forests by himself or stack up the wood into ramparts without the aid of his troops. In fact, he probably did not do any of that himself!

[57] Porph. *FGrH* 260.46.

[58] Livy 31.14.5. See also Joseph. *AJ* 12.133, Polyb. 16.39.3. See Johstono, "Defeat at Panium," 165.

south of the battlefield while Decapolis is on the route to Sidon northwards. This suggests that Scopas' army split in its efforts to escape. One group, in my view the phalanx, headed north with Scopas into the safety of the mountains. Another group, likely the cavalry, continued its flight southwards to take shelter in the two nearest citadels.

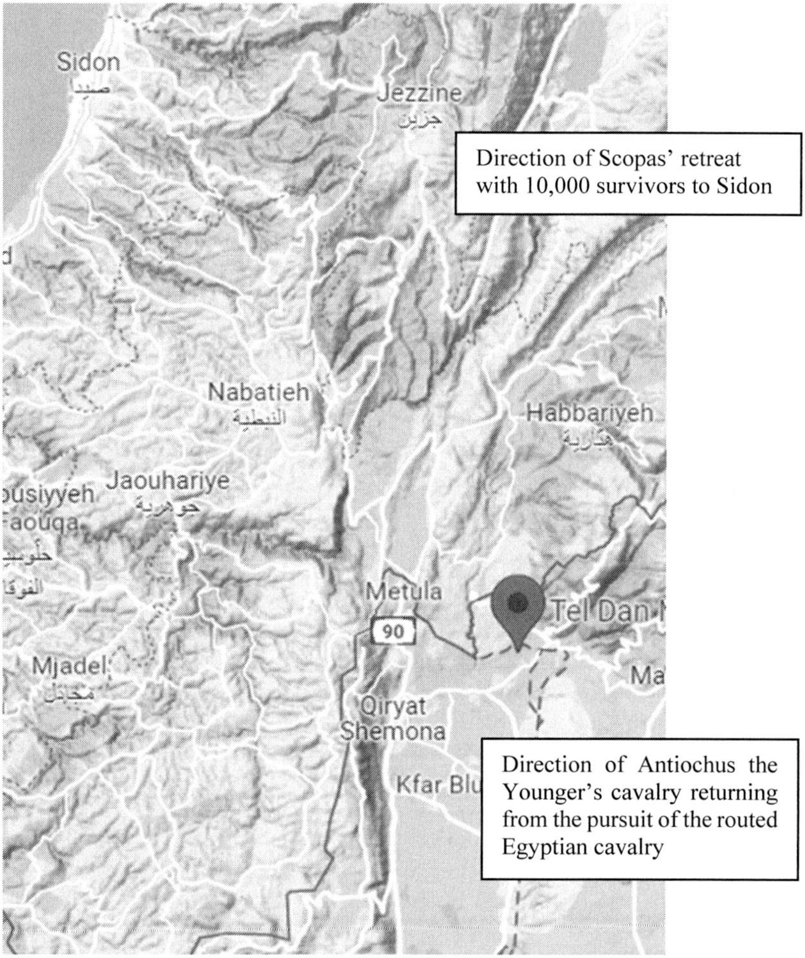

Fig. 10: Map showing route from the likely Egyptian camp at Tel Dan to their escape destination at Sidon. It also clearly shows the valley to the south into which the Egyptian cavalry fled and from which direction Antiochus the Younger returned to the battlefield forcing Scopas northwards.

In order to extract any of his army Scopas would have had to leave before Antiochus the Younger's victorious cavalry completely surrounded him. So, he gave the command to withdraw as soon as he sighted the returning cavalry and was the last to leave the field, perhaps using his personal troops as the rearguard to shield an orderly withdrawal. The battle was a disaster for the Egyptians, but that Scopas managed to extricate around 10,000 men does not suggest a headlong and chaotic pursuit with Antiochus' cavalry cutting down the fleeing infantry. Rather Scopas oversaw an orderly withdrawal perhaps with the remnants of the infantry formed in a square formation that offered protection against Antiochus' cavalry.

If this is the case, then what Polybius reports that Zeno described was Antiochus' forces from the left enveloping the right wing of the Egyptian army and it was the king's cavalry that did so. Once Scopas saw Antiochus the Younger coming with his victorious cavalry in the rear threatening to surround his other wing and thus completely encircle his army, he gave the order to withdraw. The right wing of the army that was already surrounded still had to fight its way clear, hence the description of the hottest part of the battle ensuing after the order to retreat. The center was able to escape perhaps in some semblance of order before Antiochus the Younger's cavalry cut them off completely.

Antiochus' tactics

Having made sense of Polybius' criticisms of Zeno's description we can actually go further into analyzing Antiochus' tactics at this battle. At first sight it seems strange for Antiochus to place his left wing so far in front of the central phalanx if he is not holding back his other wing to match the oblique battle line. Yet it makes perfect sense if the retreat of the Seleucid phalanx in the center was a planned move in order to draw the Egyptian forces forwards and expose their flank. This would be especially true if we are right to place the steep riverbanks shielding the Egyptian right flank. Antiochus had to find a way to draw the enemy away from the protection afforded by the river and create an opportunity to expose the flanks of the Egyptian phalanx. Antiochus used the slow retreat of the phalanx to draw thc enemy out to where he wanted them and where he was able to launch his elephants at their flank after forcing the retreat of the Aetolian cavalry.

This is a perfect use of the feigned withdrawal that was a standard part of the Seleucid phalanx's training, as discussed elsewhere.[59] Moreover, if Antiochus' elephants did not face an opposing force of Egyptian animals, then he would have been certain that nothing could prevent his elephants charging right into the enemy phalanx's vulnerable flank especially if he already knew the Aetolian cavalry would not stand up to elephants.

Though we only have Polybius' critique of Zeno's description as evidence for this battle, Egyptian elephants are noticeably absent from both. Moreover, if Zeno is right that the Aetolians fled because they were unfamiliar with the elephants, Antiochus must have guessed that this would happen. He likely knew from his scouts and deserters that Aetolians formed the core of the Egyptian army, and perhaps even knew on which wing Scopas stationed them. Calculating they would flee before an elephant charge, he stationed himself on the left, rather than the right as normal, to make sure everything on this wing went according to plan. Antiochus' knew his elephants could charge unopposed into the enemy line and so he planned accordingly. He stationed them as the most advanced part of his line protected by strong cavalry forces and missile infantry. He ordered his phalanx to feign a withdrawal to draw the enemy phalanx and cavalry down from their position on the hills with their flank shielded by the riverbanks. Antiochus' elephants chased off the Aetolian cavalry and he then launched a devastating and battle winning charge into the phalanx's flank.

In this light, it is important to note that Antiochus' son led his right-wing cavalry of heavy cataphracts in the usual assault. The king usually occupied this position of glory and led the decisive cavalry charge. It is how Antiochus fought at all his other battles, Raphia, Arius, and Magnesia.[60] Therefore, the one instance where Antiochus fought in a different location is likely of upmost significance. This is the one battle where Antiochus did not rely solely on a right-wing heavy cavalry charge for victory. He came up with a different and innovative strategy to win the battle on both wings instead. His son led the right-wing cavalry charge with the expected result of routing the enemy wing. But this time on the other wing Antiochus took command of the more complicated tactics of exploiting the Egyptian army's vulnerability to elephants.

It was a tradition in Seleucid armies for eldest sons to gain experience by commanding a cavalry squadron. Antiochus I had led the Seleucid cavalry

---

59 Wrightson, *The Battles of Antiochus*, 25-8.
60 See Wrightson, *The Battles of Antiochus*.

at Ipsus while Seleucus, his father, commanded the reserve elephant line. This battle is a perfect comparison to Panium since in both cases the eldest son commanded the heavy cavalry while the king commanded an elephant force on whose actions the decision of the battle rested. At Ipsus, Seleucus using elephants to block the return of Demetrius' victorious cavalry won that battle for the allies.[61] The role of the future Antiochus I in retreating and drawing Demetrius' force away was crucial in creating the space and time for Seleucus to deploy the elephants. At Panium, Antiochus led the elephants into a flanking attack of the Egyptian phalanx, the maneuver that decided the battle. Antiochus the Younger's cavalry charge had pushed all supporting Egyptian cavalry away thus leaving the phalanx exposed to the attack of Antiochus' elephants and cavalry. The mere sight of the cavalry's return to the battlefield was enough to cause Scopas to order the general retreat of the Egyptian army. Again, it seems, Antiochus was using tried and trusted Seleucid tactics in this battle to great effect.

Conclusions

As usual for Hellenistic armies, both generals placed cavalry on either side of the phalanx, with Scopas using fewer cavalry on one side since his flank was protected by the hills and the precipitous riverbank. Polybius is clear that Antiochus' right-wing cavalry featured a unit of cataphracts fighting in a Seleucid army for the first time at Panium. It must have joined Antiochus' army after his earlier subjugation of the Greco-Bactrian kingdom.[62] His son Antiochus the Younger commanded the right wing after occupying the hills the night before.

Polybius, clearly misunderstanding the deployment of Hellenistic armies, is confused when he criticizes Zeno's account of the battle itself. Zeno's version of events is exactly what happens in all Hellenistic and especially Seleucid battles. The Seleucid heavy cavalry force on the right wing, in this case commanded by Antiochus' son, easily routed the opposing cavalry. Zeno continues that, as you would expect, "the two phalanxes met and fought stubbornly," in the center of the line. Polybius' criticism of Zeno for "forgetting that it was impossible for them [the phalanxes] to meet as the elephants, cavalry, and light-armed troops were stationed in front of them"

---

61 Plut. *Demetr.* 28-9; App. *Syr.* 55. See note 28 on the elephant reserve.

62 Antiochus had fought against Cataphracts at the River Arius (Polyb. 10.49) and after subduing the kingdom likely incorporated cataphracts into his own army. Tarn suggests he incorporated them after his submission of Parthia (*Hellenistic Military,* 76).

is based solely on his misunderstanding of Zeno's statement of elephants deployed in front of the phalanx. In reality, the phalanxes could and certainly did meet in the center since Antiochus' force of "elephants, cavalry, and light-armed troops were stationed in front of them" but off to their left protecting their flank, as was the deployment in every single Hellenistic battle. An understanding of the usual deployment of a Hellenistic army reveals that Zeno is actually right and Polybius is completely confused.

The Aetolian cavalry that fled was almost certainly the force of cavalry assigned to shield the right flank of the advancing Aetolian phalanx in the Egyptian army. It was standard practice for units of the same origin to fight next to each other. This cavalry force, seeing the Seleucid phalanx going backwards, probably expected to pursue and attack the exposed flank of the Seleucid phalanx. However, before they could do that Antiochus' elephants charged. The elephants caused the Aetolian cavalry to retreat "because they were unaccustomed to the sight of the elephants." After routing the Aetolian cavalry flank protection, the elephants charged into the Aetolian phalanx. This unexpected retreat of the Aetolian cavalry exposed the Aetolian phalanx to a flanking attack from the elephants. As the Seleucid phalanx gave ground the elephants could attack the exposed right flank of the advancing Egyptian phalanx. This would indeed cause significant chaos in the Egyptian ranks as the large beasts plowed into the vulnerable flank of the phalanx.

Antiochus then led his elite troops to attack the rear of the Egyptian phalanx. At this point Scopas saw the cavalry forces of Antiochus the Younger returning to threaten the rear. This prompted him to sound a general withdrawal. Then "after this the hottest part of the battle began, upon the phalanx being surrounded by the elephants and cavalry," but Scopas extracted as much of the army as he could and "Scopas was the last to leave the field."

Taken in comparison to other battles, Zeno's description makes perfect sense and fits the pattern of all the other Hellenistic battles. Polybius, and all modern historians who unquestioningly follow his description, criticize poor old Zeno and reconstruct the events of the battle incorrectly. Everything stems from Polybius misunderstanding the standard deployment of Hellenistic and Seleucid armies and Zeno using a word to describe the elephants in a more forward position than the Seleucid phalanx. This clearer appreciation for the accuracy of Zeno's account reveals significant errors in Polybius' understanding of Hellenistic battles and perhaps most importantly removes the only instance in Greek warfare where elephants, light and heavy infantry and light and heavy cavalry, apparently stood in front of the phalanx. This reconstruction removes the exception from the rule and proves that all Hellenistic armies stationed elephants only as flank guards on the wings, never directly in front of the phalanx.

Therefore, if we have reconstructed the battle correctly, Antiochus' greatest tactical triumph is at the Battle of Panium. Here he successfully used an advanced left flank, feigned withdrawal, elephant charge, and a right- and left-wing heavy cavalry hammer onto an unbroken phalanx anvil in the center. It is the only occasion where Antiochus, or any later Hellenistic general, utilized to perfection in one single engagement all of the battle tactics introduced and perfected by Philip and Alexander the Great. It is his victory at Panium for which Antiochus should be lauded as a brilliant tactical general not his sojourn through the east, and it exonerates him somewhat for his spectacular failures in generalship at Raphia and Magnesia.

Bibliography

Adcock, F. E. *The Greek and Macedonian Art of War*. Berkeley and Los Angeles: California University Press, 1957.

Antela-Bernárdez, B. "Furious Wrath: Alexander's Siege of Thebes and Perdiccas' False Retreat." In G. Lee, H. Whittaker, & G. Wrightson eds. *Ancient Warfare: Introducing Current Research (IAWC vol. 1)* (Cambridge Scholars Press, 2015), 94-106.

Ashley, J. R. *The Macedonian Empire: The Era of Warfare Under Philip II and Alexander the Great, 359–323 BC*. Jefferson, NC: McFarland, 1998.

Bar Kochva, B. *The Seleucid Army.* Cambridge: Cambridge University Press, 1976.

Buckler, J., and H. Beck. *Central Greece and the Politics of Power in the Fourth Century BC*. Cambridge: Cambridge University Press, 2008.

Champion, J. *Antigonus the One-Eyed: Greatest of the Successors*. Barnsley: Pen & Sword, 2014.

Devine, A. M. "Diodorus' Account of the Battle of Paraitacene (317 B.C.)," *Ancient World* 12 (1985): 75–86.

Devine, A. M. "Diodorus' Account of the Battle of Gabiene," *Ancient World* 12 (1985): 87–96.

Errington, R. M. *A History of the Hellenistic World, 323–30 BC*. Malden, MA: Blackwell, 2008.

Eckstein, A. M. *Mediterranean Anarchy and the Rise of Rome*. Berkeley: UCP, 2006.

Gera, D. "Ptolemy son of Thraseas and the Fifth Syrian War," *AncSoc* 18 (1987): 63–73.

Grainger, J. *The Syrian Wars*. Mnemosyne suppl. 320. Leiden: Brill, 2010.

Hammond, N.G.L. "The Two Battles of Chaeronea: 338 BC and 86 BC.," *Klio* 31 (1938): 186–218.

Heckel, W., C. Willekes and G. Wrightson, "Scythed Chariots at Gaugamela." In *Philip II and Alexander the Great: Father and Son, Lives and Afterlives,* edited by E. Carney and D. Ogden, 103-113. Oxford: OUP, 2010.

Hölbl, G. *History of the Ptolemaic Empire.* Trans. T. Saavedra. New York: Routledge, 2001.

Holleaux, M. *Etudes d'epigraphie et d'histoire grecques, III: Lagides et Seleucides.* Paris: Adrien- Maisonneuve, 1942.

Huss, W. *Agypten in hellenistischer Zeit 332–30 v. Chr.* Munchen: C.H.Beck, 2001.

Johstono, P. and M. Taylor. "Reconstructing the Battle of Pydna." *Greek, Roman, and Byzantine Studies* 62 (2022): 44–76.

Johstono, P. ""No Strength To Stand": Defeat at Panium, the Macedonian Class, and Ptolemaic Decline." In *Brill's Companion to Military Defeat in Ancient Mediterranean Society*, edited by Clark, Jessica H., and Brian Turner, 162-187. Leiden: Brill, 2017.

Johstono, P. *The Army of Ptolemaic Egypt 323 to 204 BC: An Institutional and Operational History*. Barnsley: Pen and Sword, 2020.

Jouguet, P. *Macedonian Imperialism, and the Hellenization of the East.* Trans. M. R. Dobie. Reprint. New York: Routledge, 1996.

Kistler, J. War Elephants. Westport, CT: Praeger, 2006.

Matthew, C. A. *An Invincible Beast: Understanding the Hellenistic Pike Phalanx in Action*. Barnsley: Pen and Sword, 2015.

Pietrykowski, J. *Great Battles of the Hellenistic World*. Barnsley: Pen and Sword, 2009.

Rahe, P. A. "The Annihilation of the Sacred Band at Chaeronea," *AJA* 85 (1981): 84–7.

Rice, J. "Animals in Ancient Greek Warfare: A Study of the Elephant, Camel, and Dog." PhD. Diss. University of Missouri – Columbia, 2020.

Roberts, M. and B. Bennett. *Twilight of the Hellenistic World*. Barnsley: Pen and Sword, 2012.

Scolnic, B. "Second-Hand Propaganda: Polybios and Zeno on the Role of Antiochos IV at the Battle of Panion." In *Seleukid Ideology: Creation, Reception and Response. Seleukid Perspectives*, volume 1, edited by

Altay Coskun, and Richard Wenghofer, 177-200. Stuttgart: Franz Steiner Verlag, 2023.

Tarn, W. W. *Hellenistic Military and Naval Developments.* Cambridge: CUP, 1930.

Tarn, W. W. "Two Notes on Seleucid History: 1. Seleucus' 500 Elephants," *JHS* 60 (1940): 84-94.

Taylor, M. *Antiochus the Great*. Barnsley, UK: Pen and Sword, 2013.

Walbank, F. W. *A historical commentary on Polybius. Vol. 2. Oxford and New York: OUP, 1967.*

Will, E. *Histoire politique du monde hellenistique*. 2 vols. Nancy: Points, 1967.

Wrightson, G. "The Nature of Command in the Macedonian Sarissa Phalanx," *AHB* 24 (2010): 71-92.

Wrightson, G. "Macedonian Armies and the Perfection of Combined Arms," in T. Howe, E. Garvin, and G. Wrightson eds. *Greece, Macedon and Persia: Studies in the Social, Political and Military Consequences of Conquest Societies in honour of Waldemar Heckel,* 59-68. Oxford: Oxbow, 2015.

Wrightson, G. *Combined Arms Warfare in Ancient Greece from Homer to Alexander the Great and his Successors.* Oxford and New York: Routledge, 2019.

Wrightson, G. *The Battles of Antiochus the Great: The failure of Combined Arms at Magnesia that Handed the World to Rome*. Barnsley: Pen and Sword, 2022.

Wrightson, G. *The Third Macedonian War and the Battle of Pydna: Perseus' Neglect of Combined-Arms Tactics and the Real Reasons for the Roman Victory.* Barnsley: Pen and Sword, 2023.

THE UNUSUAL ENCOUNTER BETWEEN DEMETRIUS III AND ALEXANDER JANNAEUS: RELIGIOUS BELIEFS AS A MOTIVATING FACTOR IN ANCIENT WARFARE

KENNETH ATKINSON

In 90/89 BCE, the Seleucid monarch Demetrius III invaded the Hasmonean state, then ruled by Alexander Jannaeus, its king and high priest. Josephus, whose accounts provide our sole extant narratives of the battle and its aftermath (*BJ* 1.92-98; *AJ* 13.377-83), notes that Demetrius III commanded a much larger force which, remarkably, included many of Jannaeus' Jewish subjects. After being soundly defeated, Jannaeus retreated. Then something surprising occurred: many Jews fighting as mercenaries in the Seleucid army defected. Jannaeus returned home and retaliated against his subjects who had opposed him. The Dead Sea Scrolls provide a likely eyewitness account of his wrath, which included his crucifixion of eight hundred of his citizens.

This study examines the historical background of this unusual battle as described in Josephus' differing accounts compared with historical texts, numismatic evidence, and documents that shed some light on how religious beliefs precipitated this war. It highlights the religious concept known as periodization, which likely led many of Jannaeus' subjects to fight against him, and possibly motivated Demetrius III as well to invade the Hasmonean state. Periodization is found in diverse writings such as the Bible, the Dead Sea Scrolls,[1] Hesiod, Berossus, Theopompus, and Aemilius Sura. Many

[1] The Dead Sea Scrolls are identified by the number of the cave in which they were discovered followed by "Q," representing the archaeological site of Khirbet Qumran where they were found. Most texts do not have names but are simply referred to by the number randomly assigned to them after their discovery. Because many texts do not have names, and since those with names frequently change their titles as scrolls continue to be pieced together, it is customary to list both the name followed by the document number in parentheses. For discussions of periodization in ancient

Jewish texts written prior to and during Jannaeus' reign espousing periodization maintained that God would intervene in world affairs at the end of a 490-year period of history, whose termination coincided with Demetrius III's invasion. This widespread teaching offers a plausible reason for the unusual decision of Demetrius III to wage war against the Hasmonean state at a time when his kingdom was vulnerable. It also reveals that many Jews at this time believed the approaching end of the 490-year eschatological period predicted in Scripture heralded God's termination of Hasmonean rule. This investigation suggests that in antiquity, just as today, religious beliefs were sometimes significant factors that led to warfare.

1. The Historical Background: The Hasmoneans

The Hasmonean family ruled the Middle Eastern province known as Judea for nearly nine decades (152-63 BCE).[2] After a long struggle against the

---

literature with numerous examples, see John J. Collins, *Apocalypticism in the Dead Sea Scrolls* (New York: Routledge, 1997), 52-70; Joseph W. Swain, "The Theory of Four Monarchies—Opposition History Under the Roman Empire," *Classical Philology* 25 (1940): 1-21. For prominent examples, see Hdt. 1.95; Hes. *Op.* 106-21; Aemilius Sura in Vell. Pat. 1.6.6; Dion. Hal. *Ant. Rom.* 1.2.2-4 ; Polyb. 18.22; Sibylline Oracle 4:49-101; Tac. *Hist.* 5.8-9.

[2] The family is often referred to as the Maccabees and their revolt against the Seleucid ruler Antiochus IV Epiphanes is known as the Maccabean revolt. However, "Maccabee" is a nickname for Judas (Judah), the son of the Hasmonean leader Mattathias, who started the war for Judean independence. Maccabee is related to the Hebrew and Aramaic root for "Hammer-like," and was given to Judas by his followers because of his military exploits. See further Bezalel Bar-Kochva, *Judas Maccabaeus* (Cambridge: Cambridge University Press, 1989), 147-48. For studies of the sources about the Hasmonean family and the origin of their conflict with the Seleucid Empire, see further Kenneth Atkinson, *A History of the Hasmonean State: Josephus and Beyond* (London: Bloomsbury T & T Clark, 2016); Edward Dąbrowa, *The Hasmoneans and their State: A Study in History, Ideology, and the Institutions* (Kraków: Jagiellonian University Press, 2010); Hanan Eshel, *The Dead Sea Scrolls and the Hasmonean State* (Grand Rapids, MI: Eerdmans, 2008); Jonathan A. Goldstein, *1 Maccabees* (Garden City, NY: Doubleday, 1976); *ibid., 2 Maccabees* (Garden City, NY: Doubleday, 1983); Lester L. Grabbe, *A History of the Jews and Judaism in the Second Temple Period Volume 3: The Maccabean Revolt, Hasmonean Rule, and Herod the Great (175-4 BCE)* (London: T & T Clark, 2020), 248-61; Eric S. Gruen, "Hellenism and Persecution," in *Hellenistic History and Culture*, edited by Peter Green (Berkeley, CA: University of California Press, 1993), 238-74; C. Habicht, "The Seleucids and their Rivals," in *Cambridge Ancient History Vol. 8: Rome and the Mediterranean to 133 B.C.* 2d edn., edited by A. E. Austin (Cambridge: Cambridge University Press, 1989), 324-74; Sylvie Honigman, *Tales*

Seleucid Empire's rulers in the mid-second century BCE, the Hasmonean family created an independent Jewish state. At first ruling as its political leaders and high priests, they transformed their nation into a monarchy. Judas in the early stages of his family's rebellion against the Seleucid Empire made an alliance with the Roman Republic, which occurred shortly after its victory in the Achaean War and the ascension of Demetrius II. At this time, the Republic, having intervened to prevent the Seleucid monarch Antiochus IV Epiphanes in 168 BCE from conquering Egypt, viewed the Hasmonean state as a buffer between the Ptolemaic and the Seleucid Empires.[3] However, by the time the Hasmonean king and high priest Alexander Jannaeus took power, the Hasmoneans had abandoned their long-standing pro-Roman alliance. Accused of piracy by the Republic, his family lost its leading foreign supporters and eventually succumbed to the army of Pompey the Great, who terminated the Hasmonean monarchy and annexed Judea in 63 BCE.[4] Jannaeus represents a transitional figure in the history of the Hasmonaean state because of the unprecedented internal opposition to

---

of High Priests and Taxes (Berkeley, CA: University of California Press, 2014); John Ma, *Antiochos III and the Cities of Western Asia Minor* (Oxford: Oxford University Press, 1999); Eyal Regev, *The Hasmoneans: Ideology, Archaeology, Identity* (Göttingen: Vandenhoeck, & Ruprecht, 2013); Chris Seeman, *Rome and Judea in Transition: Hasmonean Relations with the Roman Republic and the Evolution of the High Priesthood* (New York: Peter Lang, 2013); Linda Zollschan, *Rome and Judea: International Law Relations, 162-100 BCE* (London: Routledge, 2017).

3 For the Republic's intervention in Egypt, see Polyb. 29.27.1-8; Diod. Sic. 31.2; Livy, *Hist.* 45.12.1-8; App. *Syr.* 66.350-2; Just. *Epit.* 34.3.1-3. Cf. Daniel 11:29. For the early Hasmonean alliance with the Republic, see 1 Maccabees 12:1-4; 14:38-40; 15:16-21. A discussion of the chronology and accuracy of the accounts of the Hasmonean alliance with the Republic is beyond the limitations of the present study. For relevant studies and issues, see further Emil Schürer, *The History of the Jewish People in the Age of Jesus Christ (175 B.C. A.D.135)*, Revised and edited by Geza Vermes, et. al. (Edinburgh: T & T Clark, 1973), 194-95; Joseph Sievers, *The Hasmoneans and Their Supporters* (Atlanta: Scholars Press, 1990), 116-19; Menahem Stern, "The Relations Between Judea and Rome During the Reign of John Hyrcanus," *Zion,* 26 (1961): 3-19 (in Hebrew).

4 The charge of piracy is alluded to in Justin's *Epitome* of the *Philippic Histories* of Pompeius Trogus. See Just. *Epit. Prol.* 39 (*Ut Syriam Iudaei et Arabes terrestribus latrociniis infestarint*); 40.2.4; cf. Strabo, 16.2.40. For Hasmonean piracy, see further Kenneth Atkinson, "Judean Piracy, Judea and Parthia, and the Roman Annexation of Judea: The Evidence of Pompeius Trogus," *Electrum* 29 (2022): 127-46.

his rule, and due to a rather unusual military episode that occurred during his reign that nearly terminated his dynasty.[5]

Alexander Jannaeus came to power in 103 BCE following the untimely death of his brother, Judah Aristobulus, due to illness.[6] Although Aristobulus was the first Hasmonean monarch to use the title king, he refrained from putting it on his coins due to immense opposition from his subjects, for whom the office was restricted, according to Scripture, to descendants of the biblical king David.[7] During the Hasmonean period, this ancient biblical expectation for a deliverer underwent a profound transformation as the Jews believed the messiah would not only be a descendant of King David, but a warrior who would gather the Jewish righteous to the Promised Land, overthrow foreign rulers and their Gentile supporters, and rule the world from Jerusalem.[8] Jannaeus departed from

---

[5] For the major historical sources for the life of Alexander Jannaeus, see Joseph. *AJ* 13.301-19; Joseph. *BJ* 1.70-84; Syncellus *Chronicle* 555.8-17 (edition Wilhelm Dindorf, ed. *Georgius Syncellus et Nicephorus Cp.* [Bonn: Weber, 1829]); Atkinson, *History of the Hasmonean State*, 80-99; Nadav Sharon, *Judea Under Roman Domination* (Atlanta: SBL Press, 2017), 209-53.

[6] Documents from the "War of Scepters" (103-101 BCE), a conflict between Cleopatra III and her son Ptolemy IX (Lathyrus) Soter II for the Ptolemaic throne, shows that the traditional chronology for the Hasmoneans must be emended: Judah Aristobulus (1 year reign from 104-103 BCE) and Alexander Jannaeus (27-year reign from 103-76 BCE). For this evidence, see Getzel Cohen, "The Beginning of the Reign of Alexander Jannaeus," in *Judean Syrian Egyptian Conflict of 103-101 B.C.*, edited by E. Van 'T Dack, W. Clarysse, G. Cohen, J. Quaegebeur, and J. K. Winnicki (Brussels: Koniniklijke Academie voor Wetenschappen, 1989), 118-21.

[7] Aristobulus' coins contain the inscription "Judah the High Priest and the Congregation of the Jews." See Dan Barag Dan and Shraga Qedar, "The Beginning of Hasmonean Coinage," *Israel Numismatic Journal* 4 (1980): 8-21 and plates 3-9; Ya'akov Meshorer, *A Treasury of Jewish Coins From the Persian Period to Bar Kokhba* (Nyack, NJ: Yad Ben-Zvi Press, 2001), 27-29. The promises that God gave King David's family eternal rule over the Jewish people and descriptions of the messiah are found in many biblical passages, most notably: Numbers 24:17; 2 Samuel 7:11-17; Isaiah 11:1-9; Jeremiah 23:5-8, 30:9, 33:14-22; Ezekiel 17:22-24, 34:23-24, 37:15-28. See further Adela Yarbro Collins and John J. Collins, *King and Messiah as Son of God* (Grand Rapids, MI: Eerdmans, 2008); Torleif Elgvin, *Warrior, King, Servant, Savior: Messianism in the Hebrew Bible and Early Jewish Texts* (Grand Rapids, MI: Eerdmans, 2022), 104-80; Kenneth E. Pomykala, *The Davidic Dynasty Tradition in Early Judaism* (Atlanta: Scholars Press, 1995).

[8] This belief in a militant Davidic messiah is found in numerous Jewish documents of the Hellenistic period, most notably the *Psalms of Solomon* and the following Dead Sea Scrolls: *4QDibHam*[a] (4Q504), *1QSb* (1Q28b), *4QCommentary on Genesis*

Jewish tradition when he took power as monarch and high priest and insisted on placing the title "king" on his currency. Because Scripture restricted the monarchy to descendants of king David, and since the Hasmoneans traced their lineage to Israel's ancient priests, he was disqualified from governing as a king. Because of his illegitimate royal pedigree, and the monarchy's traditional connection with the biblical king David and the messiah, Jannaeus faced considerable public outrage over his use of the title king. Consequently, he was forced to overstrike his currency to replace "king" with "high priest."[9] He also put the image of a lily on his currency, which was a symbol of Jerusalem and its temple, along with the anchor to represent his family's longstanding ambition of conquering the coastal territories.[10]

Jannaeus' reign was the most calamitous period of Hasmonean history. He began by attacking the coastal city of Ptolemais to fulfill his family's goal of dominating the Mediterranean coast. Its occupants appealed to the deposed Egyptian monarch Ptolemy IX Soter, at the time exiled in Cyprus, for help. His mother Cleopatra III, who had removed him from power, came to the aid of the besieged residents. She amassed her forces, assisted by her other son, Ptolemy X Alexander, to repel Jannaeus and Soter to prevent either from capturing this important maritime port. During this war, Jannaeus lost 30,000 to 50,000 soldiers fighting Soter. Cleopatra III and Ptolemy Alexander crisscrossed the Hasmonean state while pursuing Soter and forced him back to Cyprus. Jannaeus survived this conflict by becoming a Ptolemaic vassal of Cleopatra III.[11] He then began an aggressive series of campaigns to expand the Hasmonean state beginning with a siege of Gaza.

A (4Q252), *4QMidrEschat*[a] (4Q174), *4QpIsa*[a] (4Q161), *4QSefer ha-Milḥamah* (4Q285), and *4QAramaic Apocalypse* (4Q246). For an examination of these texts, see Kenneth Atkinson, *I Cried to the Lord: A Study of the Psalms of Solomon's Historical Background and Social Setting* (Leiden: Brill, 2004), 144-76; John J. Collins, *The Scepter and the Star: The Messiahs of the Dead Sea Scrolls and Other Ancient Literature*. 2nd edn. (New York: Doubleday, 2010), 20-73, 117-22; Elgvin, *Warrior, King,* 228-62.

[9] David Henden and Ilan Shachar, "The Identity of YNTN on Hasmonean Overstruck Coins and the Chronology of the Alexander Jannaeus Types," *Israel Numismatic Research* 3 (2008): 87-94.

[10] David Hendin, "Numismatic Expressions of Hasmonean Sovereignty," *Israel Numismatic Research* 17 (2007/2008): 37-38; Meshorer, *Treasury of Jewish Coins,* 37; Regev, *Hasmoneans,* 208-09, 212-14.

[11] Jannaeus' siege of Ptolemais is only described in Joseph. *AJ* 13.324-55 while the War of Scepters is documented in Joseph. *AJ* 13.348-51. Archaeological evidence and Egyptian documents provide additional materials that help us reconstruct what

Jannaeus' siege of Gaza is pivotal for understanding the invasion of Judea by Demetrius III.[12] Josephus' accounts provide no precise date for the event. In *Antiquitates Judaicae* 13.358, he implies that it took place immediately after the cessation of the War of Scepters, which would place it in approximately 101 BCE. Gaza had provided Cleopatra III's rebel son Soter shelter. Consequently, allowing Jannaeus to conquer it would have helped her by denying Soter a refuge from which to attack Egypt from the mainland. Consequently, this could explain the timing of the event as it was in the interest of Cleopatra III to allow Jannaeus to risk his life and forces to take this city, which had helped her rebellious son. However, in *Antiquitates Judaicae* 13.365, Josephus provides a more probable date when he writes that Jannaeus' siege of Gaza coincided with the murder of the Seleucid monarch Antiochus VIII Grypus, which took place in 96 BCE nearly five years after the War of Scepters.[13] This date coincides with the numismatic evidence for his death and provides a *terminus a quo* of 95/4 BCE for the conquest of Gaza by Jannaeus.[14]

Jannaeus chose an opportune time to attack Gaza. The Seleucid Empire was politically weak as its feuding rulers who had been fighting for control of Antioch, Grypus and Antiochus IX Cyzicenus, had died. For the next twelve

---

took place. See further Atkinson, *History of the Hasmonean State*, 108-21; E. Van 'T Dack, "Le conflict judéo-syro-égyptien de 103/102 av. J-C," in *Proceedings of the Sixteenth International Congress of Papyrology,* edited by Roger R. S. Bagnall, et al. (Chico, CA: Scholars Press, 1981), 303-12.

[12] The siege is documented in Joseph. *AJ* 13.372-78; Joseph. *BJ* 1.88-92.

[13] Grypus had waged a long war against his half-brother and cousin, Antiochus IX Cyzicenus, over such strategic cities as Antioch and Ptolemais. The general Heracleon murdered Grypus in 97/6 BCE. See further Kay Ehling, *Untersuchungen zur Geschichte der späten Seleukiden (164-63 v. Chr.): vom Tode des Antiochos I. bis zur Einrichtung der Provinz Syria unter Pompeius* (Stuttgart: Steiner, 2008), 231-32; Oliver D. Hoover, "A Revised Chronology for the Late Seleucids at Antioch (121/120-64 BC)," *Historia: Zeitschrift für alte Geschichte,* 56, no. 3 (2007): 285-86.

[14] Gaza became an autonomous city independent of Seleucid rule in 108/7 BCE when it stopped minting coins using the Seleucid calendar. It minted coins up to "year 14" when it was no longer an independent city-state and under Hasmonean control. Because "year 14" correlates to 95/4 BCE, it coincides with Jannaeus' siege of the city; he captured it the following year. See Oliver D. Hoover; "The Dated Coinage of Gaza in Historical Context (264/3 BC-AD 241/2)," *Schweizerische Numismatische Rundschau* 86 (2007): 63-90; Alla Kushnir-Stein, "Gaza Coinage Dated LIC—A Reappraisal," *Schweizerische Numismatische Rundschau,* 74 (1995): 49-55; *ibid.,* "Late Hellenistic Coins of Gaza and the Date of the Hasmonean Conquest of the City," *Schweizer Münzblätter,* 198 (2000): 22-24.

years the Seleucid Empire experienced almost continual civil war between the five sons of Grypus (Seleucus VI Epiphanes Nicator, Antiochus XI Epiphanes Philadelphus, Demetrius III Eucaerus, Philip I Philadelphus, and Antiochus XII Dionysus) and the son of Cyzicenus (Antiochus X Eusebes Philopator).[15] If not for this political instability, it is doubtful that Jannaeus would have followed his siege of Gaza with his campaign east of the Jordan River in the region known as the Transjordan to annex territories long claimed by the Seleucid Empire's monarchs.

Convinced he was safe from Seleucid depredations of his territory, Jannaeus undertook a series of campaigns in the Transjordan region, which had been under the control of the Seleucid Empire until the reign of Antiochus VII Sidetes (138–129 BCE).[16] This region, largely populated by Greeks and Macedonians, had made up the strength of his empire and served as a major locale from which he recruited troops for his army until he died in battle fighting the Parthian ruler Phraates II (=Arsaces VII 132-127 B.C.E.).[17] Confident that the Seleucid Empire was in a period of decline, Jannaeus fought a series of battles to annex parts of the Transjordan. After being lured into an ambush by a local tyrant named Theodorus, some of Jannaeus' Jewish subjects summoned Demetrius III to attack him.[18]

---

[15] Joseph. *AJ* 13.365-71, 384-91; App. *Syr.* 69.365-6; Alfred R. Bellinger, "The End of the Seleucids," *Transactions of the Connecticut Academy of Arts and Sciences* 38 (1949): 73-80; Ehling, *Utersuchungen zur Geschichte,*, 231-56; Jonathan D. Grainger, *The Fall of the Seleukid Empire 187-75 BC* (Barnsley: Pen & Sword Military, 2021), 166-99; Hoover, "Revised Chronology," 288-301.

[16] See further Ehling, *Untersuchungen zur Geschichte,* 178-216; Daniel R. Schwartz, "On Antiochus VII Sidetes' Parthian Expedition and the Fragmentation of Historical Research," in *The Jews in the Hellenistic-Roman World: Studies in Memory of Menahem Stern,* edited by Isaiah M. Gafni, Aharon Oppenheimer, and Daniel R. Schwartz (Jerusalem: Zalman Shazar Center for Jewish History, 1996), 83-102 (in Hebrew).

[17] Sidetes likely died in Parthia during the autumn of 129 BCE. For evidence to support this dating, see Gholamreza F. Assar "A Revised Parthian Chronology of the Period 165-91 B.C.," *Electrum* 11 (2006): 113-14; Ehling, *Untersuchungen zur Geschichte,* 201-05; Nikolaus K. Overtoom, *Reign of Arrows: The Rise of the Parthian Empire in the Hellenistic Middle East* (Oxford: Oxford University Press, 2020), 228-35. For the importance of this region for the Seleucid monarchs, see Grainger, *Fall of the Seleukid*, 88-105.

[18] The chronology of Jannaeus' Transjordan campaign (Joseph. *BJ* 1.89; Joseph. *AJ* 13.374-76) is difficult to reconstruct due to the confusing nature of his accounts as he duplicated his sources. See further Atkinson, *History of the Hasmonean State,* 121-25.

At this time, Demetrius III held the strategic city of Damascus.[19] While under Seleucid control during the second century BCE, various Seleucid rulers re-urbanized it along Hellenistic lines with the establishment of residential, commercial, agricultural, and military zones.[20] The city's coinage shows a gap for the year 93/2 BCE, suggesting that Demetrius III was briefly ejected from Damascus for that year.[21] After he departed Damascus, his enemies, likely Jannaeus or possibly the Nabatean Arabs, appear to have occupied the city, which accounts for the cessation of his coins there.[22] Demetrius III's absence likely made him vulnerable to attacks from his rival, Antiochus X Eusebes.[23] Once Demetrius III reoccupied Damascus, he minted coins there at an accelerated rate until 90/89 BCE, apparently in preparation for his invasion of Judea that took place in that year.[24] While Demetrius III was reconsolidating his rule over Damascus and preparing to invade the Hasmonean state, Jannaeus undertook his Transjordan campaign to exploit the region's political instability while Demetrius III and his brother, Philip I Philadelphus, were fighting for control of the Seleucid Empire.[25] While undertaking his war of expansion, Demetrius III invaded the Hasmonean state.

---

[19] Coins show that his reign in Damascus began in 97/6 BCE and lasted until 88/7 BCE, except for a short interruption in 93/2 BCE. See Nicholas L. Wright, "A Late Seleukid Bronze Hoard, c. 1988 (*CH* 10, 349)," in *Coin Hoards Volume X: Greek Hoards*, edited by Oliver Hoover, Andrew Meadows, and Ute Wartenberg (New York: American Numismatic Society, 2010), 254-55.

[20] For the evidence, see Getzel Cohen, *Hellenistic Settlements in Syria, the Red Sea Basin, and North Africa* (Berkeley: University of California Press, 2006), 242-45.

[21] For this evidence, see Hoover, "A Revised Chronology," 289-96.

[22] Oliver D. Hoover, Arthur Houghton, and Petr Vesley, "The Silver Mint of Damascus under Demetrius III and Antiochus XII (97/6 BC-83/2 BC)," *American Numismatic Society,* 20 (2008): 315-16.

[23] See Hoover, Houghton, and Vesley, "Silver Mint," 307. Eusebus was the son of Antiochus IX Cyzicenus.

[24] The invasion of Demetrius III is commonly dated to 88 BCE without any supporting evidence. For examples, see Ehling, *Untersuchungen zur Geschichte,* 244-5; Joseph Klausner, "Judah Aristobulus and Jannaeus Alexander," in *The World History of the Jewish People VI: The Hellenistic Age*, edited by Abraham Schalit (New Brunswick: Rutgers University Press, 1972), 232; Schürer, *History of the Jewish People,* 1.221, 223-4. The events in the Seleucid Empire favor a date of 90/89 BCE argued here. For additional support of the date adopted in the present study, see Atkinson, *History of the Hasmonean State,* 125-27; Hoover, Houghton, and Vesley, "Silver Mint," 306-7, 310.

[25] Demetrius III, Philip I, Antiochus XII Dionysus, and finally the Armenian king Tigranes II fought and occupied the city of Antioch for a time in their efforts to

II. The Encounter Between Demetrius III and Alexander Jannaeus

Josephus' accounts of Demetrius III are polemical and devoid of information, making it difficult to determine what happened when these two rulers fought one another. He provides no dates for Demetrius III's invasion of Judea, no background information, and a schematic portrayal of events in the Seleucid Empire. His use of the moniker "Akairos" ("Untimely"; Δημήτριον τὸν Ἄκαιρον) for Demetrius III (Joseph. *AJ* 13.369; Joseph. *BJ* 1.92) is clearly intended to imply that his later demise shortly after his encounter with Jannaeus was God's punishment upon him for attacking the high priest. Although Akairos would appear to presuppose the positive sobriquet Eukairos (Εὔκαιρον), which is Demetrius III's moniker in the Latin manuscripts of Josephus and the table of contents appended to his writings in Greek, there is no evidence that Eukairos was used for Demetrius III.[26] Consequently, we should have little confidence in Josephus' retelling of this dramatic encounter between Demetrius III and Jannaeus, which continues to puzzle historians.

Jannaeus confronted Demetrius III in Shechem to the north of Jerusalem, apparently to keep the Seleucid army away from his capital. According to Josephus' terse accounts, Jannaeus possessed an army of some six thousand soldiers and two hundred mercenaries; he was purportedly the first Hasmonean ruler to employ hired soldiers from Cilicia.[27] Josephus also mentions that Jannaeus hired both Cilician and Pisidian mercenaries. Although the Seleucid Empire's civil wars prevented him from employing Syrians, he must have had considerable funds to hire pagan soldiers since the inhabitants of Pisidia and Cilicia were renowned for their fighting skills and consequently highly sought by foreign rulers. He used these foreigners to quell dissent, making them an instrumental part of his military force.[28]

---

control the Seleucid Empire. See Ehling, *Untersuchungen zur Geschichte,* 231-56; Hoover, "Revised Chronology," 298-99; Arthur Houghton, "The Struggle for the Seleucid Succession, 94-92 BC: A New Tetradrachm of Antiochus XI and Philip I of Antioch," *Schwizerische numismatische Rundschau* 77 (1998): 66-68.

[26] David B. Levenson and Thomas R. Martin, "Akairos or Eukairos? The Nickname of the Seleucid King Demetrius III in the Transmission of the Texts of Josephus' *War* and *Antiquities*," *Journal for the Study of Judaism,* 40, no. 3 (2009):307-41.

[27] Joseph. *BJ* 1.88; Joseph. *AJ* 13.374.

[28] Joseph. *BJ* 1.88-89; Joseph. *AJ* 13.374. See further John M. Allegro, "*Thrakidan,* the 'Lion of Warth' and Alexander Jannaeus," *Palestine Exploration Quarterly,* 91, no. 1 (1959): 47-51; Dąbrowa, *Hasmoneans and their State,* 160-61.

Although they became essential to keeping him on the throne, his use of them posed problems for the Roman Republic that led it to annex the Hasmonean state during the reign of his younger son, Aristobulus II. This is because of the threat the Romans believed Jannaeus and his family posed to Mediterranean trade.

Cilician pirates hired themselves out as soldiers to such an extent that the name "Cilician" became synonymous in the minds of Romans with pirates, which appears to have made the Republic suspicious of the Hasmoneans.[29] Strabo (16.2.28) even writes that the principal Hasmonean port of Joppa served as a base of operations for pirates. Jannaeus' employment of Cilician mercenaries posed additional problems since it violated Rome's edict against pirates, which had included an embargo on contact with Cilicia.[30] Later, when Pompey confronted Jannaeus' two sons, Hyrcanus II and Aristobulus II, when they were fighting one another to rule the Hasmonean state, piracy became an issue. According to Diodorus Siculus (40.2), Pompey denounced Hyrcanus II for the "lawless behavior of the Jews (παρανομίας τῶν Ἰουδαίων) and the wrongs the Hasmoneans had committed against the Romans." At this meeting, according to Diodorus' account (40.2), some Jews accused the Hasmoneans of "having overthrown the ancient laws and enslaved the citizens in defiance of all justice; for it was by means of a horde of mercenaries, and by outrages and countless impious murders that they had established themselves as kings." When Pompey met the two brothers, moreover, Hyrcanus II accused Aristobulus II of raiding neighboring territories and committing acts of piracy at sea.[31]

In addition to posing a potential threat to the Republic for its apparent refusal to eliminate piracy in its territory, the Hasmoneans, largely beginning with Jannaeus, began to employ foreign mercenaries in large numbers. These men-for-hire posed a major threat to the region as they

---

[29] Atkinson, "Judean Piracy," 127-46; Zollschan, *Rome and Judea,* 267-68.

[30] Earlier, in 100 BCE, the Roman Republic passed a law (*lex de piratis persequendis*) that required the kings and states allied to Rome in Anatolia, Syria, and Egypt to stop piracy in the Mediterranean and to deny pirates the use of their ports, See further, Atkinson, "Judean Piracy," 127-46; W. R. Loader, "Pompey's Command under the Lex Gabinia," *Classical Quarterly* 54 (1940): 134-36.

[31] *AJ* 14.43. Richard Laqueur *Der jüdische Historiker Flavius Josephus* (Giessen Münchow, 1920), 145-58. proposes the reference to pirates in this passage came from Theophanes, whose work was used by Strabo and Diodorus Siculus. For the possibility that Trogus used sources also consulted by Strabo, see further Atkinson, "Judean Piracy," 127-46; Alice Borgna, "Scrivere del nemico. Pompeo Trogo e la storia de Parti (Iust., XLI-XLII)," *Dialogues d'histoire ancienne* 43 (2015): 92-94.

began to work for nearly all the surrounding kingdoms. Consequently, the Hasmoneans' continued employment of them during and after Jannaeus' reign undoubtedly worried the Republic as they could destabilize the region.[32] Earlier, in 145 BCE, the Seleucid monarch Demetrius II used mercenaries from Crete to seize Antioch.[33] Jannaeus likewise employed such contingents. He expanded his use of them when he captured the territories of the tyrant Zoilus, the ruler of Dor, who had a company of mercenaries stationed at Straton's Tower on the coast.[34] If the find of 1,735 coins found along the Dead Sea, nearly all of which were minted by Jannaeus, was used to pay mercenaries, this could suggest that his foreign armies were quite costly.[35] Because such men apparently worked for the highest bidder, the Seleucid Empire's rulers also used many of the same soldiers as mercenaries. Consequently, Demetrius III's army that confronted Jannaeus consisted of large numbers of Greek mercenaries, some of whom were possibly from the same regions as those employed in the Hasmonean army.[36]

According to Josephus, when the battle began, both armies encouraged the mercenaries fighting on the opposite side to defect.[37] However, Josephus mentions that twenty thousand Jews had joined the army of Demetrius III, who already possessed three thousand cavalry and fourteen thousand infantry.[38] Before the battle began his former Jewish subjects who had joined the Seleucid army refused Jannaeus' appeal that they change sides.

---

[32] See further Atkinson, "Judean Piracy," 127-46.

[33] See Joseph. *AJ* 13.133-42. Ehling, *Untersuchungen zur Geschichte,* 164-78; Grainger, *Fall of the Seleukid,* 79-80.

[34] Joseph. *AJ* 13.326. Josephus uses the word σύνταγμα, which can refer to a select body of troops, such as a battalion, as well as the constitution of a state. See Henry George Liddell, Robert Scott, Henry Stuart Jones, and Roderick McKenzie, *A Greek-English Lexicon,* rev. ed. (Oxford: Oxford University Press, 1996), sv. Σύνταγμα. Josephus uses this same word with the latter meaning in a citation from Strabo in *AJ* 14.116.

[35] Yizhar Hirschfeld and Donald T. Ariel, "A Coin Assemblage from the Reign of Alexander Jannaeus Found on the Shore of the Dead Sea," *Israel Exploration Journal* 55 (2005): 66-89.

[36] Joseph. *BJ* 1.95; Joseph. *AJ* 13.379. See Dąbrowa, *Hasmoneans and their State,* 81-82.

[37] Josephus' two accounts of the battle are brief, devoid of details, and contain some differences. See Joseph. *BJ* 1.93-95; Joseph. *AJ* 13.377-78.

[38] Joseph. *AJ* 13.377-78. For the events see the rest of the paragraph. Joseph. *BJ* 1.93 provides the figures: 9000 mercenaries for Jannaeus (1000 horse + 8000 infantry and 10,000 Jews) and for Demetrius 3000 cavalry and 14,000 infantry.

Those fighting in both forces were apparently unwilling to abandon their employers. Once the battle ensued, the army of Demetrius III defeated Jannaeus' forces, killing all the mercenaries fighting for the Hasmonean ruler. Nevertheless, Josephus acknowledges that many soldiers in Demetrius III's army, especially his mercenary units, perished as well, suggesting that neither side was completely victorious. Then, Josephus asserts, Jannaeus fled to the mountains. It is here where Josephus makes a startling claim that calls into question the historical accuracy of his entire account.

According to Josephus, six thousand of the Jews who had summoned Demetrius III and fought alongside him defected. Through means unspecified, they managed to leave the Seleucid army to join Jannaeus' force and join the fight against their former employer.[39] Now uncertain of his army's loyalty, and outnumbered in hostile territory, Demetrius III panicked and fled Judea. According to Josephus, once Demetrius III returned home, he besieged his brother Philip I at Beroea. Philip's ally Straton called upon Azizus, the phylarch of the Arabs, and "Mithridates Sinakes, the governor of the Parthians," for help.[40] This combined force defeated Demetrius III, took him prisoner, and exiled him to Parthia where he died in captivity.[41] The disappearance of Demetrius III meant that he was no longer a threat to the Hasmonean state. Consequently, Jannaeus was free to eradicate his religious opponents and his political rivals.

Josephus writes that Jews continued to fight against Jannaeus following the departure of Demetrius III. Jannaeus besieged some of them in Bethoma/Bemeselis.[42] He captured the city, took many prisoners to Jerusalem, and crucified 800 of them there while slaughtering their children and wives before their eyes. Josephus reports that over the next six years

---

[39] Joseph. *AJ* 13.379. The account in Joseph. *BJ* 1.95-99 is schematic and contains few parallels.

[40] Joseph. *AJ* 13.385. The name Aziz is emended from the account of Diodorus Siculus (40.1a,b). One manuscript reads "Deizus" and the others "Zizus." Concerning the enigmatic Mithridates, Strabo (*Geography*, 16.1.23) refers to a place in Northern Mesopotamia called Sinnaca, which may have been where he was from.

[41] Joseph. *AJ* 13.384-6. See further Edward Dąbrowa, "Demetrius III in Judea," *Electrum* 18 (2010): 175-181.

[42] Joseph. *AJ* 13.380. Joseph. *BJ* 1.97 reads Bemeselis. The exact location of this conflict is uncertain, but it may be located somewhere north-east of Samaria at Misilye. See further, Eshel, *Dead Sea Scrolls,* 119; Schürer, *History of the Jewish People,* 224 n. 21.

Jannaeus killed more than 50,000 of his own citizens.[43] The remainder of his reign was filled with great bloodshed and nearly constant war. But Josephus fails to explain his puzzling encounter with Demetrius III, namely how and why it abruptly ended.

A close reading of Josephus' terse account suggests that neither Demetrius III nor Jannaeus appeared to have suffered major military losses after their encounter since both continued their campaigns uninterrupted. Jannaeus resumed his foreign wars while Demetrius III besieged his brother, Philip I, at Beroea, which was the latter's base at the time.[44] This shows that Demetrius III had not been weakened by his invasion of Judea. He likely would have defeated Philip I if the Arabs and Parthians had not come to his brother's aide. However, this leaves unanswered the reason for his abrupt departure when he was about to prevail over the army of Jannaeus.

Contrary to Josephus' assertion that Demetrius III had to flee Judea because those Jews in his army who had encouraged him to invade Judea to remove Jannaeus from power had defected, a close reading of his account suggests otherwise. It was events in the Seleucid Empire that caused Demetirus III to end his campaign abruptly, on the verge of victory. The numismatic evidence shows that Antiochus X Eusebes retook Antioch in 93/2 BCE and held it until 89/8 BCE when he died fighting the Parthians to help the Arab queen Laodice.[45] At the time of the death of Eusebes, Philip I held a portion of northern Syria from his strongholds of Beroea and Cilicia. Eusebes' wife, Cleopatra Selene, acting in the name of their son, Antiochus XIII Asiaticus,

---

[43] The mass crucifixion is also mentioned in the Dead Sea Scroll text known as the *Nahum Pesher* (4Q169 3-4 I, 1-7), showing that it is a historical event and not an exaggeration of Josephus. See further Shani L. Berrin, *The Pesher Nahum Scroll from Qumran: An Exegetical Study of 4Q169* (Leiden: Brill, 2004), 165-92.

[44] Joseph. *AJ* 13.384. The toponym Beroea is similar to the Macedonian town of that name and may indicate colonists from northern Greece founded it. See further Cohen, *Hellenistic Settlements*, 153-54.

[45] Joseph. *AJ* 13.371. She is likely Laodike Thea, the Commagenian queen of Seleucid ancestry. Her father was presumably Antiochus VIII Grypus and her mother was Cleopatra Tryphaena. She married Mithridates I Callinicus. For this identification, see further Gholamreza F. Assar, "A Revised Parthian Chronology of the Period 91-55 BC," *Parthica* 8 (2006): 60; Ehling, *Untersuchungen zur Geschichte,* 230; Jonathan D. Grainger, *A Seleukid Prosopography and Gazetteer* (Leiden: Brill, 1997), 31, 48; Marek J Olbrycht, "Mithridates VI Eupator and Iran," in *Mithridates VI and the Pontic Kingdom*, edited by Jakob M. Højte (Aarhus: Aarhus University Press, 2009), 166-7.

emerged as a third claimant to the throne of the Seleucid Empire.[46] All these rulers wanted to control the strategic city of Antioch after the death of Eusebes to consolidate their hold over Syria. With Antioch potentially vulnerable, Demetrius III had to abandon his invasion of Jannaeus' realm to seize his brother's kingdom as well as the territory of his rival Eusebes.

It was luck that saved Jannaeus' kingdom. Any Jews fighting in the Seleucid army likely accompanied Demetrius III home and fought for him. But if we are to offer an explanation for Jannaeus' decision to campaign in the Transjordan to seize territory belonging to the Seleucid Empire at a time when several of its warring rulers possessed armies large enough to destroy him, we should turn to the topic of religion. Jannaeus apparently thought God was on his side as he believed he was the messiah promised in Jewish Scripture. Unfortunately, his opponents read the same Scriptural texts differently: they were confident God had predicted the demise of his kingdom which obligated them to join Demetrius III to remove Jannaeus from power and destroy his kingdom. Ironically, Jews fighting in both armies did so convinced God was on their side.

III. Jannaeus, Periodization, and Messianism

Throughout Jewish history, war and Scriptural interpretation have been closely connected. This is particularly true of the Conquest narratives, in which the ancient Israelites engaged in wars of extermination against their neighbors to conquer their lands.[47] In the case of the invasion of Demetrius III, the ancient concept known as periodization offers a possible and

---

[46] For Antiochus XIII Asiaticus, see further Ehling, *Untersuchungen zur Geschichte,* 256-77; Grainger, *Seleukid Prosopography,* 34-5.

[47] Although scholars do not believe that many of the biblical stories about the Israelite Conquest are factual since there is minimal archaeological evidence to support most of the biblical narratives, they motivated the Israelites during the post-biblical period to fight wars of conquest to take neighboring lands. See further, Jennie R. Ebeling, J. Edward Wright, Mark Elliott, and Paul Flesher, *The Old Testament in Archaeology and History* (Waco, TX: Baylor University Press, 2017). The Hasmoneans undertook their wars of conquest to fulfill the Scriptural commands of the Conquest Period that God had given the Promised Land to the Jews and that foreigners in them should be expelled. See further Kenneth Atkinson, "Josephus's Use of Scripture to Describe Hasmonean Territorial Expansion," *JSIJ—Jewish Studies, an Internet Journal*, 19 (2021): https://jewish-faculty.biu.ac.il/en/node/1602.

overlooked reason why he invaded the Hasmonean state in 90/89 BCE and why many of Jannaeus' subjects fought against him.

Periodization is the belief that time is divided into several periods, which will culminate with the end of the world or some cosmic intervention. Those who discern present signs can recognize how near we are to the end of this timetable. The concept greatly influenced ancient historians, many of whom grouped periods and empires together to result in an orderly historical sequence. Many of them presented empires as succeeding one another chronologically with the understanding that there could only be one dominant empire at a time. Herodotus, for example, divided history into three successive empires.[48] Berossus also categorized history into various periods while the fourth century BCE Greek writer Theopompus split history into successive periods of three thousand years each, which would end when the evil one is conquered.[49] A fragment of the early second century BCE Roman chronicler Aemilius Sura describes a schema of four kingdoms.[50] Sura drew upon earlier schemas of writers such as Eratosthenes and Ctesias, the latter of whom aligned the beginning of the Olympic games with the Assyrian Empire's downfall. In addition, Sura followed Ctesias' world history, which began in 2185 BCE and divided history into five parts up to the defeat of Antiochus III in 190 BCE.[51] L. Cornelius Sisenna likewise grouped events together to create an orderly sequence that accounted for the rise of the Roman Republic.[52] A similar consolidation of historical periods is found in the Hellenistic *Dynastic Prophecy* from Babylon that describes four kingdoms leading up its author's hope that the Seleucid Empire will collapse.[53] Lycophron's third century BCE poem *Alexandra* (lns. 1226-31, 1446-50) contains a prophecy in which Rome will rise at the end of time. Roman historians such as Polybius, Dionysius of

---

[48] Hdt. 1.95 (Assyrian, Median, and Persian).

[49] For Berossus, see Herman L. Jansen, *Die Henochgestalt: Eine vergleichende religionsgeschichtliche Untersuchung* (Oslo: Dybwab, 1939), 78. For Theopompus, see the testimony of Plut. *De Is. et Os.* 47.

[50] Aemilius Sura in Vell. Pat. 1.6.6. See Theodor Mommsen, "Mamilius Sura, Aemilius Sura, L. Manlius," *Rheinisches Museum für Philologie* 16 (1861): 282-87.

[51] cf. Diod. Sic. 2.22.2. For the text and discussion, see Tim J. Cornell, ed., *The Fragments of the Roman Historians* (Oxford: Oxford University Press, 2013), 1.656; 2.1144.

[52] For text, see Cornell, *Fragments,* 2.610-71.

[53] See Albert K. Grayson, *Babylon Historical-Literary Texts* (Toronto: University of Toronto Press, 1975), 14-17.

Halicarnassus, Tacitus, and Appian used periodization to portray Rome as the fifth and final empire.[54]

Many pagan and Jewish historians merged mythology and factual chronological information to derive their historical chronologies. Hesiod, for example, assigned each age of history to a different metal.[55] Thallus wrote a history from the fall of Troy to the time of Christ that offers a mythological *Urzeit*.[56] The Jewish materials that form the basis of *Sibylline Oracles* 1, 2, and 4, moreover, also contain an elaborate division of history into generations, demonstrating that the belief was both long lived and widespread.[57]

Periodization was an important component of Second Temple Period Jewish (587/6 BCE-70 CE) writings. Ancient Jewish apocalyptic literature was influenced by Scripture and frequently divided history into distinct periods with the expectation that they will come to a definitive end when God intervenes in the world to eradicate evil.[58] This model is quite ancient and is found in such varied works as the *Apocalypse of Weeks* (1 Enoch 93:1-10; 91:11-17), which divides history into periods of ten "weeks."[59] It is

---

[54] Polyb. 18.22; Dion. Hal. *Ant. Rom.*, 1.2.2-4; Tac. *Hist.* 5.8-9; App. *Hist. Rom.*, pref. 9.

[55] Hes. *Op.* 109-21. Arnaldo Momigliano ("The Origin of Universal History," *Atti della Scuole Normale Superiore di Pisa,* ser. 3.12.2 [1982]: 533-35, 553-54) notes that Hesiod adopted his chronology from earlier texts, although he believed the quadripartite scheme of world empires is a Greek peculiarity.

[56] Felix Jacoby, *Die Fragmente der griechischen Historiker. (FGrHist): Zweiter Teil: Spezialgeschichten, Autobiographien und Memoiren, Zeittafeln* (Berlin: Weidmann, 1929), 1156-58.

[57] Johannes Geffcken, *Komposition und Entstehungszeit der oracular Sibyllina* (Leipzig: De Gruyter, 1902), 47-53. Sibylline Oracle 4 describes the following succession of kingdoms, each of which will end in destruction: Assyria (4.49-53), Media (4.54-64), Persia (4.65-87), Macedonia (4.88-101), and Rome (4.102-51). For the four-kingdom motif and periodized time in the Seleucid Empire, see further Paul J. Kosmin, *Time and Its Adversaries in the Seleucid Empire* (Cambridge, MA: Belknap Press, 2018); Olivia S. Lester, "The Four Kingdoms Motif and Sibylline Temporality in Sibylline Oracles 4," in *Four Kingdom Motifs Before and Beyond the Book of Daniel*, edited by Andrew Perrin and Loren T. Stuckenbruck (Leiden: Brill, 2020), 121-41.

[58] Collins, *Apocalypticism,*, 52-70. For some of the most important Jewish texts that divide history into divisions, often tenfold and twelvefold, see 4 Ezra 14:11-12; 2 Apocalypse of Baruch 53, Apocalypse of Abraham 29:2.

[59] In this and other texts (cf. 4Q180 1:3; 1QH 9(1): 23-25) the periods are engraved on heavenly tablets to emphasize the orderliness of God's creation and the divinely

particularly evident in the "historical" apocalypses in which periodization is used to divide time into predetermined eras. Of all the biblical texts, the division of history into four periods found in Daniel 2 and 7 was the most influential.[60]

The writings of the biblical prophet Daniel interpret the earlier teachings of the prophet Jeremiah, who warned his nation that God had allowed the Babylonian king Nebuchadnezzar II in 587/6 BCE to conquer Jerusalem and exile much of its population to Babylon because of the nation's sins and its worship of pagan gods (Jeremiah 25:4-7, 9-11). Jeremiah predicted that after the end of seventy years, God would punish the Babylonians by replacing them with a more powerful kingdom (Jeremiah 25:11). At the end of this seventy-years, moreover, God would allow the exiles to return to Jerusalem and their homes and rebuild their destroyed temple (Jeremiah 29:10-14). Daniel reinterpreted and updated Jeremiah's prophecy to describe a series of empires in which the Greeks would follow the Medes and Persians (Daniel 8:20-21). In addition, Daniel 9 reinterpreted the seventy years of Jeremiah (23:11-12; 29:10) as describing a succession of empires, with the seventy-years representing seventy weeks of years, or 490 years. This figure in turn was influenced by the sabbatical theology of the biblical Book of Leviticus 25-26 which mandated that the land be allowed to rest every seventh year.[61] This same tenfold period is reflected in the *Apocalypse of Weeks* and also in *11QMelchizedek* in the Dead Sea Scrolls, showing its importance for Second Temple Period Jewish authors.[62] During the Hellenistic era following the Hasmonean revolt against Antiochus IV

---

ordained sequence of history. See George W. E. Nickelsburg, *1 Enoch: A Commentary on the Book of 1 Enoch* (Minneapolis: Fortress, 2012), 438-450.

[60] Swain ("The Theory of Four Monarchies," 1-21) traced this concept of four empires followed by a fifth found in Daniel and other Second Temple Period Jewish writings to Iran and believed that it became popular during the Greco-Roman period as part of oriental, anti-Greek propaganda. Hesiod (*Op.* 109-201) is perhaps the best Greek example as he delineates descending ages (races) that have preceded the present, fifth, race.

[61] See further Adela Yarbro Collins, "Numerical Symbolism in Apocalyptic Literature," *Aufstieg und Niedergang der Römischen Welt* II.21.2 (1984): 122-87; Klaus Koch, "Die mysteriösen Zahlen der judäischen Königc und die apokalyptischen Jahrwochen," *Vetus Testamentum* 28 (1978): 433-41. Leviticus portrays this in cosmic terms to show that the weekly Sabbath day of rest and the Sabbatical year, during which planting was forbidden, mirror the pattern of creation and God's rest following the creation. The ancient Israelites interpreted the Babylonian exile as a sabbath, which Jeremiah had predicted (2 Chronicles 36:21).

[62] See further Collins, *Apocalypticism,* 52-70.

Epiphanes, many Jews believed that the time of salvation was at hand, foreign domination would end, and that the end of Jeremiah's seventy weeks of years marking this cosmic transformation of the world was imminent.[63]

The Dead Sea Scroll document known as *4QApocryphon of Jeremiah C^e^* (4Q390) provides a plausible background for explaining the curious events surrounding the invasion of Demetrius III, namely why Jews fought in the Seleucid army against Jannaeus and continued to rebel against his rule. This document presupposes the concept of periodization based on Jeremiah's seventy weeks of years, found in many other Jewish documents, but applies this biblical prediction to the Hasmonean monarchy. In this work, the author uses the chronology of Daniel 9:24-27 to date the last historical epoch to the period of the Hasmonean revolt against the Seleucid Empire of Antiochus IV Epiphanes. It offers a plausible scenario why Demetrius III invaded Jannaeus' kingdom in in 90/89 BCE.[64] This composition divides history into the following four periods:

a) 70 years of the Babylonian exile.
b) 343 years during which the returnees from exile behaved properly.
c) 7 years of rule by Antiochus IV Epiphanes.
d) 70 years of Hasmonean rule.

The author of *4QApocryphon of Jeremiah C^e^* believed that the end-time and Israel's redemption would occur after 490 years, based on Jeremiah's (23:11-12; 29:10) seventy years prophecy, which Daniel 9 interprets as seventy weeks of years, or 490 years. The calculation in this text places the final redemption after the end of the 490-year period. Whether the writer of *4QApocryphon of Jeremiah C^e^* counted from the purge of the Temple by Judas ("Maccabee") in 164 BCE after his forces defeated the Seleucid army of Antiochus IV Epiphanes, or from 160 BCE, seven years after Antiochus defiled the Jerusalem temple when he placed a statue of Zeus in it, is unspecified. The dates respectively correspond to 94 BCE or 90 BCE.[65] If

---

[63] See further Michael Knibb, "The Exile in the Literature of the Intertestamental Period," *Heythrop Journal* 17 (1976): 253-72; Anathea E. Portier-Young, *Apocalypse Against Empire: Theologies of Resistance in Early Judaism* (Grand Rapids: Eerdmans, 2011).

[64] The extant copy of this document, based on paleographical analysis, was produced between 30-20 BCE. See Devorah Dimant, *Qumran Cave 4.XXI: Parabiblical Texts, Part 4: Pseudo-Prophetic Texts* (Oxford: Clarendon, 2001), 93, 113-16, 235-53.

[65] For the four-period timetable in *4QApocryphon of Jeremiah C^e^* (4Q390), see further Eshel, *Dead Sea Scrolls*, 22-27.

we accept the numismatic evidence that Demetrius III invaded in the latter year, then his incursion coincides with the calculation that the fulfillment of the seventy-year prophecy marks the end of Hasmonean rule. Given the dating of Demetrius III's invasion to 90/89 BCE, it is probable that the author of this text and those who accepted its interpretation believed that Jannaeus' kingdom was about to end, and that God had sent Demetrius III, as Scripture states God had previously sent the Babylonian monarch Nebuchadnezzar II to punish the Jewish nation for its sins by temporarily ending the Davidic monarchy.[66] This time, God decreed the Hasmonean monarchy's destruction and those Jews supporting it, like those in the past who had not accepted Babylonian domination, would be punished.

Given the widespread adoption of periodization in the Greco-Roman era among pagan and Jewish writers, especially among Jews following the Babylonian exile and destruction of their temple, it is plausible that the interpretation of Daniel in *4QApocryphon of Jeremiah C^e^* inspired Jannaeus' enemies to join Demetrius III, who was preparing to the invade the Hasmonaean state in 90/89 BCE. Not viewing this as a coincidence, some of Jannaeus' Jewish subjects likely joined the Seleucid army under the belief that Scripture had doomed Jannaeus and his kingdom since the completion of the predicted 490-year period had arrived in that year. Although the historian Josephus, who at first fought as a general against the Romans during the First Jewish Revolt (66-70 CE) and then joined his opponents to eventually retire in Rome as a citizen, was reluctant to document messianic fanatism, his writings suggest that he was aware of the chronology found in *4QApocryphon of Jeremiah C^e^*.[67]

In his *Antiquitates Judaicae* 13.301, Josephus states that Jannaeus' brother and predecessor, Aristobulus, became king "four hundred and eighty-one years and three months" after the return of the Jews from their Babylonian

---

[66] Nebuchadnezzar II captured the Jewish king Jehoiachin and took him into captivity in Babylon, where he died. His son and successor, Nebuchadnezzar II installed his uncle Mattanyahu as king and changed his name to Zedekiah. He too was taken to Babylon as a prisoner, where he died. See 2 Kings 24-25; Jeremiah 32, 34, 39, 52; Ezekiel 12. The loss of the Davidic monarchy stimulated reinterpretations of Scriptural texts to explain this event and to determine when the Davidic messiah would appear. These texts were constantly recycled to encourage later violence based on the belief that their prophecies were being fulfilled. See further Elgvin, *Warrior, King,* 104-42.

[67] For how Josephus' religious beliefs affected his writing of Greco-Roman history, see Atkinson, *History of the Hasmonean State,* 4-22, 172-79. In *BJ* 2.340-41, 401, Josephus states that peace is synonymous with submission to Roman rule.

captivity. This number is too large and appears to be based on the seventy weeks of years, or 490 years, in Daniel 9:24-27. If one adds to it the one-year reign of Aristobulus, then Jannaeus began his reign close to the beginning of the 70th heptad (years 483-490) as calculated in *4QApocryphon of Jeremiah C*^{e}.[68] This suggests that Josephus too knew this cosmic timetable from his sources, and likely obscured its importance in his books so as not to present the Jews as a race and religion motivated to violence based on biblical timetables.

In addition to the possibility that Jannaeus' opponents interpreted Scripture to predict the end of his reign, and therefore joined the army of Demetrius III convinced that they were fighting on God's side, numismatic evidence suggests that another Scriptural interpretation motivated Jannaeus to seize lands belonging to the Seleucid Empire despite having inferior military forces. His coins reveal that he had messianic ambitions. He minted a coin with a star surrounded by a diadem, which was a messianic symbol based on the biblical prophecy of Numbers 24:17 ("I see him but not now; I behold him but not near—a star shall come out of Jacob, and a scepter shall rise out of Israel; it shall crush the foreheads[of Moab and the heads[of all the Shethites").[69] This biblical oracle, attributed to the gentile prophet Balaam regarding Israel's future, was widely understood during the Hellenistic period by Jews as a messianic prophecy.[70] The passage was interpreted to mean that the messiah would lead the nation's armies in a successful battle against the unrighteous, represented by ancient Israel's foes in Moab in the Transjordan, since God would be on his side. Jannaeus' public appeal to this ancient biblical interpretation by placing the star on his currency made him unique among the Hasmonean monarchs, and likely contributed to his hubris that led him to start a series of foreign wars that nearly destroyed his kingdom, and which weakened his nation, making it vulnerable to Demetrius III's invasion. It likely prompted many of Jannaeus' Jewish subjects to wage a series of revolts to overthrow him, and

---

68 Lester L. Grabbe, "The Seventy-Weeks Prophecy (Daniel 9:24-27) in Early Jewish Interpretation," in *The Quest for Context and Meaning*, edited by Craig A. Evans and Shemaryahu Talmon. (Leiden: Brill, 1997), 595-611.

69 Meshorer, *Treasury of Jewish Coins,* 37-38.

70 See Collins, *Scepter and the Star*, 73-91; Elgvin, *Warrior, King,* 65-68. The most famous use of this biblical passage is in connection with the star that foretold the birth of Jesus in Matthew 2:1-12. Also, Rabbi Akiva used the same passage to proclaim the Jewish rebel leader Bar Kokhba, who led the Second Jewish Revolt of 132-35 CE, as the messiah. See Jerusalem Talmud, *tractate Ta'anit* 4.5; *Lamentations Rabbah* 2.4.

for some to join the Seleucid army, convinced that the messianic era and the end of the Hasmonean dynasty had arrived. This would not be the only time that messianism led to wars between Jews and foreigners.

Messianism was a potent force in ancient Judaism that led to several wars. Although Josephus tended to avoid discussing messianism in connection with violence, largely to present the Jews in a favorable light to the Romans, his account of the First Jewish Revolt against the Romans (66-70 CE) shows that messianism ignited the conflict:

> What incited them to the war was an ambiguous oracle in their sacred scriptures that at that time one from their country would become ruler of the world. They understood this to refer to someone of their own race, and many of their wise men went astray in their interpretation of it (*BJ* 6.312-13).

This oracle is based on Numbers 24:17-19, which describes a messiah who will subdue enemy nations. The interpretation that this oracle was being fulfilled led the Jews to revolt during the emperor Nero's reign to reestablish an independent religious state under the leadership of a messianic ruler.[71] During the latter days of the Jews' ill-fated revolt, Josephus mentions that a messianic prophet urged those in Jerusalem during the 70 CE Roman siege to go up to the temple courts to await divine deliverance.[72] The rebel leader Menahem entered the temple adorned with the robes of a king and an entourage of armed fighters, suggesting that he too believed he was the messiah and expected divine deliverance and God's army to arrive and subdue the Roman forces.[73]

Earlier, during the reign of Herod the Great, whom the Roman Senate appointed king of Judea and who fought in 37 BCE with the Roman general Sossius to capture Jerusalem, a spiritual leader urged the population to resist his rule, saying that Jerusalem could never be conquered (Joseph. *BJ* 1.347). Likewise, the author of the *Oracle of Hystaspes*—a Jewish apocalyptic text dating to around 37 BCE—predicts that the city will never be conquered.[74]

---

[71] See Collins, *Scepter,* 220-21; Elgvin, *Warrior, King,* 262-65. Anthony J. Tomasino ("Oracles of Insurrection: The Prophetic Catalyst of the Great Revolt," *Journal of Jewish Studies* 59 [2008]: 86-111) believes the oracle Josephus refers to is Daniel 9:24-27.

[72] Joseph. *BJ* 6.285-86, 288.

[73] Elgvin, *Warrior, King,* 264-65.

[74] David Flusser, *Jerusalem and the Origins of Christianity* (Jerusalem: Magness, 1988), 392.

In 132-35 CE, the Jewish community of ancient Palestine revolted against Roman rule in a conflict known as the Second Jewish Revolt (132-35 CE). The leader of this revolt, Simon Bar Kosiba, took the name Bar Kokhba (son of the star) to show he was the messianic leader predicted in Numbers 24. He also minted coins with a depiction of the temple which the Romans had destroyed in 70 CE during the First Jewish Revolt alongside a star—a messianic symbol—and signed his letters using the biblical messianic title "Prince of Israel."[75] The rebels were motivated by religion, as evident by their overstriking Roman coins with the depiction of the reconstructed Jerusalem temple and the legends "Jerusalem is holy," "Freedom of Zion," and "For the redemption of Zion."[76] This revolt was likely inspired by the messianism that ignited the First Jewish Revolt and events that took place during Trajan's reign.

The Diaspora Revolt of 115-17 CE also was rooted in biblical messianic prophecies and the belief that the Roman Empire's end was near and that the Jews would return to their ancestral land. The approaching seventy-year anniversary of the 70 CE Roman destruction of the temple stimulated further eschatological hopes and led Jews to massacre pagans during Trajan's reign.[77] To prevent future violence, Trajan made Judea a regular consular province and stationed an additional legion there, and in nearby Arabia.[78]

Conclusion

Ancient pagan and Jewish historians shared a belief in periodization, which they used to construct their historical chronologies and to divide empires into successive phases. This belief, and religious interpretations of Scripture, particularly messianic passages, appear to have motivated Jews

---

[75] Elgvin, *Warrior, King,* 277-89; Justin Martyr, *First Apology*, 31.6

[76] Meshorer, *Ancient Jewish Coinage,* 259-63.

[77] For the start of this revolt in the spring of 116 CE, in contrast to Eusebius' placing it in 115 CE, see the papyrological evidence cited in Miriam Pucci Ben Zeev, *Diaspora Judaism in Turmoil, 116/17 CE: Ancient Sources and Modern Insights* (Leuven: Peeters, 2005), 145-56. For the messianic factors behind this revolt, see further Shimon Applebaum, *Jews and Greeks in Ancient Cyrene* (Leiden: Brill, 1979), 251-60; Katell Berthelot, *Jews and Their Roman Rivals: Pagan Rome's Challenge to Israel* (Princeton: Princeton University Press, 2021), 150-54. The massacres of Greeks by Jews mentioned in Cassius Dio (68.32.1-3) provide evidence for the violence of the revolt.

[78] Elgvin, *Warrior, King,* 275-92.

to war. In the perplexing case of the encounter between Demetrius III and Alexander Jannaeus, those fighting on both sides appealed to Scripture. Those Jews who joined Demetrius III believed they were not traitors, but warriors in God's army fighting against a doomed ruler whose demise had been predicted in Scripture. Whether Demetrius III knew this belief, and took advantage of it, is unknown, but probable. He likely used it to motivate Jews to join his forces. Jannaeus, moreover, interpreted Scripture to claim he was the messiah, which led him to wage costly wars to annex his neighbors to expand the biblical Promised Land in a new Conquest. In the end, both rulers ended their reigns in disaster: Demetrius III in exile and Jannaeus facing constant rebellions until his death.

The invasion of Demetrius III provides an ancient example of the importance of considering religious beliefs as a factor that led rulers to go to war, and which motivated their citizens to join the fighting in the belief that they were siding with God. It remained so for centuries as witnessed by three successive Jewish revolts against the Roman Empire (66-70 115-17, and 132-35 CE) that were based on messianic prophecies, and which motivated ordinary citizens to take up arms against a foreign power, confident they were fighting in God's army.

Bibliography

Allegro, John M. "*Thrakidan,* the 'Lion of Warth' and Alexander Jannaeus." *Palestine Exploration Quarterly,* 91, no. 1 (1959): 47-51.

Applebaum, Shimon. *Jews and Greeks in Ancient Cyrene.* Leiden: Brill, 1979.

Assar, Gholamreza F. "A Revised Parthian Chronology of the Period 91-55 BC." *Parthica* 8 (2006): 55-104.

—. "A Revised Parthian Chronology of the Period 91-55 BC." *Parthica* 8 (2006): 55-104.

Atkinson, Kenneth. *I Cried to the Lord: A Study of the Psalms of Solomon's Historical Background and Social Setting*. Leiden: Brill, 2004.

—. *A History of the Hasmonean State: Josephus and Beyond.* London: Bloomsbury T & T Clark, 2016.

—. "Josephus's Use of Scripture to Describe Hasmonean Territorial Expansion." *JSIJ—Jewish Studies, an Internet Journal*, 19 (2021): https://jewish-faculty.biu.ac.il/en/node/1602

—. "Judean Piracy, Judea and Parthia, and the Roman Annexation of Judea: The Evidence of Pompeius Trogus." *Electrum* 29 (2022): 127-46.

Bar-Kochva, Bezalel. *Judas Maccabaeus.* Cambridge: Cambridge University Press, 1989.

Barag, Dan and Shraga Qedar. "The Beginning of Hasmonean Coinage." *Israel Numismatic Journal* 4 (1980): 8-21.

Bellinger, Alfred R. "The End of the Seleucids." *Transactions of the Connecticut Academy of Arts and Sciences* 38 (1949): 51-102.

Berrin, Shani L. *The Pesher Nahum Scroll from Qumran: An Exegetical Study of 4Q169.* Leiden: Brill, 2004.

Berthelot, Katell. *Jews and Their Roman Rivals: Pagan Rome's Challenge to Israel.* Princeton: Princeton University Press, 2021.

Borgna, Alice. "Scrivere del nemico. Pompeo Trogo e la storia de Parti (Iust., XLI-XLII)." *Dialogues d'histoire ancienne* 43 (2015): 87-116.

Cohen, Getzel. "The Beginning of the Reign of Alexander Jannaeus." In *Judean Syrian Egyptian Conflict of 103-101 B.C.*, edited by E. Van 'T Dack, W. Clarysse, G. Cohen, J. Quaegebeur, and J. K. Winnicki, 118-21. Brussels: Koniniklijke Academie voor Wetenschappen, 1989.

—. *Hellenistic Settlements in Syria, the Red Sea Basin, and North Africa.* Berkeley: University of California Press, 2006.

Collins, John J. *Apocalypticism in the Dead Sea Scrolls.* New York: Routledge, 1997.

—. *The Scepter and the Star: The Messiahs of the Dead Sea Scrolls and Other Ancient Literature*. 2nd edn. New York: Doubleday, 2010.

Cornell, Tim J. ed. *The Fragments of the Roman Historians.* 3 vols. Oxford: Oxford University Press, 2013.

Dąbrowa, Edward. "Demetrius III in Judea." *Electrum* 18 (2010): 175–181.

—. *The Hasmoneans and their State: A Study in History, Ideology, and the Institutions.* Kraków: Jagiellonian University Press, 2010.

Dimant, Devorah. *Qumran Cave 4.XXI: Parabiblical Texts, Part 4: Pseudo-Prophetic Texts*, Oxford: Clarendon, 2001.

Dindorf, Wilhelm, ed. *Georgius Syncellus et Nicephorus Cp.*, Bonn: Weber, 1829

Ebeling, Jennie R., J. Edward Wright, Mark Elliott, and Paul Flesher. *The Old Testament in Archaeology and History.* Waco, TX: Baylor University Press. 2017.

Ehling, Kay. *Untersuchungen zur Geschichte der späten Seleukiden (164-63 v. Chr.): vom Tode des Antiochos I. bis zur Einrichtung der Provinz Syria unter Pompeius*. Stuttgart: Steiner, 2008.

Elgvin, Torleif. *Warrior, King, Servant, Savior: Messianism in the Hebrew Bible and Early Jewish Texts.* Grand Rapids, MI: Eerdmans, 2022.

Eshel, Hanan, *The Dead Sea Scrolls and the Hasmonean State*. Grand Rapids, MI: Eerdmans, 2008.

Flusser, David. *Jerusalem and the Origins of Christianity.* Jerusalem: Magness, 1988.

Geffcken, Johannes. *Komposition und Entstehungszeit der oracular Sibyllina.* Leipzig: De Gruyter, 1902.

Goldstein, Jonathan A. *1 Maccabees.* Garden City, NY: Doubleday, 1976.

—. *2 Maccabees.* Garden City, NY: Doubleday, 1983.

Grabbe, Lester L. "The Seventy-Weeks Prophecy (Daniel 9:24-27) in Early Jewish Interpretation." In *The Quest for Context and Meaning*, edited by Craig A. Evans and Shemaryahu Talmon, 595-611. Leiden: Brill, 1997.

—. *A History of the Jews and Judaism in the Second Temple Period Volume 3: The*

Maccabean Revolt, Hasmonean Rule, and Herod the Great (175-4 BCE). London: T & T Clark, 2020.

Grainger, Jonathan D. *A Seleukid Prosopography and Gazetteer*. Leiden: Brill, 1997.

—. *The Fall of the Seleukid Empire 187-75 BC.* Barnsley: Pen & Sword Military, 2021.

Grayson, Albert K. *Babylon Historical-Literary Texts.* Toronto: University of Toronto Press, 1975.

Gruen, Eric S, "Hellenism and Persecution." In *Hellenistic History and Culture*, edited by Peter Green, 238-74. Berkeley, CA: University of California Press, 1993.

Habicht, C. "The Seleucids and their Rivals." In *Cambridge Ancient History Vol. 8: Rome and the Mediterranean to 133 B.C.* 2d edn. edited by A. E. Astin, 324-74. Cambridge: Cambridge University Press, 1989.

Hendin, David. "Numismatic Expressions of Hasmonean Sovereignty." *Israel Numismatic Research* 17 (2007/2008): 37-38.

Hendin, David and Ilan Shachar. "The Identity of YNTN on Hasmonean Overstruck Coins and the Chronology of the Alexander Jannaeus Types." *Israel Numismatic Research* 3 (2008): 87-94.

Hirschfeld, Yizhar and Donald T. Ariel. "A Coin Assemblage from the Reign of Alexander Jannaeus Found on the Shore of the Dead Sea." *Israel Exploration Journal,* 55 (2005): 66-89.

Honigman, Sylvie. *Tales of High Priests and Taxes.* Berkeley, CA: University of California Press, 2014.

Hoover, Oliver D. "The Dated Coinage of Gaza in Historical Context (264/3 BC-AD 241/2)." *Schweizerische Numismatische Rundschau* 86 (2007): 63-90.

—. "A Revised Chronology for the Late Seleucids at Antioch (121/120-64 BC)." *Historia: Zeitschrift für alte Geschichte,* 56, no. 3 (2007): 280-301.

Hoover, Oliver D., Arthur Houghton, and Petr Vesley. "The Silver Mint of Damascus under Demetrius III and Antiochus XII (97/6 BC-83/2 BC)." *American Numismatic Society,* 20, (2008): 280-301.

Houghton, Arthur. "The Struggle for the Seleucid Succession, 94-92 BC: A New Tetradrachm of Antiochus XI and Philip I of Antioch." *Schwizerische numismatische Rundschau* 77 (1998): 66-68.

Jacoby, Felix. *Die Fragmente der griechischen Historiker.* (FGrHist): Zweiter Teil:

Spezialgeschichten, Autobiographien und Memoiren, Zeittafeln. Berlin: Weidmann, 1926.

Jansen, Herman L. *Die Henochgestalt: Eine vergleichende religionsgeschichtliche Untersuchung.* Oslo: Dybwab, 1939.

Klausner, Joseph. "Judah Aristobulus and Jannaeus Alexander." In *The World History of the Jewish People VI: The Hellenistic Age*, edited by Abraham Schalit, 222-41. New Brunswick: Rutgers University Press, 1972.

Knibb, Michael. "The Exile in the Literature of the Intertestamental Period." *Heythrop Journal* 17 (1976): 253-72.

Koch, Klaus. "Die mysteriösen Zahlen der judäischen Könige und die apokalyptischen Jahrwochen." *Vetus Testamentum* 28 (1978): 433-41.

Kosmin, Paul J. *Time and Its Adversaries in the Seleucid Empire.* Cambridge, MA: Belknap Press, 2018.

Kushnir-Stein, Alla. "Gaza Coinage Dated LIC—A Reappraisal." *Schweizerische Numismatische Rundschau,* 74 (1995): 49-55.

—. "Late Hellenistic Coins of Gaza and the Date of the Hasmonean Conquest of the City." *Schweizer Münzblätter,* 198 (2000): 22-24.

Laqueur, Richard. *Der jüdische Historiker Flavius Josephus.* Giessen Münchow, 1920.

Lester, Olivia S. "The Four Kingdoms Motif and Sibylline Temporality in Sibylline Oracles 4." In *Four Kingdom Motifs Before and Beyond the Book of Daniel*, edited by Andrew Perrin and Loren T. Stuckenbruck, 121-41. Leiden: Brill, 2020.

Levenson, David B. and Thomas R. Martin. "Akairos or Eukairos? The Nickname of the Seleucid King Demetrius III in the Transmission of the Texts of Josephus' *War* and *Antiquities*." *Journal for the Study of Judaism,* 40, no. 3 (2009): 307-41.

Liddell, Henry George, Robert Scott, Henry Stuart Jones, and Roderick McKenzie, *A Greek English Lexicon,* rev. ed. Oxford: Oxford University Press, 1996.

Loader, W. R. "Pompey's Command under the Lex Gabinia." *Classical Quarterly* 54 (1940): 134-36.

Ma, John. *Antiochos III and the Cities of Western Asia Minor.* Oxford: Oxford University Press, 1999.

Meshorer, Ya'akov. *Ancient Jewish Coinage. Volume II. Herod the Great through Bar Cochba.* New York: Amphora Books, 1982.

—. *A Treasury of Jewish Coins from the Persian Period to Bar Kokhba.* Nyack, NJ: Yad Ben-Zvi Press, 2001.

Momigliano, Arnaldo. "The Origin of Universal History." *Atti della Scuole Normale Superiore di Pisa,* ser. 3.12.2 (1982): 533-60.

Mommsen, Theodor. "Mamilius Sura, Aemilius Sura, L. Manlius." *Rheinisches Museum für Philologie* 16 (1861): 282-87.

Nickelsburg, George W. E. *1 Enoch: A Commentary on the Book of 1 Enoch*. Minneapolis: Fortress, 2012.

Olbrycht, Marek J. "Mithridates VI Eupator and Iran." In *Mithridates VI and the Pontic Kingdom*, edited by Jakob M. Højte, 163-90. Aarhus: Aarhus University Press, 2009.

Overtoom, Nikolaus K. *Reign of Arrows: The Rise of the Parthian Empire in the Hellenistic Middle East.* Oxford: Oxford University Press, 2020.

Pomykala, Kenneth E. *The Davidic Dynasty Tradition in Early Judaism.* Atlanta: Scholars Press, 1995.

Portier-Young, Anathea E. *Apocalypse Against Empire: Theologies of Resistance in Early Judaism.* Grand Rapids: Eerdmans, 2011.

Pucci Ben Zeev, Miriam. *Diaspora Judaism in Turmoil, 116/17 CE: Ancient Sources and Modern Insights.* Leuven: Peeters, 2005.

Regev, Eyal. *The Hasmoneans: Ideology, Archaeology, Identity*. Göttingen: Vandenhoec, & Ruprecht, 2013.

Schürer, Emil. *The History of the Jewish People in the Age of Jesus Christ (175 B.C. A.D.135)*, Revised and edited by Geza Vermes, et. al. Edinburgh: T & T Clark, 1973.

Schwartz, Daniel R. "On Antiochus VII Sidetes' Parthian Expedition and the Fragmentation of Historical Research." In *The Jews in the Hellenistic-Roman World: Studies in Memory of Menahem Stern,* edited by Isaiah M. Gafni, Aharon Oppenheimer, and Daniel R. Schwartz, 83-102. Jerusalem: Zalman Shazar Center for Jewish History, 1996 (in Hebrew).

Seeman, Chris. *Rome and Judea in Transition: Hasmonean Relations with the Roman Republic and the Evolution of the High Priesthood*. New York: Peter Lang, 2013.

Sharon, Nadav. *Judea Under Roman Domination.* Atlanta: SBL Press, 2017.

Sievers, Joseph. *The Hasmoneans and Their Supporters.* Atlanta: Scholars Press, 1990.

Stern, Menahem. "The Relations Between Judea and Rome During the Reign of John Hyrcanus." *Zion,* 26 (1961): 3-19 (in Hebrew).

Swain, Joseph W. "The Theory of Four Monarchies—Opposition History Under the Roman Empire." *Classical Philology* 25 (1940): 1-21.

Tomasino, Anthony J. "Oracles of Insurrection: The Prophetic Catalyst of the Great Revolt." *Journal of Jewish Studies* 59 (2008): 86-111.

Van 'T Dack, E. "Le conflict judéo-syro-égyptien de 103/102 av. J-C." In *Proceedings of the Sixteenth International Congress of Papyrology,* edited by Roger R. S. Bagnall, et al., 303-12. Chico, CA: Scholars Press, 1981.

Wright, Nicholas L. "A Late Seleukid Bronze Hoard, c. 1988 (*CH* 10, 349)." In *Coin Hoards Volume X: Greek Hoards*, edited by Oliver Hoover, Andrew Meadows, and Ute Wartenberg, 245-64. New York: American Numismatic Society, 2010.

Yarbro Collins, Adela. "Numerical Symbolism in Apocalyptic Literature." *Aufstieg und Niedergang der Römischen Welt* II.21.2 (1984): 122-87.

Yarbro Collins, Adela and John J. Collins. *King and Messiah as Son of God.* Grand Rapids, MI: Eerdmans, 2008.

Zollschan, Linda. *Rome and Judea: International Law Relations, 162-100 BCE.* London: Routledge, 2017.

DETERRENCE AND SELF REPRESENTATION: THE ROMAN HEADQUARTERS AT SYRACUSE[1]

BORJA VERTEDOR BALLESTEROS[2]

1. Introduction

In the context of the Republican expansion, the Roman camp - understood not only in its material sense of castra, but also as the base of the imperium-wielding magistrate, whether it was the military enclosure itself, the *hibernae* or a city - was articulated as a political headquarters and administrative center.[3] This feature determined the important role of the

[1] This article has been made possible thanks to the activity carried out at the Escuela Española de Historia y Arqueología de Roma (EEHAR), attached to the CSIC (Consejo Superior de Investigaciones Científicas).
[2] This paper has been written in the context of the FPI Predoctoral Fellowship (PRE-2019-089903), and the Research Project "Diplomatic Expression in the Central and Eastern Mediterranean under Roman Expansion: The Gift in its Political and Institutional Context" (PGC2018-096415-B-C22), funded by MCIN/AEI/ 10.313039/501100011033/ and FEDER Una manera de hacer Europa.
[3] On the political role of Roman republican camps, see Borja Vertedor Ballesteros "Buying loyalties, grating rewards: the Roman headquarters as a space of diplomatic Interaction," in *The materiality of diplomacy in the Hellenistic-Roman Mediterranean: Gifts, Bribes, Offerings,* edited by E. Sánchez Moreno and Enrique García Riaza (Edinburgh: Edinburgh University Press, forthcoming). For imperial times we have some references, Lorenz Baumer, Tonio Hölscher, and Lorenz Winkler, "Narrative Systematik und politisches Konzept in des Reliefs der Trajanssäule. Drei Fallstudien," *Jahrbuch des Deutschen archäologischen Institut* 106 (1991): 261-295. See also, on various aspects of provincial command, Arthur Eckstein, *Senate and General: Individual Decision-making and Roman Foreign Relations, 264-194 B.C.* (Berkeley: The University of North Carolina Press, 1987); Alejandro Díaz Fernández, "*¿Dum populus Senatusque Romanus vellet*? La capacidad de decisión de los mandos provinciales en el marco de la política romana (227-49 a.C.)," in *Poder central y poder local. Dos realidades paralelas en la órbita política romana,* edited by Raul González Salinero (Madrid: Signifer Libros, 2015),

camp as a valid space for interaction at the highest level with local authorities and their representatives.

Various diplomatic missions, legations, heralds and foreign representatives were received in the camp. The military base and its immediate vicinity also served as a venue for certain meetings, particularly bilateral conferences between dignitaries from different political entities and the Roman commander. At the same time, in a proactive manner, it was configured as a dispatching center for representatives of the Roman provincial authority to cities and kingdoms within the commander's demarcation or in more remote areas.[4]

The Roman camp or headquarters was an epicenter of interaction, in which dialogue played an important role. This fact was emphasized by the important symbolic role of the headquarters as the seat of Roman provincial authority. Monarchs, ambassadors, political officials, heralds, converged systematically in the camp, an itinerant communication node which, in spite of its military character, facilitated interaction. This is not to deny that, in many cases, the decisions taken were conditioned by the military environment and the objective and subjective conditions of the war context[5].

The unifying factor in these dynamics, both political and diplomatic, is the commander-in-chief, the highest authority at the headquarters. As the Roman representative deployed in the field,[6] he played a preponderant role in the diplomatic protocol held in the military enclosure.[7] Alongside the

---

135-152; Fred Drogula, *Commanders and Command in the Roman Republic and Early Empire* (Chapel Hill: The University of North Carolina Press, 2015).

[4] For the reception of embassies see the case mentioned by Appian in the context of the end of the Third Punic War, when the Punic delegation comes to the Roman camp, App. *Pun.* 78. For the sending of *legati* from Roman headquarters, note the case of the dispatch of three centurions as ambassadors to Syphax by the Scipio brothers, Liv. 24.48.1-4.

[5] This chapter follows Díaz Fernández, *Provincia et Imperium. El mando provincial en la República romana (227-44 a.C.)* (Sevilla: Universidad de Sevilla, 2015), especially p. 34: *provincia* understood as the jurisdiction mandated to a holder of *imperium*.

[6] On the exemplarity of ambassadors, see Elena Torregaray Pagola, "*Legatorum facta:* la ejemplaridad de los embajadores romanos," *VELEIA* 26 (2009): 127-152.

[7] Enrique García Riaza, "Le protocole diplomatique entre particularisme romain et universalisme: quelques réflexions sur l'Occident républicain," in *La diplomatie romaine sous la République: réflexions sur una pratique*, edited by Barthélémy

magistrate, in the ceremonial audience, the general's *consilium* also performed important roles at the political,[8] legal and symbolic levels.

In this sense, we propose to study and analyze the political-diplomatic communication in the process that led to the final capture of Syracuse by Marcellus in 211. In the period between the beginning of the siege and the final fall of the city, a whole series of exchanges, meetings and diplomatic negotiations took place around the Roman camp, the city, and the island of Sicily itself. We will therefore examine the types of meetings that took place, the different roles played by the interlocutors involved and the more immediate contexts.

We have divided the process of the Roman conquest into three phases, which correspond to different stages in the siege and the capture of the city. Firstly, we have grouped together those references that are confined to the moments prior to the siege; secondly, we have collected data regarding the activities that took place during the height of the conflict; and finally, we have evaluated the evidence on the process of the cessation of hostilities. Through these three groups, linked together, we will try to reconstruct the sequence of diplomatic developments at Roman headquarters.

2. The Roman camp as a political-administrative center

Epigraphy, through a series of republican examples, constitutes a testimony of primary importance for the knowledge of the political-administrative function of the Roman camp. The Bronze of Lascuta,[9] an edict issued by Aemilius Paulus, praetor in 191, mentions a change of status initiative for the inhabitants of the Turris Lascutana in Hispania Ulterior. The release of the Lascutans from their relationship of dependence with the city of Hasta Regia was carried out under the *imperium* of the Roman magistrate. The last

---

Grass & Ghislaine Stouder (Génova: Institut des Sciences et Techniques de l'Antiquité, 2015), 15-41.

[8] On the *consilium* as a permanent body in the Roman camp, see Pamela Johnston, *The Military Consilium in Republican Rome* (Piscataway, New Jersey: De Gruyter, 2008).

[9] Luis García Moreno, "Sobre el decreto de Emilio Paulo y la Turris Lascutana (*CIL*.12, 614)," *Epigrafía hispánica de época romano-republicana* (1986): 195-218; María Hidalgo de la Vega, "El bronce de Lascuta. Un balance historiográfico," *Studia historica. Historia Antigua* 7 (1989): 59-66.

line of the epigraphic document, which mentions the place of issue, is worth emphasizing: *Act(um) in castreis*.

This allusion to the Roman camp is repeated in a later bronze, the Bronze of Ascoli,[10] showing the importance by explicitly stating that the camp was the place where decrees were issued and implemented. This document dates to the Social War, specifically in the year 89, during the siege of the city of Ascoli by Pompeius Strabo. This bronze accumulates a significant amount of information not only regarding the formula, which is reproduced twice: *[fecit in cast]reis apud Ausculum* and *donavit in castreis apud Asculum*.[11] It also attests to the function of the general's *consilium*, which acted as witness.

Finally, and although the reference to the camp is not explicitly mentioned, a third inscription, the Alcántara Bronze,[12] is of interest in relation to the functions of the general's *consilium*. This document is dated to the year 104 and consists of the legal articulation of the *deditio in fidem* of the people of the *Seano* [---] to the *imperator L. Caesius C.f.* As already stated, the tablet includes no mention of the above-mentioned sentence, but it alludes to the active role of the *consilium* in the process: *de consili sententia.*[13] On the

---

[10] *CIL* I^2 709.

[11] Lines 2 and 57, Nicola Criniti, *L'Epigrafe di Asculum di Gn. Pompeo Strabone* (Milano: Vita e Pensiero, 1979), 17; 26.

[12] Raquel López Merelo, García Jiménez Santiago, & Jose Sánchez Abal, "El bronce de Alcántara. Una deditio del 104 a.C." *Gerión. Revista De Historia Antigua*, 2 (1984): 495; *HEp* 1, 1989, 151; Dieter Nörr, *Aspete des römischen Völkerrechts. Die Bronzetafel von Alcántara.* (München: Bayerische Akademie der Wissenschaften, 1989); Borja Díaz Ariño, *Epigrafía latina republicana de Hispania (ELRH)* (Barcelona: Publicacions i Edicions, Universitat de Barcelona, 2008), 191-194; Díaz Fernández, "*Deditio, restitutio* y cláusulas revocatorias en el Bronce de Alcántara (AE, 1984, no. 495)," In *In fidem venerunt. Expresiones de sometimiento a la República romana en Occidente*, edited by E. García Riaza and A. M. Sanz (Madrid: Dykinson, 2019), 168.

[13] The role of the *consilium* is recorded on the Alcántara Bronze, which intervenes directly in the resolution of the *deditio*, see Estela García Fernández, "La *deditio in fidem populi Romani* y la construcción de un orden exterior en la República Romana," in *Hispania et Roma: Estudios en homenaje al profesor Narciso Santos Yangua,* edited by Jose San Vicente González de Aspuru, Carolina Cortés Bárcena, and Emma González González (Madrid: Ediciones de la Universidad de Oviedo, 2019), 99-108; Eduardo Sánchez-Moreno and Tomás Aguilera Durán, "Lusitanos frente a Galba (151-150 a.C.) ¿una deditio in fidem?" in *In fidem venerunt: expresiones de sometimiento a la República romana en Occidente,* edited by Enrique García Riaza and Anthony-Marc Sanz (Madrid: Dykinson, 2019), 139-166; Ines

other hand, there is abundant literary evidence of the role of the Roman camp not only as a venue for incoming embassies at the highest level, but also as dispatching center of Roman representatives with the capacity for dialogue.

We will now look at some cases that illustrate this role of the Roman headquarters as the location for political-administrative and diplomatic activities[14]. After the Roman victory over Macedonia at the Battle of Pydna (168), the Roman troops moved to a camp established near the city of Pella. In this new location, the military enclosure was visited by several *legationes* to express their thanks to the Roman people for the liberation of Macedonia[15].

A second instance involves noble horsemen from Capua, in the context of the Second Punic War, in this case discussing surrender or submission (*in fidem*). The group, seeking to join the Roman side before the fall of the city, which was in Punic hands, went to the Roman camp at Suessula, commanded by the praetor Cnaeus Fulvius. After having been accepted for entry, ten of them were received by the commander-in-chief, with whom they negotiated the terms of their submission to the Roman people.[16] The headquarters, as discussed above, was also active in dispatching Roman representatives to local populations. For example, Scipio Africanus sent three centurions as emissaries to Syphax in 213 (*ad eum centuriones tres legatos miserunt*), with the aim of securing allies for the landing at

---

Sastra and Domingo Plácido Suárez, *Deditio in fidem* and Peasant Forms of Dependence in the Roman Provincial System: the Case of Northwestern Iberia," *XXX colloque du GIREA* (2005): 501-509.

[14] On the capacity of the Roman camp as a reception center for embassies, see Jean-François Claudon, "Les ambassades des cites greques d'Asie Mineure auprès des autorités romaines: de la libération des Grecs à la fin du Haut-Empire (196 av. J.-C. – 235 apr. J.-C.)." (PhD diss., EPHE Paris, 2015); John D. Great Power Diplomacy in the Hellenistic World. (New York: Routledge, 2017). Cf. Torregaray Pagola, "Fracaso y oportunidad en la diplomacia de la República romana", In Algunas sombras en la diplomacia romana, edited by E. Torregary Pagola & J. Lanz Betelu, 31-47. =Vitoria-Gasteiz: Universidad del País Vasco = Euskal Herriko Unibertsitatea, 2020)

[15] Liv. 44.46.9.

[16] Liv. 24.47.12-13: (…) *equites centum duodecim nobiles Campani per speciem praedandi ex hostium agro permissu magistratuum ab Capua profecti ad castra Romana* (…).

Carthage.[17] Another episode occurs in the context of the Indibilis uprising (205), specifically Lucius Lentulus' and Lucius Manlius' actions against the Ausetans. Once the Roman generals were established and encamped in the territory of the enemy, an attempt was made through emissaries to get them to lay down their arms. When this diplomatic approach failed, the armed conflict resumed (*primo per legatos nequiquam temptatum, ut discederetur ab armis*).[18] In both cases, officials were sent as emissaries. Considering the military context, it is possible that their functions were not only diplomatic but also to include gathering tactical and military information from the enemy.[19]

Diplomatic interaction took place in certain areas of the camp or headquarters,[20] but mainly at the *praetorium*, understood both as the tent itself of the commanding general as well as the space used to house his rooms. This space, together with the *forum* of the camp and the *principia*, were the main areas where diplomatic activity took place. Once they had entered the camp, either the Roman magistrate alone or he and his *consilium* received the visiting embassies, legations, foreign monarchs, and local representatives depending on the needs and dispositions of the retinue.

The arrangement of these sites, denoting a coercive or merely receptive environment, highlights the adaptation of military compounds for negotiations and administration.[21] Furthermore, we can identify in the military enclosure a duality between public and the private spheres. As we

---

[17] Liv. 24.48.3. On the role of Roman *legati*, see M. M, Salomonsson, "Roman Legates in the Republic," *Opuscula Romana* 25-26, (2000-2001): 79-88.
[18] Liv. 29.2.2-3.
[19] See Rose Mary Sheldon, *Intelligence Activities in Ancient Rome. Trust in the Gods, but Verify* (London and New York: Routledge, 2005), 57-58; N. J. E. Austin and N. B. Rankov, *Exploratio. Military and political intelligence in the Roman world from the Second Punic War to the battle of Adrianople* (London and New York: Routledge, 2006).
[20] On the internal layout of the camp see in the first instance Polyb. 6.40-42, also Catherine M. Gilliver, "The *de munitionibus castrorum*: Text and Translation," *JRMES* 4 (1993): 33-48; on the archaeological remains: Michael Dobson, M. *The Army of the Roman Republic. The Second Century BC, Polybius and the Camps at Numantia, Spain* (Oxford: Oxbow Books, 2008); Ángel Morillo Cerdán, "Criterios arqueológicos de identificación de los campamentos romanos en Hispania," *SALDVIE* 8 (2008): 73-93; Alicia Jiménez *et al.*, "Renewed Work at the Roman Camps at Renieblas near Numancia (2nd-1st c. B.C.)," *JRA* 33 (2020): 4-35.".
[21] On spaces for diplomatic interaction, particularly in the city, see Torregaray Pagola, "Los espacios para la diplomacia," *Caesarodunum* XL (2006): 223-258.

shall see below, private conversations took place in the context of diplomatic negotiations, as well as clandestine or secret meetings. In short, a whole series of contacts and connections were woven into the provincial governors' camps.

3. Preliminary moments

The death of Hieron II of Syracuse, a staunch ally of Rome, was followed by a turbulent and unstable period. His grandson, Hieronymus, heir to the throne, inexperienced in royal and political affairs, allowed himself to be influenced by Punic agents and signed an alliance with Carthage.[22] The agreement guaranteed, in case of victory, the complete annexation of Sicily.[23] However, in 214 he perished in a popular uprising, and, in this context, which was exploitable from the Roman perspective, Rome sent in the troops already on the island, together with reinforcements commanded by Claudius Marcellus.[24]

In these circumstances, Livy informs us that an embassy had previously been sent to the propraetor Appius Claudius.[25] The aim of this diplomatic mission was to obtain a ten-day truce (*indutiis dierum decem legatos isse ad Appium Claudium*). Once this demand was met, talks continued at a higher stage, and another delegation was sent to negotiate the renewal of the old treaty of alliance between Syracuse and Rome (*de foedere antique renovando agerent missos*).[26]

Once Marcellus arrived in the province as consul, Appius Claudius forwarded the Syracusan ambassadors to him. The new magistrate gave the *legati* an audience, paid attention to their proposals, and after weighing a possible peace in view of the suggested terms, sent ambassadors to Syracuse (*et ipse legatos Syracusas*). It is arguable, therefore, that the emissaries of the magistrates were empowered to undertake negotiations with local rulers

---

[22] Gabriel Rosselló Calafell, *Relaciones exteriores y praxis diplomática cartaginesa. El período de las guerras púnicas* (Sevilla: Universidad de Sevilla), 46-47.
[23] Liv. 24.4-9.
[24] Liv. 24.21.1-3.
[25] Legate of Marcellus in charge of the fleet and the camp at Leontini, Liv. 24.39.
[26] Liv. 24.27.4. See Paul J. Burton, *Friendship and Empire. Roman Diplomacy and Imperialism in the Middle Republic (353-146 BC)* (Cambridge: Cambridge University Press, 2011).

on an issue as sensitive as the renewal of the alliance with Rome (*qui coram cum praetoribus de renovando foedere agerent misit*).[27]

At this point we can already visualize several of the aspects we have previously discussed. The capacity of Roman generals to manage complex diplomatic processes is attested, including conversations on the renewal of treaties on behalf of Rome.[28] Secondly, it should be stressed that the negotiating capacities of this kind of "military embassies" were obviously subordinated to the will of the magistrate and his consilium.

The final phase of the treaty renewal process was implemented by the Syracusans by sending a delegation to Marcellus. The purpose of this delegation was to draw up the terms of the treaty (*legatos ad rem cum eis confirmandam*).[29] At this point, however, Hippocrates, whose objective was to boycott the agreement, undertook a series of attacks and raids on the boundaries of the area controlled by Rome. He even went so far as to confront a detachment sent by Appius Claudius to protect the allied lands. Marcellus was informed of these Syracusan actions and proceeded to send an embassy to Syracuse to report that the peace agreement previously reached had been broken.[30] Moreover, the Roman general added a request: the expulsion of Hippocrates and Epicydes not only from Syracuse, but also from Sicily, as a guarantee against the outbreak of an imminent conflict.[31]

Not satisfied with their previous actions, Hippocrates and Epicydes took advantage of the assault and capture of the city of Leontini by the Romans to create a *casus belli*. Thus, they spread the rumor of the near total annihilation of the population of this city by the Roman troops, as well as of the cruelty of the Roman troops, and they claimed that the destiny of Syracuse would be the same if they did not act. The turmoil created in the

---

[27] Liv. 24.27.6. On Rome's alliances and interaction see John Percy Vyvian Dacre Balsdon, *Romans and Aliens* (London: The University of North Carolina Press), 82-84; David Braund, *Rome and the Friendly King* (New York: Routledge), 25, 63; García Riaza, "Derecho de guerra en Occidente durante la expansión romano-republicana. Planteamientos metodológicos," In *De fronteras a provincias. Interacción e integración en Occidente (ss. III-I a.C.),* edited by E. García Riaza, 31-65 (Palma de Mallorca: Edicions Universitat de les Illes Balears, 2011), 31-38.

[28] On provincial command and the magistrate's ability to act, see Eckstein, *Senate and General*; Drogula, *Commanders and Command*.

[29] Liv. 24.28.9.

[30] Liv. 24.29.4-5; Liv. 24.30.1, the aggravating factor of the massacre of the Roman guard post during peace negotiations is emphasized.

[31] Liv. 24.29.5.

city led to a massacre between those loyal to Rome and the supporters of the Punic side, resulting in the election of Hippocrates and Epicydes as praetors by the latter.[32]

As soon as the Romans became aware of these events, they moved their camp from Leontini to Syracuse (*castra ad Syracusas*), to establish themselves in a position closer to the future theatre of operations.[33] Furthermore, Livy informs us that an embassy sent by Appius Claudius narrowly escaped capture.[34] Thus, continues Livy, not only the rights of war, but also the rights of peace were no longer respected because of this flagrant attack on the sacredness and inviolability of embassies.[35]

The Roman military enclosure was located next to the Olympium, a temple of Jupiter, some fifteen hundred paces from Syracuse. A *legatio* was dispatched from that camp but was not allowed in the city. Instead, Hippocrates and Epicydes, together with their friends, went out to the gates.[36] In that liminal location, *ad portas*, the Roman spokesman (*romanus orator*) informed the Syracusans that Rome's intention was not to wage war, but to help and support those who had been enslaved and massacred, i.e., the Syracusan people allied with the Romans.[37] If they did not return to a previous situation with Rome, hostilities would be undertaken with all the military capacity the Roman army had at its disposal. The local praetors ironically begged them to return when the people they had alluded to were in power in Syracuse, and challenged them to take the city, arguing that any comparison with Leontini did not match their city's defensive capabilities. He then left the Roman ambassadors and closed the gates.[38] The failure of the diplomatic option inevitably gave way to the military alternative.

---

[32] Liv. 24.30-33.
[33] Liv. 24.33.1.
[34] Liv. 24.33.2-3.
[35] García Riaza 2020, 7-9.
[36] Liv. 24.33.3-6, *mille et quingentos passus ab urbe castra posuit.*
[37] See: Torregaray Pagola, "The Roman Ambassador's Speech: Public Oratory on the Diplomatic Stage," In *Community and Communication Oratory and Politics in Republican Rome*, edited by C. Steel & H. van der Blom, 229-245 (Oxford: Oxford University Press, 2013), 235; Pierre Sánchez, "Quand Rome se cherchait de nouveaux alliés: les accords de coopération militaire négociés à l'initiative des Romains sur le théâtre des operations (IVe- IIIe s. av. n.è.)," *Ktéma* 41 (2016): 165-190.
[38] Liv. 24.33.3-8.

4. Open Hostilities

From the time of the outbreak of the conflict until the beginning of the new negotiations, the Roman attack unsuccessfully faced Archimedes' inventions during the siege of the city. In this new phase, certain political-diplomatic meetings took place, and once again the Roman camp was a key locale.

In the last phases of the siege, prior to the final assault on the city, a series of interviews with some of the Syracusans who had committed defection took place. While it is true that they can not be considered *de iure* as representatives of the new Syracusan power, they acted on behalf of the internal resistance and of the pro-Roman faction. From this perspective, they must be considered as valid agents. Communication between the Roman camp and the city was carried out by means of a slave, who managed to bring several Syracusan defectors from the city to the Roman camp.[39]

A plan was devised to strike a coup d'état in the city. However, after several days of back and forth, the plan came to an end due to the tip-off of one of the participants, which led to the torture and death of the members of the plot.[40] As we can see, the camp was also the scenario of several informal interviews, since they were not carried out through the usual channels, nor through "official" representatives of the government, in this case, Syracusan.[41] The Roman compound provided the ideal conditions for the organization of counter-intelligence activities. We will later discuss how the pro-Roman resistance was also sheltered inside the camp to jointly organize the siege and final assault on Syracuse.

The failure of this venture did not mean the end of Roman attempts to penetrate the city. A new plot was hatched as follows. Roman troops had captured one Damippus, a Lacedaemonian sent by Syracuse to King Philip. Epicydes, the Syracusan praetor, wanted at all costs to ransom him; Marcellus was not opposed to this idea, as he was interested in connecting the Romans with the Aetolians. Thus, they arranged a meeting between the two sides. The location chosen was equidistant between the city and the Roman camp. Once there, the Roman emissaries, over successive days,

---

[39] Liv. 25.23.5-6.

[40] Liv. 25.23.7.

[41] In the same vein, the case of the horsemen of Capua, who came to the Roman camp clandestinely while their city was under Punic rule, has been discussed above, Liv. 24.47.12-13.

observed a weak point in the wall, from which they would later assault the city.[42] Negotiations carried out in this type of high-level meeting (*conloquia*) could indeed last for several days, generating tactical advantages (such as resupply, or redeployment). At the same time, *legationes* were used as an advance party to gather information about the enemy capacities and position.

The new Roman assault was successful, and finally managed to penetrate the city. However, the initial resistance pushed Marcellus in a different direction, and as it was impossible to take Acredina, he headed towards Euryalus. In that area was located a fortress commanded by Philodemus of Argos, whom Epicydes had appointed. The Roman magistrate took the diplomatic initiative, sending Sosis to meet the Syracusan. The latter, in order to delay the situation and await reinforcements, postponed replying to the Romans.[43]

Then, as the situation remained unchanged, and once he understood that it was not possible to seize the fortress, Marcellus chose to take by force the space between Tica and Naples, two districts of the city of Syracuse.[44] The Roman headquarters pivoted from the outside of the urban area to almost the heart of the city. This location was intended not only to intensify the threat, but also to make the camp visible as a point of confluence to accept communications of surrender and allegiance to Rome.

From this moment on, the political reference point for the inhabitants of the surrounding area became the new headquarters. It is indeed the place where they came with ribbons and veils of supplicants,[45] praying that no massacres

---

[42] Liv. 25.23.8-10.

[43] Liv. 25.25.1-4.

[44] Liv. 25.25.6.

[45] Torregaray Pagola, "Embajadas y embajadores entre Hispania y Roma en la obra de Tito Livio," In *Diplomacia y autorrepresentación en la Roma antigua*, edited by E. Torregaray Pagola, J. Santos Yanguas, 25-62 (Vitoria/Gasteiz: Universidad del País Vasco = Euskal Herriko Unibertsitatea, 2005), 57. For other examples of diplomatic practices in the context of surrender, see Anthony-Marc Sanz, "La deditio: un acte diplomatique au cœur de la conquête romaine," In *La diplomatie romaine sous la République: réflexions sur una pratique,* edited by Barthélémy Grass and Ghislaine Stouder, 87-105 (Génova: Institut des Sciences et Techniques de l'Antiquité, 2015). García Riaza, "In Conloqvivm venire: Interviews between Roman Commanders and Western Leaders in the Age of Republican Expansion," In *Provinces and Provincial Command in Republican Rome: Genesis, Development*

or fires be committed (*legati eo ab Tycha et Neapoli cum infulis et velamentis venerunt, precantes ut a caedibus et ab incendiis parceretur*).[46] This initiative was motivated by the knowledge of the Roman *ius belli* and the dire consequences of a renewed armed assault on the city.[47] Marcellus, as part of the mechanism of the diplomatic reception ritual, convened his *consilium* to determine a course of action. In line with the general feeling of the magistrate's council (*de quorum precibus quam postulatis magis consilio*), he issued an order to the soldiers that no violence should be committed against free persons, although the rest would be subject to pillage (*habito Marcellus ex omnium sententia edixit militibus ne quis liberum corpus violaret: cetera praedae futura*).[48]

The seizure of these areas and the establishment among them of the headquarters fortified with the rubble of the houses, came to the knowledge of Philodemus. After receiving assurances (*fide accepta ut inviolatus ad Epicyden rediret*) from the Roman magistrate that he could return safely to Epicydes, he handed over to him the fortress and the mountain.[49]

5. Surrender, end of hostilities

Finally, after the withdrawal of the Carthaginian troops, the Sicilians encamped in the vicinity of Syracuse sent a delegation to Marcellus to discuss the terms of the surrender of the city, having previously canvassed the opinions of the besieged (*legatos de condicionibus dedendae urbis explorata prius per conloquia voluntate eorum qui obsidebantur ad Marcellum mittunt*).[50] An agreement was reached whereby the Roman general guaranteed the minimum concessions for the Syracusan population: those assets that had belonged to the kings would be handed over to the Romans, but the Sicilians would keep everything else along with their freedom and their own laws.[51]

After accepting the Roman demands, a revolt and purge arose within Syracuse against the pro-Punic elements, particularly against Hippocrates

---

and Governance, edited by A. Díaz Fernández (Sevilla: Universidad de Sevilla, 2021), 127-144.
[46] Liv. 25.25.6.
[47] García Riaza, "Derecho de Guerra," 54-56.
[48] Liv. 25.25.5-7. Diod. 26.20.1.
[49] Liv. 25.25.10.
[50] Liv. 25.28.2.
[51] Liv. 25.28.3.

and Epicydes and their henchmen. In this context of renewal of local power, it was decided, before appointing ambassadors, to elect praetors. Some of these were then sent as delegates to Marcellus (*numero missi oratores ad Marcellum*). The spokesman for the Syracusans informed the commander-in-chief of their intention to surrender unconditionally, i.e., a *deditio.*[52]

The last stronghold in the city, Acradina, was next. In this strategic location the mercenary forces had concentrated. The delegates from Syracuse previously sent to Marcellus returned to the Roman general in order to act as mediators. A plan was hatched with the help of Mericus, a prefect of Acradina, who turned out to be Hispanus, in such a way that among the delegates sent to him to surrender there was an auxiliary of the same provenance. The soldier's aim was to hold a private interview with Mericus to induce him to surrender the enclave after explaining to him the diplomatic and military sensitivities of the situation. Later on, the ambassadors left Acradina for Marcellus' headquarters. This new delegation included a brother of Mericus. The private meeting was thus replicated, and they agreed on the conditions and mechanisms by which the surrender and seizure of the neighborhood was to take place.[53]

Finally, after the diplomatic process and the military offensive, the Syracusans themselves opened the gates of Acradia. They immediately sent representatives (*oratores*) to Marcellus with the sole request that he safeguard their lives and those of their children. Again, the Roman magistrate summoned his *consilium*, which was also attended by those Syracusans who had remained faithful to Rome from the beginning.[54]

The answer of the commander-in-chief was harsh, in response to the perfidious behavior of the Syracusans against Rome, staining with perjury them and their ancestors. Marcellus proceeded as follows: first, he sent a quaestor to Naso along with a contingent to take charge of the royal treasury and take it into custody. Secondly, the whole city was handed over to the troops to carry out looting; while plunder was going on, Archimedes was killed by the hand of a soldier.[55]

---

[52] Liv. 25.29.1-10. About the mechanism of *deditio* see Sanz, "La *deditio*"; García Riaza and Sanz (eds.), *In Fidem Venerunt, expresiones de sometimiento a la República Romana en Occidente* (Madrid: Dykinson, 2019).

[53] Liv. 25.30.1-12.

[54] Liv. 25.31.1-3.

[55] Liv. 25.31.8-11.

Once he fully captured the city, Marcellus committed to the complete organisation of Sicily. At this time, he was still receiving embassies from many cities on the island. The aim for the delegations was to confirm peace and achieve a new status with Rome. According to the role played by each city during the war, the Roman general punished those that had gone over to the Punic side or those that had maintained a neutral or inactive role, and rewarded those that had remained loyal to Roman interests.[56]

6. Conclusions

Roman headquarters provided a privileged setting for diplomatic interaction before, during, and after the conclusion of conflicts. Despite its purely military nature, the Roman camp was not only the point of arrival for surrender emissaries, but also the venue for political negotiations. It worked as a diplomatic node between the local representatives and Rome. At the same time, it played a key role as a center for dispatching embassies to undertake political-diplomatic actions. The emissaries sent by the commanding general had significant negotiating skills.

The itinerant nature of the camp was not a handicap in terms of its political-administrative capacities, but on the contrary, mobility provided the creation of ties and relationships in the magistrate's *provincia*. Moreover, the internal spatial articulation of the camp together with the scenography of the diplomatic ritual did not only create mechanisms of pressure and coercion, but also of negotiation and strengthening of ties, both personal and institutional.

The existence of non-official diplomacy at the headquarters is also attested. The conversations held between the Roman representative and foreign interlocutors in a confidential, quasi-clandestine manner reveal a remarkable offensive and coercive facet of Rome's power. In a complementary way, the camp housed pro-Roman exiles seeking refuge and acted as a base for double agents, helping Rome to achieve its more immediate ends.

The *consilium*, as we have seen on several occasions, was articulated alongside the general, providing him with support and advice at extremely complex moments, particularly in decisions that could entail repercussions on key aspects of the development of the campaign. Likewise, the *consilium*

---

[56] Liv. 24.40.5

provided the Roman commander with an assessment of the circumstances, and offered him a solution to possible political-diplomatic conflicts, particularly in the context of punitive actions against the local populations. The *consilium* used to accompany the commanding general in his tent or outside, in the praetorium, most probably in the court.

High-level interviews of the *conloquia* type, crucial in political-diplomatic negotiations, took place outside the camp. Such bilateral meetings not only occurred between local agents and the Roman magistrate but also by representatives of Rome. Their location was also used as a tactical-military outpost, acting in parallel as a source of information on the terrain surrounding the camp or headquarters. Diplomacy and the military thus went hand in hand, creating not only a political but also a coercive formula through the multiple contacts undertaken both within the camp and in the vicinity of foreign cities or locations. During all stages of the Roman-Syracusan conflict, diplomatic activity in the camp did not cease, and became more intense as the conflict evolved. This process highlights the general's ability to orchestrate and coordinate the political life of his province from his headquarters or camp.

We must understand the Roman camp in all its richness and complexity. We should examine it from a military point of view, but also from a diplomatic perspective. The camp was not only a perfect machine for intimidation and violence, but it also acted, systematically, as a point of reference and representation of Rome's political and diplomatic power.

Bibliography

Austin, N. J. E. and N.B. Rankov. *Exploratio. Military and political intelligence in the Roman world from the Second Punic War to the battle of Adrianople.* London and New York: Routledge, 2006.

Balsdon, J. P. V. D. *Romans and Aliens.* London: The University of North Carolina Press, 1979.

Baumer, L., T. Hölscher, and L. Winkler. "Narrative Systematik und politisches Konzept in des Reliefs der Trajanssäule. Drei Fallstudien." *Jahrbuch des Deutschen archäologischen Institut* 106 (1991): 261-295.

Braund, D. *Rome and the Friendly King*. New York: Routledge, 1984.

Burton, P. J. *Friendship and Empire. Roman Diplomacy and Imperialism in the Middle Republic (353-146 BC).* Cambridge: Cambridge University Press, 2011.

Claudon, J.-F. "Les ambassades des cites greques d'Asie Mineure auprès des autorités romaines: de la libération des Grecs à la fin du Haut-Empire (196 av. J.-C. – 235 apr. J.-C.)." PhD diss., EPHE Paris, 2015.

Criniti, N. *L'Epigrafe di Asculum di Gn. Pompeo Strabone.* Milano: Vita e Pensiero , 1979.

Díaz Ariño, B. *Epigrafía latina republicana de Hispania (ELRH).* Barcelona: Publicacions i Edicions, Universitat de Barcelona, 2008.

Díaz Fernández, A. "*¿Dum populus Senatusque Romanus vellet*? La capacidad de decisión de los mandos provincials en elm arco de la política romana (227-49 a.C.)." In *Poder central y poder local. Dos realidades paralelas en la órbita política romana,* edited by R. González Salinero, 135-152. Madrid: Signifer Libros, 2015.

Díaz Fernández, A. *Provincia et Imperium. El mando provincial en la República romana (227-44 a.C.).* Sevilla: Universidad de Sevilla, 2015.

Díaz Fernández, A. "*Deditio, restitutio* y cláusulas revocatorias en el Bronce de Alcántara (AE, 1984, no. 495)." In *In fidem venerunt. Expresiones de sometimiento a la República romana en Occidente,* edited by E. García Riaza and A. M. Sanz, 167-193. Madrid: Dykinson, 2019.

Dobson, M. *The Army of the Roman Republic. The Second Century BC, Polybius and the Camps at Numantia, Spain.* Oxford: Oxbow Books, 2008.

Drogula, F. K. *Commanders and Command in the Roman Republic and Early Empire.* Chapel Hill: The University of North Carolina Press, 2015.

Eckstein, A. M. *Senate and General: Individual Decision-making and Roman Foreign Relations, 264-194 B.C.* Berkeley: The University of North Carolina Press, 1987.

Eckstein, A. M. *Rome Enters the Greek East: from Anarchy to Hierarchy in the Hellenistic Mediterranean, 230-170 BC.* Oxford: Wiley-Blackwell, 2008.

Eckstein, A. M. "Ancient International Law, the Aetolian League, and the Ritual of Surrender during the Roman Republic: a Realist View." *The International History Review* 31 (2) (2009): 253-267.

García Fernández, E. "La *deditio in fidem populi Romani* y la construcción de un orden exterior en la República Romana." In *Hispania et Roma: Estudios en homenaje al profesor Narciso Santos Yangua,* edited by J. I. San Vicente González de Aspuru, C. Cortés Bárcena, E. González González, 99-108. Madrid: Ediciones de la Universidad de Oviedo, 2019.

García Moreno, L. "Sobre el decreto de Emilio Paulo y la Turris Lascutana (CIL.12, 614)," *Epigrafía hispánica de época romano-republicana* (1986): 195-218.

García Riaza, E. "Derecho de guerra en Occidente durante la expansi,ón romano-republicana. Planteamientos metodológicos." In *De fronteras a provincias. Interacción e integración en Occidente (ss. III-I a.C.),* edited by E. García Riaza, 31-65. Palma de Mallorca: Edicions Universitat de les Illes Balears, 2011.

García Riaza, E. "Le protocole diplomatique entre particularisme romain et universalisme: quelques réflexions sur l'Occident républicain." In *La diplomatie romaine sous la République: réflexions sur una pratique*, edited by Barthélémy Grass & Ghislaine Stouder, 15-41. Génova: Institut des Sciences et Techniques de l'Antiquité, 2015.

García Riaza, E. "Guerra, agresiones a embajadas y prejuicios de género en la expansión romano-republicana: el caso de Teuta de Iliria," *Gladius* 40 (2020): 7-17.

García Riaza, E. "*In Conloqvivm venire:* Interviews between Roman Commanders and Western Leaders in the Age of Republican Expansion." In *Provinces and Provincial Command in Republican Rome: Genesis, Development and Governance*, edited by A. Díaz Fernández, 127-144. Sevilla: Universidad de Sevilla, 2021.

García Riaza, E. and Sanz, A. M. (eds.). *In Fidem Venerunt, expresiones de sometimiento a la República Romana en Occidente.* Madrid: Dykinson, 2019.

Gilliver, C. M. "The *de munitionibus castrorum:* Text and Translation." *JRMES* 4 (1993): 33-48.

Grainger, J. D. *Great Power Diplomacy in the Hellenistic World.* New York: Routledge, 2017.

Gruen, E. S. *The Hellenistic World and the Coming of Rome.* Berkeley: The University of North Carolina Press, 1986.

Hidalgo de la Vega, Mª. J. "El bronce de Lascuta. Un balance historiográfico," *Studia historica. Historia Antigua* 7 (1989): 59-66.

Jiménez, A., Bermejo, J., Valdés, P., Moreno, F., Tardio, K. "Renewed Work at the Roman Camps at Renieblas near Numancia (2nd-1st c. B.C.)," *JRA* 33 (2020): 4-35.

Johnston, P. D. *The Military Consilium in Republican Rome.* Piscataway, New Jersey: De Gruyter, 2008.

López Merelo, R., García Jiménez Santiago, G. J. S., & Sánchez Abal, J. "El bronce de Alcántara. Una deditio del 104 a.C.." *Gerión. Revista De Historia Antigua*, 2 (1984): 265-323.

Morillo Cerdán, Á. "Criterios arqueológicos de identificación de los campamentos romanos en Hispania." *SALDVIE* 8 (2008): 73-93.

Nörr, D. *Aspete des römischen Völkerrechts. Die Bronzetafel von Alcántara.* München: Bayerische Akademie der Wissenschaften, 1989.

Rosselló Calafell, G. *Relaciones exteriors y praxis diplomática cartaginesa. El período de las guerras púnicas.* Sevilla: Universidad de Sevilla, 2023.

Salomonsson, M. M. "Roman Legates in the Republic." *Opuscula Romana* 25-26 (2000-2001): 79-88.

Sánchez Moreno, E. and Aguilera Durán, T. "Lusitanos frente a Galba (151-150 a.C.) ¿una deditio in fidem?" In *In fidem venerunt: expresiones de sometimiento a la Repúblicana romana en Occidente,* edited by E. García Riaza y Sanz, A. M., 139-166. Madrid: Dykinson, 2019.

Sánchez, P. "Quand Rome se cherchait de nouveaux alliés: les accords de coopération militaire négociés à l'initiative des Romains sur le théâtre des operations (IVe-IIIe s. av. n.è.)," *Ktéma* 41 (2016): 165-190.

Sanz, A. M. "La *deditio*: un acte diplomatique au cœur de la conquête romaine." In *La diplomatie romaine sous la République: réflexions sur una pratique*, edited by Barthélémy Grass and Ghislaine Stouder, 87-105. Génova: Institut des Sciences et Techniques de l'Antiquité, 2015.

Sastra, I. & Plácido Suárez, D. "*Deditio in fidem* and Peasant Forms of Dependence in the Roman Provincial System: the Case of Northwestern Iberia," *XXX colloque du GIREA* (2005): 501-509.

Sheldon, R. Mary. *Intelligence Activities in Ancient Rome. Trust in the Gods, but Verify.* London and New York: Routledge, 2005.

Torregaray Pagola, E. "Embajadas y embajadores entre Hispania y Roma en la obra de Tito Livio." In *Diplomacia y autorrepresentación en la Roma antigua*, edited by E. Torregaray Pagola, J. Santos Yanguas, 25-62. Vitoria/Gasteiz: Universidad del País Vasco = Euskal Herriko Unibertsitatea, 2005.

Torregaray Pagola, E. "Los espacios de la diplomacia en la Roma republicana," *Caesarodunum X*L (2006): 223-258.

Torregaray Pagola, E. "*Legatorum facta:* la ejemplaridad de los embajadores romanos," *VELEIA* 26 (2009): 127-152.

Torregaray Pagola, E. "The Roman Ambassador's Speech: Public Oratory on the Diplomatic Stage." In *Community and Communication Oratory and Politics in Republican Rome,* edited by C. Steel & H. van der Blom, 229-245. Oxford: Oxford University Press, 2013.

Torregaray Pagola, E. "Fracaso y oportunidad en la diplomacia de la República romana." In *Algunas sombras en la diplomacia romana,* edited by E. Torregary Pagola & J. Lanz Betelu, 31-47. Vitoria-Gasteiz: Universidad del País Vasco = Euskal Herriko Unibertsitatea. 2020.

Vertedor Ballesteros, B. "Buying loyalties, grating rewards: the Roman headquarters as a space of diplomatic Interaction." In *The materiality of diplomacy in the Hellenistic-Roman Mediterranean: Gifts, Bribes, Offerings,* edited by E. Sánchez Moreno and E. García Riaza, Edinburgh: Edinburgh University Press, *Forthcoming.*

ILLA VERO PARS CALLIDITATIS EGREGIA ET AB OMNI REPREHENSIONE PROCUL REMOTA... HANNIBAL, THE CARTHAGINIANS AND THE STRATAGEMS IN VALERIUS MAXIMUS

GABRIELE BRUSA

Introduction

Modern-day scholarship has debated at length the employment of stratagems by ancient armies, and their ethical consideration in the Greek and Roman world. In the Greek scenario, this problem is connected to discussions about how the phalanxes worked and fought, as well as the relative importance of the other sections of the armies.[1] In the Roman

[1] This text will not take into account a comparison between the Greek and the Roman world, as this would make it way too long. For an attempt to draw such a comparison, see Giovanni Brizzi, *I sistemi informativi dei Romani: principi e realtà nell'età delle conquiste oltremare (218-168 a.C.)* (Wiesbaden, 1982), 7-26. Brizzi offers a clear-cut distinction between the Roman refusal of tricks and the common Greek reliance on such practices. Everett L. Wheeler, *Stratagem and the Vocabulary of Military Trickery* (Leiden, 1988a) studies the strategematic vocabulary of both cultures, and theorizes a much more similar approach to the issue by Greeks and Romans. Specifically on Greek stratagems in relation to the so-called Greek way of war, see, among the most important works, Josiah Ober, "The Rules of War in Classical Greece," in *The Athenian Revolution: Essays on Ancient Greek Democracy and Political Theory*, ed. Josiah Ober (Princeton, 1999), 57-62, and Victor D. Hanson, *The Western Way of War: Infantry Battle in Classical Greece* (Oxford, 1990), for the classical view linking the monolithic nature of the phalanx to the absence of stratagems and elaborate tactical devices in Greek battles, and Peter M. Krentz and Everett L. Wheeler, *Stratagems of War: Polyaenus* (Chicago, 1994), vi-ix; Peter M. Krentz, "Deception in Archaic and Classical Greek Warfare," in *War and Violence in Ancient Greece*, ed. Hans van Wees (London, 2000), 167-200; Hans van Wees, *Greek Warfare: Myths and Realities* (London, 2004), 131-145; and Roel Konijnendijk, "Mardonius' Senseless Greeks," *CQ* 66 (2016): 3-4, for a newer theory which refutes both the simplistic nature of Greek battle and the lack of stratagems.

context, technical and tactical reconstructions have been less relevant to this problem than strictly cultural and moral considerations. It is fruitless to retrace here in detail every step of the debate; it will suffice to recall some particularly important works and to outline the general trends in historiography. In 1982, a much-discussed book by Brizzi systematically reframed theories already sketched by previous scholarship.[2] The author suggested that the Romans refused, on moral grounds, to make use of any stratagem prior to the Hannibalic war. The concept of a military ruse itself was unknown to them, and, when they had to describe it, they were compelled to borrow a Greek word, στρατήγημα, the Latin language lacking a corresponding noun. Hannibal, the shrewd general *par excellence*, put these ethical preconceptions to the test. He forced the Romans to oppose him with *mens*, a sort of defensive version of the strategematic approach to war. Scipio the Elder learned the employment of military tricks from Hannibal, and, in time, the Romans became as skilled as the Greeks and the Carthaginians. However, they retained a cultural distaste for any un-open way to wage war, and always felt compelled to justify their stratagems as a defensive response to the enemy's *perfidia*.[3] The concept of *fides* is central to Brizzi's reconstruction:[4] it represents a chivalrous approach to warfare, sanctioned by religion, both *in indicendo* and *in gerendo*,[5] and forbids any trick. Battles had to be decided only by *virtus*, through an open clash between the two armies; anything else was described as *perfidia*. All of Rome's enemies were represented as *perfidi hostes*.

---

theory which refutes both the simplistic nature of Greek battle and the lack of stratagems.

[2] See esp. Jean-Paul Brisson, "Les mutations de la seconde guerre punique," in *Problèmes de la guerre à Rome*, ed. Jean-Paul Brisson (Paris-La Haye, 1969); Enrico Montanari, "*Mens*," *R&C* 2 (1976): 172-235, and above all Gérard Freyburger, *Fides: étude sémantique et religieuse depuis les origines jusqu'à l'époque augustéenne* (Paris, 1986), 105 ff.: "la morale militaire romaine exigeait du combattant qu'il fit preuve de *fides*, c'est-à-dire en somme d'esprit chevaleresque"; "la morale aristocratique traditionclle des Romains méprisait sans nul doute profondément les manœuvres tortueuses que sont les embuscades et l'utilisation des trahisons."

[3] Brizzi 1982.

[4] Brizzi 1982, 6-30.

[5] The passage he cites is Cic. *Leg*. 2.14.34. However, Cicero does not specify what this *ius in bello* entails.

Brizzi's view is still widespread especially among Italian and French authors.[6] However, it was challenged by Wheeler in a monograph devoted to the Greek and Latin vocabulary of military trickery and in an article in the same year.[7] Wheeler contended not only that the Romans did in fact have a range of words to describe the stratagems, but also that these expressions did not necessarily carry a negative characterization. On the contrary, they were *voces mediae*, and any possible reproach depended on the context in which they were used.[8] The Romans made use of military tricks from the very start of their history and could take pride in them.[9] According to the author, however, from the late Republic they started to choose to portray Rome as the "people of *fides*". This led to the creation of the stereotype of the *perfidus hostis*, and to a new emphasis on Roman *fides*, with a denial of previous Roman reliance on stratagems: these, in a word, became un-ethical and un-Roman.[10] It must be noted that, despite the marked differences in their reconstructions, Brizzi and Wheeler shared a similar view about the characteristics of *fides*. They both saw it as a value that – whether in reality or just in propaganda – forbade the employment of ruses in battle.[11]

Wheeler's views have gained credit: many researchers nowadays agree about the fact that stratagems were widely employed, despite the

[6] See, for example, Yann Le Bohec, *Histoire militaire des guerres puniques: 264-146 avant J.-C.* (Monaco, 1996), 150-154; Guy Achard, "César et la ruse," in *Ruses, secrets et mensonges chez les historiens grecs et latins*, eds. François Bérard, Pascale Giovannelli-Jouanna, and Hélène Olivier (Lyon 2006), 85-97; Giambattista Cairo, "La guerra nel mondo eroico greco-romano: tra propaganda, ideologia e religione," in *Domi forisque: omaggio a Giovanni Brizzi*, ed. Stefano Magnani (Bologna, 2018), 49-58. Brizzi himself revisited the issue in several other publications: see, in English, Giovanni Brizzi, "Carthage and Hannibal in Greek and Roman Memory," in *A Companion to the Punic Wars*, ed. Dexter Hoyos (Malden, 2011), 484-485.

[7] Wheeler 1988a and Everett L. Wheeler, "Sapiens and Stratagems: The Neglected Meaning of a Cognomen," *Historia* 37 (1988b): 166-195.

[8] Wheeler 1988a, 93-102.

[9] Wheeler 1988b convincingly shows that some Roman *cognomina* were derived from the world of military trickery and were intended as marks of honor.

[10] Wheeler 1988b, 193-195: "Romans as the people of *fides* wished to disassociate themselves from stratagem in the second century. Stratagem had become inconsistent with Roman propaganda."

[11] This notwithstanding, Wheeler recognizes that *fides* does not coincide with the lack of stratagems (Wheeler 1988a, 17); however, he considers it a rather fluid concept (pp. 63-64) and ties it, at least potentially, with the world of military trickery (p. 65).

"propaganda of openness" and the construction of an ideology based on *fides*.[12] The main point that must be stressed for the purposes of this text is that much of the scholarship detects in the late Republic and early Empire an ideological and moral distaste for the military ruse. This was described as *perfidia*, attached as a charge of barbarism to Rome's enemies, and opposed to Roman *fides*. The Carthaginians in particular have been seen as the incarnation of *perfidia* in Roman eyes.

In this text I address these issues through the case study of the *Facta et Dicta Memorabilia* of Valerius Maximus, particularly his depiction of Hannibal and the Carthaginians, and their comparison and opposition to the Romans. While this narrow focus will not allow extending my conclusions to Roman authors and culture at large, Valerius remains an interesting object of research for various reasons. In recent years, research on his work has led to a vindication of his "authorship," that is, to him being recognized as an author with personal ideas, which are expressed through the *exempla* he reports.[13] It is interesting, therefore, to verify whether he had any particular

[12] Rose M. Sheldon, *Ambush: Surprise Attack in Ancient Greek Warfare* (Barnsley, 2008) accepts Wheeler's theories and applies them to the context of military espionage; Jean-Vincent Holeindre, *La ruse et la force: une autre histoire de la stratégie* (Paris, 2017), 114-127 thinks that the Romans always criticized any ruse in battle, but made use of it themselves, taking care to represent it as "defensive" in nature and directed against *perfidi hostes*.

[13] Clive Skidmore, *Practical Ethics for Roman Gentlemen: The Work of Valerius Maximus* (Exeter, 1996) vindicated the *Facta et Dicta Memorabilia* as a work with specific moral purposes, and not just as a collection of passages that could be of use to rhetoricians. From this starting point, the ethical aims of Valerius have been the subject of a vigorous scholarship. Some authors must be singled out as particularly important for this paper: Yves Lehmann, "Les rivendications morales et politiques de Valère Maxime," in *Valeurs et mémoire à Rome: Valère Maxime, ou la vertu recomposé* , ed. Jean-Michel David (Paris, 1998), 19-26; Hans-Friedrich Mueller, *Roman Religion in Valerius Maximus* (London-New York, 2002); Sarah J. Lawrence, "Inside Out: The Depiction of Externality in Valerius Maximus" (PhD diss., University of Sydney, 2006); Rebecca Langlands, "Reading for the Moral in Valerius Maximus: The Case of *Severitas*," *Cambridge Classical Journal* 54 (2008): 160-187; Simon R. Matravers, "Commentary on Valerius Maximus' Book IX.1-10. A Discussion on *Vitia*: An Apotreptic Approach" (PhD diss., University of Birmingham, 2016); Jeffrey Murray, "Valerius Maximus on Vice: A Commentary on *Facta et Dicta Memorabilia* 9.1-11" (PhD diss., University of Cape Town, 2016); and Jorg Rüpke, "Knowledge of Religion in Valerius Maximus' *Exempla*: Roman Historiography and Tiberian Memory Culture," in *Memory in Ancient Rome and Early Christianity*, ed. Karl Galinsky (Oxford, 2016), 89-111. Some features of all of these will be discussed later.

opinions about stratagems, tricks, and Carthaginians. On the other hand, Valerius is undeniably the recipient of a tradition, which he re-shapes and re-organizes,[14] and could thus serve well as a starting point for further studies. Finally, despite the poverty of specific studies about his approach to Punic ruses,[15] one could detect a whole wealth of *en passant* comments, in modern scholarship, about how poorly he portrays the Carthaginians and Hannibal from this point of view. There is an extremely wide consensus about his depiction of Hannibal as the antithesis of Roman virtues: deceitful and cunning, without any consideration for religion or justice, whose only ability in war was to trick his opponents instead of facing them honorably; in short, *perfidia* incarnated.[16] Only one author, to my knowledge, has denied Valerius' tendency to contrapose Roman and Carthaginian values sharply. This theory, however, is inserted into the context of a general work about his depiction of the universality of human virtues and vices, and does not focus

---

[14] Mueller 2002, 2. Lehmann 1998, 19 sees Valerius as building, drawing on his sources, "une étique officielle du people romain." About his relationship with his sources, see Eva Valvo, "La rappresentazione di Annibale in Valerio Massimo," *Millennium* 5 (2008), 50 (on his alterations of Livy), and, more generally, G. Maslakov, "Valerius Maximus and Roman Historiography: A Study of the *Exempla* Tradition," *ANRW* II.32.1 (1984), 437-496, and Martine Chassignet, "La deuxième guerre punique dans l'historiographie romaine: fixation et évolution d'une tradition," in, *Valeurs et mémoire à Rome: Valère Maxime, ou la vertu recomposé* , ed. Jean-Michel David (Paris, 1998), 55-72.

[15] An exception is Valvo 2008, who studies Valerius' representation of Hannibal, with a particular focus on trickery and deceit (esp. pp. 47-50).

[16] Once again, Valvo 2008. See also Paolo Desideri, "Greci, barbari, Cartaginesi in Valerio Massimo", in *Costruzione e uso del passato storico nella cultura antica*, ed. Anna M. Biraschi, Paolo Desideri, Sergio Roda (Alessandria, 2007), 305-312, and Murray 2016, 190-191 and 199-202. Cf., in less detail, William J. Watts, "Race Prejudice in the Satires of Juvenal," *Aclass* 19 (1976): 92; Brizzi 1982, 64-65 and 2011; Federico Mazza, "L'immagine dei Fenici nel mondo antico," in *I Fenici*, ed. Sabatino Moscati (Milano, 1988), 563; Wheeler 1988a, 16-17; Arthur J. Pomeroy, "Hannibal at Nuceria," *Historia* 38 (1989): 162-163 (the author, however, focuses more on treachery, and less on stratagems); Georges Devallet, "*Perfidia plus quam Punica*: l'image des Carthaginois dans la littérature latine de la fin de la république à l'époque des Flaviens," *Lalies* 16 (1996): 26; Le Bohec 1996, 13-14; Maria T. Schettino, *Introduzione a Polieno* (Pisa, 1998), 83-84; Lehmann 1998, 23; Claire Stocks, *The Roman Hannibal: Remembering the Enemy in Silius Italicus'* Punica (Liverpool, 2014), 29-32; Matravers 2016, 255-258. Note, however, the more nuanced remarks by Ben Tipping, *Exemplary Epic: Silius Italicus'* Punica (Oxford, 2010), 53-54, who sees Hannibal in Valerius Maximus as a complex character, who is not condemned in a clear-cut way.

specifically on stratagems.[17] Overall, scholars agree to present the *Facta et Dicta Memorabilia* as a typical example of Roman pre-conceptions about the Carthaginians. This, once again, makes Valerius' text an ideal starting point for a reappraisal of these problems.

Hannibal and the stratagems in Valerius Maximus

In his seventh book, Valerius devotes a chapter to Roman and foreign *exempla* of stratagems. The introduction to this chapter is well known, but it is worth quoting once more:

> *Illa vero pars calliditatis egregia et ab omni reprehensione procul remota, cuius opera, quia appellatione† vix apte exprimi possunt, Graeca pronuntiatione strategemata dicantur.*[18]

The text is partially corrupt (in the lost words the author was probably saying that the Latin language lacked a suitable term to translate the Greek στρατήγημα), but the general meaning is clear. A stratagem is a witty action, mainly in war, which cannot incur any reproach. Valerius' moral clarification seems to imply that there were indeed, in his opinion, other *partes calliditatis* which should be criticized. Stratagems, however, should not. Among the Romans, the author lists Tullus Hostilius, Sextus Tarquinius (the son of Superbus), the Roman defenders during the Gallic siege, Nero and Salinator against Hasdrubal and Hannibal, and Metellus Macedonicus;[19] among foreigners, just

---

[17] Lawrence 2006. The author argues that there's no sharp distinction between Romans and strangers in the *Facta et Dicta*.

[18] Val. Max. 7.4.1.

[19] Val. Max. 7.4.1 (Tullus Hostilius, informed of the treachery of the Albans on the battlefield, pretends to have ordered them to retreat and to attack the enemy's flank to restore his army's morale: cf. Liv. 1.27 and Dion. Hal. *Ant. Rom.* 3.24-25); 7.4.2 (Sex. Tarquinius, pretending to be at variance with his father, manages to be accepted as a citizen and leader of Gabii, which he subsequently delivers to his father: cf. Liv. 1.53 and Dion. Hal. *Ant. Rom.* 4.55-58); 7.4.3 (the Romans, to convince the Gauls that they have an abundance of food, throw bread off the walls; cf. Liv. 5.48); 7.4.4 (Salinator and Nero manage to trick Hannibal and Hasdrubal, making the former believe that Nero was still encamped in front of him, and attacking the latter with two combined armies; cf. Liv. 27.46 and Polyb. 11.1-3); 7.4.5 (Metellus Macedonicus, in Spain, manages to conquer Contrebia by pretending to lead his army away and then quickly turning back). To these passages, one could add 3.3.1, Mucius Scaevola's attempt to murder King Porsena.

Agathocles[20] and Hannibal[21] are mentioned. About the latter, Valerius only mentions the Battle of Cannae, which he takes as the supreme example of Punic *astutiae*, of which he counts three during the battle itself.[22] At the end of the *exemplum*, however, the author adds a personal consideration, to justify the Roman defeat:

> *Haec fuit Punica fortitudo, dolis et insidiis et fallacia instructa. Quae nunc certissima circumventae virtutis nostrae excusatio est, quoniam decepti magis quam victi sumus.*[23]

This passage is often taken to reflect a sort of a "double standard" employed by Valerius: when stratagems are employed by the Romans, these are good examples of *providentia*; when they are directed against the Romans, they are criticized as unworthy of *virtus*.[24] Something more about the juxtaposition of *dolus* and *virtus* will be said in the next few pages. For now, one just has to note that this passage cannot be taken as an instance of criticism directed towards Punic stratagems. The *exemplum* is part of the chapter about stratagems, a "section of shrewdness,"[25] and it is quite natural to find an instance of *dolus*, *insidia* and *fallacia*. As Wheeler showed, these terms are

---

[20] Val. Max. 7.4.ext.1 (Agathocles, while Syracuse is besieged by the Carthaginians, decides to invade Africa, compelling the Carthaginians to abandon the siege to defend their homeland).

[21] Val. Max. 7.4.ext.2.

[22] Livy, on the other hand, considers just the false surrender of the Numidians as a stratagem (Liv. 22.48); of the other two (the Punic retreat to lure the Romans into a trap, and the disposition of the army to compel the Romans to fight with sun and dust in their eyes), the former is mentioned, but not emphasized as a ruse, and the latter is partly denied (the sun equally affected the armies) and partly considered the outcome of mere chance (Liv. 22.46.8-9). On these changes, see Valvo 2008, 50, who thinks that they are functional to the author's presentation of Hannibal as a bad, deceitful commander.

[23] Val. Max. 7.4.ext.2.

[24] See the authors quoted at n. 16. Concerning Valerius' "double standard," see Le Bohec 1996, 13-14; Matravers 2016, 256; and especially Wheeler 1988a, 16-17 ("trickery against the Romans is base, but trickery by the Romans is prudence"); and Schettino 1998, 84 ("in Valerio Massimo l'astuzia è atto di prudenza e saggezza quando è opera dei Romani, appare moralmente riprovevole se è a spese di questi ultimi").

[25] Tipping 2010, 53-54 is, to my knowledge, the only author to note very briefly that the collocation of the Hannibalic stratagems in this chapter is a hint of their acceptability.

voces mediae, not necessarily implying any reproach.[26] Valerius himself considered these actions *ab omni reprehensione procul remota*. The fact that the Romans were *circumventi* and not *victi* does not mean that the Carthaginians acted in a bad way by tricking them, just that they did not display any *virtus* comparable to that of the Romans.[27]

The absence of reproach is also shown by the fact that, throughout the whole work, Carthaginian victories over the Romans are never criticized because of Hannibal's stratagems, and Roman generals are never excused because of it. On the contrary, they are censured either because of their lack of *pietas*, which resulted in the gods' anger and therefore in defeat, or because of their rashness and lack of *prudentia* and *consilium*.[28] The enemy's tricks are never really an excuse for defeat.[29] On the other hand, Claudius Nero and Livius Salinator, who managed to outwit Hannibal and defeat his brother Hasdrubal in the Battle of Metaurus, are praised for their ability to "deceive the deceiver." In this case, *Punica calliditas* was *elusa Romana prudentia*.[30] One must note that Punic shrewdness is here characterized as *toto terrarum orbe infamis*. Valerius does not specify what gave it this bad repute, and this appears to be the only hint of a real moral reproach to their *calliditas*. While this cannot be ruled out, the Carthaginians do not appear to be explicitly censured: Valerius may be just saying that their predisposition towards stratagems made their enemies always suspicious of them and required employing an extra dose of *prudentia* against an enemy who always fought through deceit, and never through open *virtus*.

Even in the chapter about *vafritia* (a more general concept, describing any kind of trick) Hannibal is not criticized. *Vafritia*, unlike *strategema*, was not considered always positive by Valerius. The fact that he takes pains to

---

[26] Wheeler 1988a, 58-62 (on *dolus*), 75-76 (on *fallacia*), 81-82 (on *insidia*), 93-110 (on the importance of the concept of *vox media*).

[27] On the distinction between *dolus* and *virtus* (and Romans and Carthaginians in this context) see the following pages.

[28] Val. Max. 1.1.16 (on Cannae, with focus on the divine anger); 1.6.6 (on Flaminius' *impietas* and *temeritas*); 1.6.9 (the same, about Marcellus); 4.5.2 (on Varro's *temeritas* at Cannae, and on the subsequent explanations through divine anger); 5.2.4 (on Minucius' *temeritas*); 5.4.2 (on the Battle of the Ticino River, fought *adversis auspiciis*); 9.3.1 (even Livius Salinator is too anxious to fight against Hannibal).

[29] Note that, according to Polybius (3.75.1), Tib. Sempronius Longus tried to justify himself after the defeat near the Trebia through the adverse meteorological conditions, and not through Hannibal's trick. In Polybius' reconstruction, he was well aware of his own fault.

[30] Val. Max. 7.4.4.

underline the moral acceptability of the latter in the chapter immediately following that on *vafritia* points to the fact that the former could be seen in a more negative way.[31] Among the Romans, Q. Fabius Labeo, M. Antonius the orator, and M. Volusius are criticized;[32] among foreigners, Tullus, the chief of the Volsci,[33] and partially Bias of Priene.[34] Hannibal, on the other hand, is not. His attempt to cast suspicion onto Fabius Maximus by sparing only his properties while devastating the rest of Italy is cited in a neutral way as an example of his *vafri mores*.[35] The moral focus is just on Fabius: the author underlines his *pietas*, which was so well-known that no-one at Rome fell into Hannibal's trap.[36]

Perhaps the most important passage, however, comes from the chapter on self-confidence. Here, Valerius records the answer given by Hannibal to Prusias when the latter told the former that the result of the sacrifices he had performed forbade him to fight a battle, as Hannibal instead suggested.[37] Hannibal asked the king whether he preferred to put his trust in pieces of meat or in the opinion of a seasoned general. What is most important is that Valerius Maximus takes this sentence as the starting point to briefly sketch a picture of Hannibal as a general. First, he cites his triumphs: Spain, Gaul and Liguria subdued, the way across the Alps opened, the battles of the

---

[31] On the concept of *vafritia* in Valerius, see Wheeler 1988a, 71 and Skidmore 1996, 69.

[32] Val. Max. 7.3.4; 5; 8. The first man, who tricked the Nolans and Neapolitans to adjudicate to the Romans a disputed stretch of land, is called *improbus*; Antonius' refusal to write any of his orations to be able later to deny having said what he had said is considered shameful (*facti vix pudentis*), although partially justified; the last man saved his own life through an indecent disguise.

[33] Val. Max. 7.3.ext.10. Note that, according to the tradition (Liv. 2.37.2; Dion. Hal. *Ant. Rom.* 8.2; Plut. *Coriol.* 26.2), the treacherous plan to draw the Romans into a war with the Volsci and Coriolanus was either devised by Coriolanus alone (Dionysius) or by Coriolanus and Tullus together (Livy).

[34] Val. Max. 7.3.ext.3. The author describes his principle of considering friends as potential enemies as too cynical, but admits that, if applied with moderation, it is acceptable.

[35] Val. Max. 7.3.ext.8. The preceding example refers to another Hannibal, the commander of the Punic fleet defeated by C. Duilius at Mylae, who was able to avoid condemnation by the Carthaginian Senate through a trick, which is not reproached.

[36] Here Valerius distances himself from Livy once again: according to Liv. 22.23.4, Hannibal's plan was effective in ruining the dictator's popularity.

[37] Val. Max. 3.7.ext.6. The same anecdote is quoted also by Cic. *Div.* 2.24.52 (who agrees with Hannibal) and Plut. *De exil.* 16 (who wants to show that Hannibal preserved his freedom of speech even while in exile). See, very briefly, Stocks 2014, 30-31.

Trasimene and Cannae (*Punicae victoriae clarissumum monumentum*), Italy devastated, and Capua convinced to side with him. These are all proofs of his *gloria*, which is *longo experimento testata*. The final phrase is even more important:

> *Et sane, quod ad exploranda bellica artificia aestimandosque militaris ductus adtinebat, omnis foculos, omnis aras Bythiniae Marte ipso iudice pectus Hannibalis praegravasset.*

Even Mars himself would have recognized Hannibal's greatness as a commander. This greatness is at least in part related to the employment of stratagems, *bellica artificia*, for which Hannibal was famous. This is not the only passage in which the author speaks so highly of Hannibal: later, he mentions his *insignem nomini sui memoriam*,[38] and previously he had touched on his *magna gloria*.[39] This seems to be sufficient to cast doubt on the usual reconstruction of Valerius' portrayal of the Carthaginian and his stratagems. Hannibal, to summarize the evidence, is present, as are the Romans, in the section about stratagems. At the start of this section Valerius pointed out that no one could blame their employment. Hannibal is seen as a particularly shrewd commander, and he is not criticized for this. On the contrary, his ability makes him a great general, for whom Valerius only has words of praise in the only passage in which the author tries to reassess his generalship.

Romans and Carthaginians

It would not be fair to maintain that Roman and Punic practices in warfare are not contrasted in the *Facta et Dicta Memorabilia*. The accusations of *crudelitas* and *perfidia* will be dealt with in the next section. Here, I would like to show that the main difference between the two peoples does not lie in the employment of stratagems, but in the concept of *virtus*.[40] As is customary in Latin literature, the Romans always display an impressive amount

---

[38] Val. Max. 9.6.ext.2; the author criticizes only his *perfidia*, which, as will be argued below, has nothing to do with his stratagems.

[39] Val. Max. 5.3.ext.1.

[40] On the concept of *virtus*, central to the self-representation of the Romans, see Miles McDonnell, *Roman Manliness: Virtus and the Roman Republic* (Cambridge, 2006), esp. pp. 59-71, and Catalina Balmaceda Errazuriz, *Virtus Romana: Politics and Morality in the Roman Historians* (Chapel Hill, 2007), esp. pp. 15-18, with bibliography. About the importance of *virtus* in Valerius' work, see Steven J. Simon, "Some Concepts of Republican Virtue as Expressed by Valerius Maximus," *AncW* 33 (2002): 147-152.

of military virtue. For example, to show his own courage, Romulus prefers single combat against the chief of Caenina to a regular battle.[41] M. Marcellus attacks the king of the Gauls and his army almost alone.[42] Cato flings himself in the thick of a battle, unarmed, to recover his lost sword.[43] According to Valerius, commanders were even expected not to let themselves be besieged, as this would deprive their men of the occasion to display their valor.[44] On the same lines, those who are taken prisoners are blamed as cowards: they should have died heroically, as "real Romans" do.[45] Hannibal's invasion itself is considered a trigger to awaken Roman *virtus*.[46] The Romans, in short, are the people of *virtus par excellence*. Even when they employ stratagems, Valerius takes care to underline that this *calliditas* is just an addition to the regular *virtus*. When he depicts the defenders of the Capitol throwing bread against the Gauls to dishearten them by pretending to have an abundance of food, he says that Roman courage was finding refuge in shrewdness (*Romana virtus, praesidium ab astutia mutuans*).[47] This is implied to be unusual, as Jupiter pities Roman courage, which, for once, could not fight by its forces alone. When they outwit Hannibal and Hasdrubal, Salinator and Nero give proof of *providentia* and *consilium*: they lie and they deceive (*mentitus*; *decipiendum*), but in the end Hasdrubal was defeated by their valor (*utriusque virtute*).[48]

Exactly the opposite is true for the Carthaginians: just as the Romans are the people who always fight with *virtus*, they are the people who never do. They always win through ambushes, for example against Regulus,[49] against Marcellus,[50] at the Trasimene or at Cannae. When they defeat the Romans, they do not do so in a clash of courage, but through deceit. This appears to be the key to interpreting the passage quoted about Cannae, where Valerius writes about the *Punica fortitudo, dolis et insidiis et fallacia instructa*, and

---

41 Val. Max. 3.2.3. This chapter is full of Roman examples along these same lines.
42 Val. Max. 3.2.5.
43 Val. Max. 3.2.16. Here the author confuses Cato the Elder with his son (Plut. *Aem.* 21.1-5).
44 Val. Max. 2.7.7: Cincinnatus compels L. Minucius to resign his consulship because of this.
45 Val. Max. 2.7.15 (Hannibal's Roman prisoners); 3.5.1 (Scipio Africanus' son against Antiochus). At 3.3.8, one of Octavian's centurions is taken prisoner by Antony, but immediately asks to be killed.
46 Val. Max. 7.2.3; cf. 3.2.11 and 3.7.10.
47 Val. Max. 7.4.3; cf. Mueller 2002, 89-90.
48 Val. Max. 7.4.4.
49 Val. Max. 1.1.14.
50 Val. Max. 1.6.9.

states that the Romans were not *victi*, but *decepti*.[51] Hannibal wins thanks to his shrewdness, but this – despite not being dishonorable in itself – is not paired with *virtus*, which allows the Romans to retain their primacy in this field. When Hannibal is praised as a commander, it is due to his cunning and intelligence; but neither he nor his soldiers are ever said to be courageous. Mars would have given to him the first place only *ad exploranda bellica artificia aestimandosque militaris ductus*.[52]

In the *Facta et Dicta Memorabilia*, in short, Romans and Carthaginians are sharply juxtaposed. On the one side, one reads of the exploits of the Romans, always daring and courageous. They can make use of stratagems, without incurring any reproach, but even when they do so, *virtus* remains their supreme value. On the other hand, the reader finds that the Carthaginians always rely on stratagems, and on nothing else. Their shrewdness is not criticized; what really sets them apart is their inability to imitate Roman *virtus*.

Valerius' charges against Hannibal: *crudelitas* and *perfidia*

So far, very little attention in this text has been devoted to the concept of *perfidia*. This may come as a surprise, given that many authors connected it to the employment of stratagems and considered this as the main charge levelled at Hannibal. While it is true that Hannibal and the Carthaginians are in the *Facta et Dicta* (but also in most of the Latin tradition)[53] described as *perfidi hostes*, *perfidia* does not necessarily relate to the world of stratagems. In his work about the vocabulary of military trickery, Wheeler rightly identified the violation of good faith as the key element of *perfidia*. He proceeded to ask whether it should be supposed that a normal stratagem violates *fides*: "does not any stratagem violate the sense of good faith implicit in a code of honor in warfare?"[54] From the passages analyzed in the previous pages, it appears that Valerius does not have such a conception as a "code

---

[51] Val. Max. 7.4.ext. 2.

[52] Val. Max. 3.7.ext.6; cf. 9.1.ext.1, where Hannibal is described as *vigilantissimus dux*.

[53] About Roman stereotypes against the Carthaginians (and about their origin in the Greek world), see Luisa Prandi, "La *fides Punica* e il pregiudizio anticartaginese," *CISA* 6 (1979): 90-97; Mazza 1988; Devallet 1996. Particularly famous is Livy's description of Hannibal's *perfidia* as *plus quam Punica* (Liv. 21.4.9).

[54] Wheeler 1988a, 17. Matravers 2016 simplistically writes that *perfidia* belongs "to a more severe group on V.'s moral barometer" than a stratagem.

of honor" implicit in warfare which involves stratagems. Reviewing the occurrences of the word *fides* in the work, it is possible to establish its meaning as "fidelity" or "trust," both in a social context[55] and as a concept pertaining to the world of foreign relationships. In this latter case, it refers to the loyalty that everyone must show towards one's own allies, and to the respect of oaths and agreements.[56] The religious value of *fides* is evident, especially in this latter case: a general cannot break a sworn treaty, because it has been sanctioned by the gods.[57] As is well-known, at least during the Republic the Romans took care to justify the wars they started as *bella iusta*, stating that the enemies had been the first to break a treaty or their *fides* in some way.[58] A military trick that is directed against the enemies, and does not violate an oath, on the other hand, is never defined by Valerius Maximus as an act of *perfidia*. When he accuses Hannibal and the Carthaginians (but also Romans, or other foreigners)[59] of being *perfidi*, he always shows them breaking oaths, attacking their own friends, and not caring about the gods. Let us review the evidence.

The first instance refers to the treacherous murder of Tib. Gracchus by his own *hospes* Flavus in accord with the Carthaginian Mago; Flavus is therefore *perfidus*, while the author does not focus on the Carthaginians'

[55] Ute Lucarelli, *Exemplarische Vergangenheit: Valerius Maximus und die Konstruktion des sozialen Raumes in der frühen Kaiserzeit* (Göttingen, 2007) studies this concept in a social context, with a focus on *fides coniugalis* (esp. pp. 168-177) and *fides servorum* (esp. pp. 225-228).

[56] See Val. Max. 6.6, on public *fides*. The Roman examples refer to the loyalty shown towards the Egyptian allies, the impunity granted to the Carthaginians because of a truce and the respect shown to foreign ambassadors in three instances. See also, among the many occurrences scattered throughout the whole work, 2.9.8 (on the respect of the oaths sworn to the Carthaginians); 4.8.1 (Fabius Maximus' respect of the agreement he had with Hannibal); and 6.6.ext.1-ext.2 and 7.6.2 (on the loyalty of the allies towards Rome).

[57] About the religious importance of *fides*, see Pierre Boyancé, "Les Romains, peuple de la *fides*", in *Études sur la religion romaine*, ed. Pierre Boyancé (Roma, 1972), 135-152, and Gérard Freyburger, "La *fides* et les *mores* dans les conceptions religieuses anciennes de Rome," in *Hommages à Carl Deroux*, ed. Pol Defosse (Bruxelles, 2003, vol. 4), 378-386. In Liv. 21.4.9, Hannibal's absence of *fides* is immediately paired with *impietas*: *nihil veri nihil sancti, nullus deum metus nullum ius iurandum nulla religio*.

[58] On the debated issue of the characteristics of Roman *bellum iustum* see Giovanni Turelli, "*Fetialis religio*: una riflessione su religione e diritto nell'esperienza romana," in *Religione e diritto romano: la cogenza del rito*, ed. Salvo Randazzo (Tricase, 2014), 449-493 with earlier bibliography.

[59] See the appendix.

responsibility.[60] The Carthaginians are also involved in the treason of Capua.[61] Hannibal is criticized for accepting the surrender of Nuceria and Acerra, in exchange for the promise to let the inhabitants go unharmed, and then killing them anyway: he acted *adversus ipsam fidem*, and the author suggested that he rejoiced in this course of action (*mendaciis et fallacia gaudens*).[62] The harshest criticism, however, concerns the murder of Xanthippus: in this case, the Carthaginians killed their benefactor without any reason, and therefore they earn the appellative of *fons perfidiae*.[63] Even the only instance of Carthaginian courage throughout the entire work is stained by treachery. According to Valerius, Carthage and Cyrene had decided to define the exact border between the two cities by sending two envoys; the place where they would meet would become the border.[64] The Carthaginian envoy started its journey ahead of the pre-arranged time. When they met, the Cyrenaic embassy accepted the recognition of this new demarcation, but only if the Carthaginians accepted being buried alive on the spot. They consented, dying for the enlargement of their fatherland: they are praised for their courage, but nevertheless censured as *perfidi*, as they did not honor the terms.

To be sure, despite the characterization of the Carthaginians as "the source of *perfidia* itself," this vice is not their own monopoly. As shown in the table in the appendix, other peoples and individuals are accused of it, including not a few Romans, especially during the civil wars. The Romans, however, may be *perfidi* as individuals; they are not so as a people.[65] For the Carthaginians, on the contrary, this is a national feature: in an *exemplum*, Valerius depicts Hamilcar and Hanno debating about the opportunity to accept a Roman offer to come to a meeting. Hamilcar feared that he could be arrested despite the truce, as the Carthaginians had previously done exactly this with Cornelius Asina. Hanno, instead, decided to go, knowing that the Romans were different from the Carthaginians: through his mouth, Valerius expresses his convictions about the opposition between Roman *fides* and Punic

---

[60] Val. Max. 1.6.8.
[61] Hannibal instigated *fallacibus promissis*: Val. Max. 3.8.1.
[62] Val. Max. 9.6.ext.2. It is interesting to note that, in this passage, "for the rest" (*alioqui*) Hannibal is praised as a great man.
[63] Val. Max. 9.6.ext.1.
[64] Val. Max. 5.6.ext.4. Sall. *Iug.* 79 records the same story, but says that the Cyrenaic envoys arrived late, and therefore invented an accusation against the Carthaginians.
[65] Skidmore 1996, 63-68; Mueller 2002, 110-117; Rüpke 2016, 100-103.

perfidia.[66] This, of course, stemmed from a different attitude towards religion: while the Romans tend to provide good examples of *pietas*, Valerius relates with surprise of the display of *pietas* by Masinissa, commenting that it was *magis Masinissae animo quam Punico sanguini conveniens*.[67] The presentation of the Carthaginians partially serves to build, through a contrast, an opposite image of the Roman people.[68]

The other common accusation levelled at the Carthaginians is that of cruelty.[69] This is less important here, but one must notice that even in this case the Punic vice originates from their lack of *pietas*. While discussing their brutal killing of Regulus (who is, of course, praised for his *fides* and *pietas*), Valerius says that the Carthaginians' inhumane cruelty was so distasteful to the gods, that they decided to punish the city with its destruction during the third Punic war.[70] Once again, the general picture is that of a people without

---

[66] Val. Max. 6.6.2. Apparently, a military tribune threatened to arrest Hanno, but was immediately reproached by the consuls, who reassured the Carthaginian: *isto te metu, Hanno, fides civitatis nostrae liberat.* At the start of the chapter, Valerius had already stated that *fides* was a distinctively Roman value: *quam [fidem] semper in nostra civitate viguisse et omnes gentes senserunt et nos paucis exemplis recognoscamus*. Significantly, the only two foreign examples are related to the allies' loyalty towards Rome, which remains the fulcrum of the chapter.

[67] Val. Max. 1.1.ext.2. The author's confusion of the Numidians and Carthaginians is not relevant.

[68] At least in the context of the themes treated here, the traditional view of the portrayal of foreigners in the *Facta et Dicta* as intended to underline, by contrast, Roman virtues, seems preferable: see Watts 1976, 92-93; Lehmann 1998, 23 and especially Desideri 2008 (cf. Valvo 2007, 39); *contra* Lawrence 2006; Murray 2016, 41-43; Matravers 2016, 44-46. Pomeroy 1989 studied how Valerius changes Livy's text to create a picture of Hannibal deeply characterized by *perfidia*.

[69] Val. Max. 1.1.14; 9.2.1 (Hannibal is here the *crudelis hostis par excellence* and becomes the touchstone for Sulla's massacres); 9.2, ext.1-ext.2; 9.8.ext.1; at 2.7.ext.1 their brutal conduct towards their own defeated commanders is not censured. Even when Hannibal's *clementia* is shown, the author is careful to underline the fact that this is a Roman value, exceptionally borrowed by a normally *crudelis* enemy (Val. Max. 5.1.ext.6; cf. Desideri 2007, 312; Valvo 2008, 50-53; Stocks 2014, 31; Matravers 2016, 149-152 and especially Brizzi 2011).

[70] Val. Max. 1.1.14. In accord with his conceptions about *fides*, here Valerius does not accuse the Carthaginians of *perfidia*: Regulus had sworn to return (which he did), while his enemies had not promised to let him go unharmed: by killing him, they did not break an oath.

gods or rules, without the good moral values which are at home in Rome,[71] whose skill with lies and deceit is not restrained by any religious fear.

Conclusion: *fides* and *perfidia*, *virtus* and *ars*

Two pairs of opposite words represent Valerius' moral conceptions about deceit in warfare very well, as suggested by this text: *fides* and *perfidia* on the one hand, *virtus* and *ars* on the other. Throughout his work, Valerius Maximus seems to provide the reader with two separate ethical categories. *Perfidia* appears to be the art of lying and deceiving through a breach of *fides*. This attitude is impious and wrong, as *fides* always must be kept both towards the allies and in cases of a sworn agreement. The Carthaginians are not the only ones to incur the author's reproaches from this point of view, but Valerius suggests that a lack of both *fides* and *pietas* is common among them; as a corollary, they also tend to be cruel. Punic commanders also show their shrewdness in the field of stratagems, which can be described with a Latin word as *ars*, or *consilium*.[72] In this case, they are not criticized, because they are not breaking *fides*.[73] However, their constant reliance on trickery shows them as lacking *virtus*: they can only win through stratagems, unlike the Romans, who can be defeated in a contest based on *ars*, but not in one involving *virtus*. The Romans themselves make use of shrewd actions in war, and, when they do so successfully, they are praised by the author. However, their fundamental values remain *virtus* and *fides*, and *ars*, when employed, is only a complement to *virtus*. In Valerius' construction,[74] these two virtues define the Romans as a people and have nothing to do with the Carthaginians.

To this summary, I would like to add two brief clarifications. The first is that one should be careful not to compel every single piece of information

---

[71] See Val. Max. 5.1.6, where Scipio, after the conquest of Carthage, gives back to the Sicilians the sacred objects previously stolen by the Carthaginians.

[72] As noted, at the start of his chapter about stratagems Valerius appears to say (despite a lacuna in the text) that the Romans did not have a word as meaningful as the Greek στρατήγημα to describe this concept. However, the author himself employs the words *ars* and *consilium* at the end of his first *exemplum*. See once again Wheeler 1988a, 50-92 for a review of the Latin strategematic vocabulary.

[73] See Val. Max. 9.6.ext.2, where the author reproaches Hannibal for his *perfidia*, but also comments that, for the rest (*alioqui*), he left *insignem nominis sui memoriam*. Here the author must be referring to his generalship and, consequently, to his stratagems.

[74] Simon 2002.

found in the author to neatly fit inside a superimposed scheme. In any ethical construction there is room for inconsistency, and Valerius is no exception.[75] The second Roman example in the chapter about stratagems relates how Sex. Tarquinius, the son of King Tarquinius Superbus, pretended to be hated by his father in order to be accepted as a leader of Gabii. He then proceeded to kill many members of the aristocracy, so helping King Tarquinius conquer the city.[76] If the distinction proposed in this work is accepted, one is inevitably puzzled to find this breach of *fides* (although not explicitly described as such) in the chapter about stratagems. It may be argued that there was no treaty between Rome and Gabii (but Sextus was welcomed in Gabii as a citizen and a town leader, and by betraying the city he broke the relation of *fides* he had with its inhabitants), or that the author carelessly reported an *exemplum* he found in his sources (but there is no ground to suppose this). It is better to honestly acknowledge that the reconstruction proposed here does not account for all the evidence which can be found in the work. I would suggest that this is because the moral construction which emerges from the *Facta et Dicta* may have its occasional inconsistencies. The case quoted appears to me to be the only serious counterexample for the arguments dealt with in this text.

The second note concerns a point already raised in the introduction. Because this is a study with a narrow focus – namely just on the work of Valerius Maximus – it cannot be taken for granted that the conclusions reached can be extended to Roman authors and Roman culture in general. To be sure, Valerius provides an excellent starting point for a discussion about these themes in other thinkers, but one should keep in mind the possibility both of a change over time in Roman cultural values and of a simple divergence of opinion among different authors, even in the same period. With this text, I hope to have cast light on Valerius Maximus' position, and at the same time to have provided a stimulus to further research about Roman attitudes towards stratagems, through an analysis of the difference between *perfidia* and stratagems.

Appendix: instances of the word "*perfidia*" and its cognates in the *Facta et Dicta Memorabilia*		
Passage	**Culprit**	**Situation**

[75] Moreover, as Valvo 2006, 40-41 notes, an example can be slightly bent by the author to fit the context of the chapter into which it is fitted. Langlands 2008 reads Valerius' collection of examples as an attempt to lead his readers to think critically about moral issues, which can be presented in different ways throughout the work.
[76] Val. Max. 7.4.2.

1.6.8	Flavus	Treason of his *hospes* Gracchus
1.8.9	Ptolemy	Murder of Pompey
2.3.3	Capuans	Treason of Rome, siding with Hannibal
2.7.12	Latins	Desertion, treason of Rome
2.9.8	Roman prisoners	Attempt not to return to Hannibal, despite the oath they had sworn
3.2.ext.1	Campanians	Treason of Rome, siding with Hannibal
3.2.ext.3	Leonida's allies	Treason of the Spartans, letting Xerxes pass
5.1.10	Egyptians (Ptolemy)	Murder of Pompey
5.3.3	Sextilius	Treason of his own patron (C. Caesar), handed to the proscriptors
5.6.ext.4	Carthaginians	Treachery of the Punic envoys towards the Cyrenaic ones
6.2.8	Pompey	Killing of M. Brutus the Elder, whom he had promised safety
6.5.4	L. Cotta	Tries to use his tribunate to gain impunity for not paying a debt
6.8.4	C. Cassius	Murder of Caesar
9.6.1*	Tarpeia	Treason of the citadel
9.6.2	Ser. Galba	Murder of 8,000 enemies, after promising them safety
9.6.3	Cn. Domitius	Imprisonment of Bituitus, during a truce and under cover of hospitality
9.6.4	Q. Servilius Caepio	Killing of Viriathus, by bribing Viriathus' friends
9.6.ext.1	Carthaginians	Killing of Xanthippus, their benefactor, without reason
9.6.ext.2*	Hannibal	Murder of the inhabitants of Nuceria and Acerra, despite his promises
Note: the two *exempla* marked by * do not contain the word *perfidia*, but are listed in the chapter about it		

Bibliography

Achard, Guy. "César et la ruse." In *Ruses, secrets et mensonges chez les historiens grecs et latins*, edited by François Bérard, Pascale Giovannelli-Jouanna, and Hélène Olivier, 85-97. Lyon: De Boccard, 2006.

Balmaceda Errazuriz, Catalina. *Virtus Romana: Politics and Morality in the Roman Historians*. Chapel Hill: University of North Carolina Press, 2007.

Boyancé, Pierre. "Les Romains, peuple de la *fides*." In *Études sur la religion romaine*, edited by Pierre Boyancé, 135-152. Roma: École Française de Rome, 1972.

Brisson, Jean-Paul. "Les mutations de la seconde guerre punique." In *Problèmes de la guerre à Rome*, edited by Jean-Paul Brisson, 33-59. Paris-La Haye: Editions de l'EHESS, 1969.

Brizzi, Giovanni. *I sistemi informativi dei Romani: principi e realtà nell'età delle conquiste oltremare (218-168 a.C.)*. Wiesbaden: Franz Steiner Verlag, 1982.

—. "Carthage and Hannibal in Greek and Roman Memory." In *A Companion to the Punic Wars*, edited by Dexter Hoyos, 483-498. Malden: Wiley-Blackwell, 2011.

Cairo, Giambattista. "La guerra nel mondo eroico greco-romano: tra propaganda, ideologia e religione." In *Domi forisque: omaggio a Giovanni Brizzi*, edited by Stefano Magnani, 49-58. Bologna: Il Mulino, 2018.

Chassignet, Martine. "La deuxième guerre punique dans l'historiographie romaine: fixation et évolution d'une tradition." In *Valeurs et mémoire à Rome: Valère Maxime, ou la vertu recomposé*, edited by Jean-Michel David, 55-72. Paris: De Boccard, 1998.

Desideri, Paolo. "Greci, barbari, Cartaginesi in Valerio Massimo." In *Costruzione e uso del passato storico nella cultura antica*, edited by Anna M. Biraschi, Paolo Desideri, and Sergio Roda, 305-312. Alessandria: Edizioni dell'Orso, 2007.

Devallet, Georges. "*Perfidia plus quam Punica*: l'image des Carthaginois dans la littérature latine de la fin de la république à l'époque des Flaviens." *Lalies* 16 (1996): 17-28.

Freyburger, Gérard. *Fides: étude sémantique et religieuse depuis les origines jusqu'à l'époque augustéenne*. Paris: Les Belles Lettres, 1986.

—. "La *fides* et les *mores* dans les conceptions religieuses anciennes de Rome." In *Hommages à Carl Deroux.*, edited by Pol Defosse, vol. 4, 378-386. Bruxelles: Latomus, 2003.

Hanson, Victor D. *The Western Way of War: Infantry Battle in Classical Greece*. Oxford: Oxford University Press, 1990.

Holeindre, Jean-Vincent. *La ruse et la force: une autre histoire de la stratégie*. Paris: Perrin, 2017.

Konijnendijk, Roel. "Mardonius' Senseless Greeks." *CQ* 66 (2016): 1-12.

Krentz, Peter M. "Deception in Archaic and Classical Greek Warfare." In *War and Violence in Ancient Greece*, edited by Hans van Wees, 167-200. London: Duckworth, 2000.

Krentz, Peter M. and Everett L. Wheeler. *Stratagems of War: Polyaenus*. Chicago: Ares Publishers, 1994.

Langlands, Rebecca. "Reading for the Moral in Valerius Maximus: The Case of *Severitas*." *Cambridge Classical Journal* 54 (2008): 160-187.

Lawrence, Sarah J. 2006. "Inside Out: The Depiction of Externality in Valerius Maximus." PhD diss., University of Sydney, 2006.

Le Bohec, Yann. *Histoire militaire des guerres puniques: 264-146 avant J.-C*. Monaco: Éditions du Rocher, 2006.

Lehmann, Yves. "Les rivendications morales et politiques de Valère Maxime." In *Valeurs et mémoire à Rome: Valère Maxime, ou la vertu recomposé*, edited by Jean-Michel David, 19-26. Paris: De Boccard, 1998.

Lucarelli, Ute. *Exemplarische Vergangenheit: Valerius Maximus und die Konstruktion des sozialen Raumes in der frühen Kaiserzeit*. Göttingen: Vandenhoeck & Ruprecht, 2007.

Maslakov, G. "Valerius Maximus and Roman Historiography: A Study of the *Exempla* Tradition." *ANRW* II.32.1 (1984): 437-496.

Matravers, Simon R. "Commentary on Valerius Maximus' Book IX.1-10. A Discussion on *Vitia*: An Apotreptic Approach." PhD diss., University of Birmingham, 2016.

Mazza, Federico. "L'immagine dei Fenici nel mondo antico." In *I Fenici*, edited by Sabatino Moscati, 548-567. Milano: Bompiani, 1988.

McDonnell, Miles. *Roman Manliness: Virtus and the Roman Republic*. Cambridge: Cambridge University Press, 2006.

Montanari, Enrico. "*Mens*." *R&C* 2 (1976): 172-235.

Mueller, Hans-Friedrich. *Roman Religion in Valerius Maximus*. London-New York: Routledge, 2002.

Murray, Jeffrey. "Valerius Maximus on Vice: A Commentary on *Facta et Dicta Memorabilia* 9.1-11." PhD diss., University of Cape Town, 2016.

Ober, Josiah. "The Rules of War in Classical Greece." In *The Athenian Revolution: Essays on Ancient Greek Democracy and Political Theory*, published by Josiah Ober, 53-71. Princeton: Princeton University Press, 1999.

Pomeroy, Arthur J. "Hannibal at Nuceria." *Historia* 38 (1989): 162-176.

Prandi, Luisa. "La *fides Punica* e il pregiudizio anticartaginese." *CISA* 6 (1979): 90-97.

Rüpke, Jorg. "Knowledge of Religion in Valerius Maximus' *Exempla*: Roman Historiography and Tiberian Memory Culture." In *Memory in*

Ancient Rome and Early Christianity, edited by Karl Galinsky, 89-111. Oxford: Oxford University Press, 2016.

Schettino, Maria T. *Introduzione a Polieno*. Pisa: ETS, 1998.

Sheldon, Rose M. *Ambush: Surprise Attack in Ancient Greek Warfare*. Barnsley: Frontline Books, 2008.

Simon, Steven J. "Some Concepts of Republican Virtue as Expressed by Valerius Maximus." *AncW* 33 (2002): 147-152.

Skidmore, Clive. *Practical Ethics for Roman Gentlemen: The Work of Valerius Maximus*. Exeter: Exeter University Press, 1996.

Stocks, Claire. *The Roman Hannibal: Remembering the Enemy in Silius Italicus' Punica*. Liverpool: Liverpool University Press, 2014.

Tipping, Ben. 2010. *Exemplary Epic: Silius Italicus' Punica*. Oxford: Oxford University Press, 2010.

Turelli, Giovanni. "*Fetialis religio*: una riflessione su religione e diritto nell'esperienza romana." In *Religione e diritto romano: la cogenza del rito*, edited by Salvo Randazzo, 449-493. Tricase: Libellula, 2014.

Valvo, Eva. "La rappresentazione di Annibale in Valerio Massimo." *Millennium* 5 (2008): 37-55.

Van Wees, Hans. *Greek Warfare: Myths and Realities*. London: Duckworth, 2004.

Watts, William J. "Race Prejudice in the Satires of Juvenal." *Aclass* 19 (1976): 83-104.

Wheeler, Everett L. *Stratagem and the Vocabulary of Military Trickery*. Leiden: Brill, 1988a.

—. "*Sapiens* and Stratagems: The Neglected Meaning of a *Cognomen.*" *Historia* 37 (1988b): 166-195.

The Suspension of the War-Tax (*Tributum*) in 167 BCE: A Short-Sighted Decision?

François Gauthier

Recent studies have made it clear that the backbone of Roman military financing until 167 was provided by the *tributum*, the war-tax paid by Roman citizens eligible for military service.[1] It was the most reliable and often most important source of income that the senate could count on when it came to planning a campaign. Other sources of income resulting from victorious wars such as plunder were of course only available after the fighting was done and could thus not be counted on to finance a campaign.

Over the course of the late third century and the beginning of the second, Rome defeated most of the major powers of the Mediterranean, notably including Carthage, Macedon, and the Seleucid kingdom. As a result of their defeat, these polities had to pay important war indemnities to Rome (see Table 1 below). Furthermore, the appropriation of the Macedonian treasury in 168 after the Third Macedonian War provided Rome with an influx of money: some 75, 50 or 30 million denarii as the sources give conflicting figures.[2] This prompted the senate to stop collecting *tributum* the following year. Despite the fact that this money and that provided by war indemnities represented great wealth flowing into the Republic's treasury, it hardly ensured Rome's perennial fortune. Yet, for a long time, modern historians

[1] Nathan Rosenstein, "*Bellum Se Ipsum Alet*? Financing Mid-Republican Imperialism," in *Money and Power in the Roman Republic*, eds. Hans Beck, Martin Jehne, and John Serrati (Brussels: Latomus, 2016), 117; Michael Taylor, "State finance in the Middle Roman Republic: a reevaluation," *AJPh* 138 (2017): 143–180; Michael Taylor, *Soldiers and Silver. Mobilizing Resources in the Age of Roman Conquest* (Austin: University of Texas Press, 2020), 111-138. On the *tributum*, see Claude Nicolet, *Tributum. Recherches sur la fiscalité directe et indirecte sous la république romaine* (Bonn: Rudolf Habelt, 1976).

[2] Plin. *HN* 33.56: 75 million; Vell. Pat. 1.11: 50 million; Liv. 45.40.2: 30 million It would only be reintroduced in 43 in the context o civil war: Cic. *ad Fam.* 12.30.

have argued that this was the culmination of the trend of shifting the burden of financing Rome's armies on the shoulders of the defeated and the provinces.[3] Whereas the provinces were certainly contributing to the Roman war effort, a wide-ranging and systematic implementation of taxation to finance the army *before* the suspension if *tributum* is not clearly discernable in the sources.[4]

Recent studies on Republican army financing have focused on the period roughly between 200 and 157.[5] The loss of Livy's informative narrative after the Third Macedonian War probably helps to explain this focus since the sources available for the late second century do not allow for the same compilation of data. Yet, the Roman Republic continued to field relatively large armies even after the abrogation of *tributum*.[6] How did the state manage to finance them without the war-tax? The aim of this paper is to try to address that question. It will argue that the suspension of *tributum* was not caused by an increase of provincial revenues but by the influx of wealth caused by the appropriation of Macedonian wealth. Rather, the senate actually took steps to intensify mining activities in the provinces *after* the suspension of *tributum*, which points to a need to increase revenues.

---

[3] William Thomas Arnold, *The Roman System of Provincial Administration to the Accession of Constantine the Great* (Oxford: Blackwell, 1914), 194: "When in the year 167 B.C., the tribute was taken off Italy, the expenses of administration and public works were undisguisedly supported by the taxation of the provinces." More recently, James Tan, *Power and Public Finance at Rome*, 264–49 BCE (Oxford: Oxford University Press, 2017), 145: "The Romans decided that mining profits, indirect taxes, and (at least in the early second century) indemnities would have to suffice."

[4] Tan, *Power*, 145.

[5] Rosenstein, "*Bellum*," 114-130; Taylor, "State Finance," 143-180; *idem*, *Soldiers and Silver*.

[6] Peter Astbury Brunt, *Italian Manpower, 225 BC – AD 14* (Oxford: Oxford University Press, 1971), 426-434.

Table 1: War Indemnity Payments, 263-188[7]

Payor	Lump sum (talents)	Annual Payment	Year	Source
Syracuse	100 or 200	-	263	Polyb. 1.16.9; Eutr. 2.19.2; Oros. 4.7.3
Carthage	1,000	2,200 over 10 years	241	Polyb. 1.63.3; 3.27.5
Carthage	1,200	-	237	Polyb. 1.88.12
Carthage	-	10,000 over 50 years	201	Polyb. 15.18.7; Liv. 30.37.5; App. *Pun.* 54.235.
Macedon	500	500 over 10 years	196	Polyb. 18.44.7
Boeotia	30		196	Liv. 33.29.12.
Sparta	100	400 over 8 years	195	Liv. 34.35.
Aetolians	200	300 over 6 years	189	Polyb. 21.32.8
Seleucid Kingdom	3,000	12,000 over 10 years	188	Polyb. 21.17.5; 21.41.8; 21.43.19
Cappadocia	300	-	188	Polyb. 21.45
Ambracia	150	-	188	Polyb. 21.30.10; Liv. 38.9.13.

1 - Military Commitment and Finances in the Second Century

According to the calculations of T. Frank and M. Taylor, *tributum* represented between 10% and 20% of Rome's income in the first half of the second century.[8] Whereas N. Rosenstein argued that it perhaps made up half of Rome's total military expenditure.[9] Following the estimates of P. Brunt, Rome fielded an average of 6.48 legions per year between 167 and 91 (sometimes with important variations from year to year).[10] This means

[7] Similar tables can be found in Philip Kay, "What Did the Attalids Ever Do for Us? The View from the Aerarium," in *Attalid Asia Minor: Money, International Relations, and the State*, ed. Peter Thonemann (Oxford: Oxford University Press, 2013), 138; Philip Kay, *Rome's Economic Revolution* (Oxford: Oxford University Press, 2014), 39; Taylor, "State Finance," 173; *idem*, *Soldiers and Silver*, 135.

[8] Tenney Frank, *An Economic Survey of Ancient Rome* (Baltimore: Johns Hopkins University Press, vol. 1, 1933.), 141; Taylor "State finance," 169.

[9] Rosenstein, "*Bellum*," 117.

[10] Brunt, *Manpower*, 422-434.

an average of roughly 4 million denarii in *stipendium*, in addition to about half a million denarii in fleet expenses, according to Frank.[11]

Now, by the mid-second century, war indemnity payments had run out and so did the supply of rich enemies whose treasury could be ransacked at will. According to Pliny the Elder, the fall of Carthage in 146 only brought 4,370 pounds of silver (about 367,080 denarii) to the treasury.[12] Only after the war with Jugurtha in 104 do we hear of more substantial plunder: 3,297,980 denarii, a sum that probably did not cover the expenses required in a war spread over six years fought in North Africa and which entailed heavy losses.[13] Another passage in Pliny the Elder provides a rare snapshot at the state of the Roman treasury in the mid second century. This author reports that ten years after the suspension of *tributum*, in 157, a survey of the *aerarium* revealed it contained 22.6 million denarii in coin and bullion.[14] This was enough to pay the *stipendium* of some thirty-seven legions, or roughly 5.79 years of *stipendium* with an average establishment of 6.48 legions. This was not an insignificant sum, but it was also definitely not the bonanza that one could be tempted to imagine only ten years after the decision to suspend the collection of *tributum*. In terms of ships, the loss of Livy after 167 is a serious hindrance in trying to reconstruct Roman naval strength in the late second century. Fifty quinqueremes were involved in the last war against Carthage and it can be assumed that these ships were then used to patrol the western Mediterranean.[15] Following the end of the Macedonian monarchy, Rome mostly relied on its allies, notably Pergamum and Rhodes, to control the eastern Mediterranean.[16] This hypothesis is supported by the fact that, when Mithridates invaded Asia in 88, Rome had a fleet at the mouth of the Pontus, but since it surrendered without a fight, it was probably not very significant.[17] Pay for the crew of fifty quinqueremes could have amounted to a little over one million denarii.[18]

---

[11] Frank, *Survey*, 1.226; 1.228; Brunt, *Manpower*, 424-425. The annual *stipendium* for a legion was around 600,000 denarii, cf. François Cadiou, *Hibera in terra miles. Les armées romaines et la conquête de l'Hispanie sous la république (218–45 av. J.-C.)* (Madrid: Casa de Velázquez, 2008), 512.

[12] Plin. *HN* 33.141

[13] Plut. *Mar*. 12.4: 3,007 pounds of gold, 5,775 pounds of silver, 287,000 denarii.

[14] Plin. *HN* 33.55.

[15] App. *Lib*. 75.350.

[16] Frank, *Survey*, 1.142.

[17] App. *Mith*. 17.61; 19.75.

[18] According to Polybius 1.26.7, each quinquereme had a crew of 300 sailors and 120 marines. However, modern scholars think that the figure of 120 marines is too

Did Pliny's figures include the funds in the *aerarium sanctius*? His wording does not explicitly mention it since he speaks of the money available in the *aerarium populi*. The *aerarium sanctius* was a reserve fund financed by a five percent tax on the manumission of slaves. It has not attracted a lot of scholarly attention. Yet, it is the only reserve fund attested in Republican Rome. According to Livy, the five percent tax was created in 357 and the purpose of the money was to act as an emergency stock of money (*ad ultimos casus servaretur*).[19] It was notably used in 209, during the Hannibalic War, when it was decided to withdraw 4,000 pounds of gold to finance armies in the field (3.36 million denarii measured in sextantal asses).[20] This is a significant sum, enough to provide the *stipendium* of about 5 and a half legions, but we have no information concerning how long it took to amass. The next and only other time we hear about the *aerarium sanctius* is in 49, when Caesar appropriated the treasury. On this occasion, Cicero reports that the consuls had been ordered to take the money from the *aerarium sanctius* and to give it to Pompey. However, the consul who opened the treasury fled after hearing a (false) rumor that stated Caesar was

high since Polybius simply states that four legions embarked on the fleet, indicating stronger numbers than usual. They point out that the number of marines frequently changed and was probably closer to 40-60, cf. Johannes Kromayer, "Die Entwicklung der römischen Flotte vom Seeräuberkriege des Pompeius bis zur Schlacht von Actium," *Philologus* 56 (1897): 485-491; Brunt, *Manpower*, 669; Taylor "State Finance," 153. It is likely that at least half the crews were provided by the Italian allies who funded them themselves, which greatly reduced the cost of the Roman fleet. On the *socii nauales*, see Johannes Hendrik Thiel, *Studies on the History of Roman Sea Power in Republican Times* (Amsterdam: North-Holland Publishing, 1946), 195-196; Brunt, *Manpower*, 50; Alessandro Milan, "I socii navales di Roma," *Critica storica* 10 (1973): 193-221; Christa Steinby, *The Roman Republican Navy, from the sixth century to 167 B.C.* (Helsinki: Societas Scientiarum Fennica, 2007), 59.

[19] Liv. 7.16.7: "The other consul accomplished nothing worth recording, except that without precedent he got a law passed in his camp before Sutrium—the men voting by tribes—which levied a tax of one-twentieth on manumissions" (*Ab altero consule nihil memorabile gestum, nisi quod legem novo exemplo ad Sutrium in castris tributim de vicensima eorum qui manumitterentur tulit*); 27.10.11: "As the consuls were making all the other necessary preparations for war, a decision was taken to withdraw gold raised by the five percent tax that was being kept in reserve in the inner treasury to meet emergencies." (*Cetera expedientibus quae ad bellum opus erant consulibus, aurum vicensimarium quod in sanctiore aerario ad ultimos casus servaretur promi placuit*); Cic. *Att.* 2.16.1. See also Nicolet, *Tributum*, 13-14.

[20] Liv. 27.10.12-13.

approaching. He left the money there in his haste.[21] Thus Caesar was able to seize the funds in the *aerarium sanctius*. Yet, not all sources refer to the *aerarium sanctius*, some simply mention the public treasury.[22] However, since the episode is famous enough to have been recorded by a variety of sources, it can be reasonably inferred that authors who do not refer explicitly to the *aerarium sanctius* abridged their text by simply referring to *aerarium* or ταμιεῖον.[23]

Thus, since Pliny reports the result of the inventory of the treasury for 157, his figure may well include the reserve funds in the *aerarium sanctius*. However, since this was considered an emergency treasury, the money it contained was unlikely to be used in the second half of the second century when no such situation required it (with the possible exception of the war against the Teutons and the Cimbri). It is impossible to know what part of the money reported by Pliny was in the *aerarium sanctius*. Whatever its worth, it should be subtracted from the total amount as it could not be used for regular military expenditure (i.e. non-extreme emergency).

As highlighted by N. Rosenstein and M. Taylor, a few exceptionally lucrative campaigns kept Rome's finances afloat for a while, but this was not sustainable, nor was it accompanied by a swift implementation of new wide-ranging financing mechanisms.[24] Rather, the Romans were fairly slow into organizing and systematically tapping into the financial resources of their new provinces. The senate's approach seems to have been to maintain whatever fiscal regime was in place when one existed, or to find quick ways to tap into the wealth of the provinces where none existed. In Spain and Macedon, this took the form of mining revenues, whereas the Gracchan era

---

[21] Cic. *Att.* 7.21.2; Caes. *BCiu.*1.14.1: "When this news reached Rome the panic that suddenly hit was so great that, although the consul Lentulus had come to open the treasury to provide money for Pompey as per the senate's decree, he fled the city directly after opening the treasury reserve" (*Quibus rebus Romam nuntiatis tantus repente terror invasit ut, cum Lentulus consul ad aperiendum aerarium venisset ad pecuniamque Pompeio ex senatus consulto proferendam, protinus aperto sanctiore aerario ex urbe profugeret*).

[22] Explicit reference to the *aerarium sanctius* in 49: App. *B. Ciu.* 2.41.164; Flor. 2.13.21; Plut. *Caes.* 35.3-4. No explicit reference to the *aerarium sanctius*: Cass. Dio 41.17.1; Luc. *Phar.* 3.114-68; Oros. 6.15.5; Plin. *NH* 33.56; Plut. *Pomp.* 62.1.

[23] Brunt, *Manpower*, 549.

[24] Kay, *Revolution*, 26-27; Tan, *Power*, 150; Rosenstein, "*Bellum*," 114-130; Taylor, *Silver*, 111-168.

saw the first attempt at organizing a systematic approach to a province's fiscality.

2 - New Sources of Income in the Second Century

I will argue that the initial measures taken by the senate after the suspension of *tributum* were pragmatic. They were designed to quickly bring about the necessary funds needed to finance the Roman army and fleet, without having to implement wide-ranging reforms in military fiscality to compensate for the end of indemnity payments and the abrogation of *tributum*. The senate had no desire to increase state revenue just for the sake of it. In fact, attempts at augmenting public income could even motivate accusations of demagoguery, as the example of Gaius Gracchus would show.[25]

2.1 - Macedonian Mines

The Romans originally chose to close the gold and silver mines of Macedon after the Third Macedonian War, which supports the idea that the senate thought the *aerarium* well stocked enough for the time being. Moreover, the senate was initially content with half the tribute that the Macedonians used to pay to their king (100 talents, equivalent to 600,000 denarii, enough for the *stipendium* of single legion).[26]

However, the late antique chronicler Cassiodorus records that the mines were once again exploited in 158.[27] Cassiodorus' *Chronicle* contains few events, besides lists of kings, consuls, and emperors, so it is reasonable to believe that this was an important event.[28] According to Livy, these mines used to yield important revenues to the Macedonian monarchy.[29] Additionally, Diodorus claims that at the time of Philip II in the mid-fourth century, the mines provided Macedon with 1,000 talents (6 million denarii)

[25] Tan, "Fiscality." 267.
[26] Liv. 45.18.7.
[27] Cassiod. *Chron. sub anno* 158 BCE: "Under these consuls mines were established in Macedonia." (*His conss. metalla in Macedonia instituta*). See also Michael Crawford, *Coinage and Money under the Roman Republic. Italy and the Mediterranean Economy* (Berkeley and Los Angeles: University of California Press, 1985),143.
[28] Kay, *Revolution*, 57.
[29] Liv. 45.18.3.

annually.[30] Even halved, this figure would still make a significant contribution to the Roman treasury, allowing the state to finance 75% of the 4 million denarii *stipendium* necessary to fund the average of 6.48 legions fielded each year between 167 and 91. P. Kay argued that the mines were reopened in response to a need for new revenues.[31] Indeed, the inventory of the money contained in the *aerarium* conducted in 157 and the reopening of the Macedonian mines the previous year may have been part of a concern over the Republic's flow of revenues after the gradual end of war indemnity payments and the depletion of the funds gained from the Macedonian treasury in 168. The decision to exploit the mines was also probably motivated by the expediency with which they could contribute to the treasury. Increasing tribute could have been dangerous, especially in an area that had been recently conquered (and that would once again rise up in rebellion in 150). Focusing on the revenues provided by the mines was a safer solution to increase the revenues from Macedon without exacerbating resentment.

2.2 Spanish Mines

According to Polybius, the exploitation of mines in Spain provided 25,000 drachmae per day to the Republic (9 million denarii annually).[32] However, scholars now believe that this figure only applied to the second half of the second century, or even after the end of the Numantine War in 133, when a relative period of peace allowed for more intensive operations of extraction.[33] Moreover, it is also only in the late second century that Roman

[30] Diod. Sic. 16.8.6–8

[31] Kay, *Revolution*, 58: "However, by 158, with no further indemnities or booty arriving in the aerarium, there was a requirement for new sources of revenue. Both the closure and the reopening of the Macedonian mines can therefore be attributed in large part to the varying financial requirements of the Roman state."

[32] Polyb. 34.9.8; Strab. 3.147-148. See also Diod. Sic. 5.36.3-4: "but at a later time, after the Romans had made themselves masters of Iberia, a multitude of Italians have swarmed to the mines and taken great wealth away with them, such was their greed." (ὕστερον δὲ τῶν Ῥωμαίων κρατησάντων τῆς Ἰβηρίας, πλῆθος Ἰταλῶν ἐπεπόλασε τοῖς μετάλλοις, καὶ μεγάλους ἀπεφέροντο πλούτους διὰ τὴν φιλοκερδίαν). However, it is unlikely that there was a massive arrival of Italians until the first century, cf. Cadio, *Hibera*, 631-635.

[33] Toni Ñaco del Hoyo, "The Late Republican West: Imperial Taxation in the making?," in *Crises and the Roman Empire. VII Workshop Impact of Empire (Roman Empire, 200 BC – AD 476)*, eds. Olivier Hekster, Gerda de Kleijn, and Daniëlle Slootjes (Leiden and Boston: Brill, 2007), 229; Taylor, "State Finance," 166; *Silver*, 128-129.

silver issues start to match the isotope signature of Spanish mines.[34] M. Taylor proposed that Polybius' high figure was perhaps the result of a temporary silver mining boom that did not reflect the income of the mines over a long period.[35] Although the latter remains speculation, it is quite certain that the contribution of the Spanish mines for most of the second century was far lower than what Polybius claims. Otherwise, Rome would have been able to fund the *stipendium* of all the legions mobilized during the period 167-91 (and even more) with the income of the Spanish mines alone. This is not credible as this would have made the collection of other types of revenues in the provinces redundant if the mines provided that many denarii.

The mines were of course not the only resources exploited by Rome in the Spanish provinces but it is difficult to assess what part of the various levies and requisitions reached the *aerarium*, if any. It might well be the case that the money, food, and supplies gathered in the peninsula served primarily to provide for the troops stationed there.[36]

2.3 Gaius Gracchus and the *Lex Sempronia de prouincia Asia*

Although the intention of Gaius Gracchus was not specifically to provide new sources of funding for the army, he did nevertheless increase state funding through the *lex Sempronia de provincia Asia*, and the taxes brought to Rome could certainly be used to finance the army following Gaius' murder in 121. Despite its importance, the *lex Sempronia de provincia Asia* is actually not mentioned explicitly by Plutarch or Appian, the authors providing the most detailed accounts of the Gracchi.

A speech in Appian credited to Mark Antony mentions that the tribute which used to be collected under the Attalids was not raised by Rome until the appearance of populist politicians in Rome (*δημοκόπων ἀνδρῶν*), something that probably refers to the Gracchi, more precisely Gaius.[37] Cicero explicitly mentions the *lex Sempronia* in the *Verrines*, when he

[34] Wilhelm Hollstein (ed.), *Metallanalytische Untersuchungen an Münzen der Römischen Republik* (Berlin: Gebr. Mann Verlag, 2000), 114–117; 122–123.
[35] Taylor, *Silver*, 129. See also Frank, *Survey*, 1.138-139.
[36] Toni Ñaco del Hoyo, *Vectigal incertum. Economía de guerra y fiscalidad republicana en el occidente romano: su impacto histórico en el territorio, (218-133 a.C.)* (Oxford: BAR, 2003).
[37] App. *B. Ciu.* 5.4.17-20.

discusses taxation contracts in the province of Asia.[38] Such tax farming contracts in Asia established by Gaius Gracchus are also attested in Diodorus and Fronto.[39]

However, the publication of the inscription of the *Senatus Consultum de Agro Pergameno* in the late nineteenth century seemed to suggest that Asia continued to be taxed after the Roman takeover of the Attalid kingdom. This view was based on the fact that the document mentions a Roman official resolving a conflict between Pergamum and some *publicani*, which was dated to around 129. This was interpreted as evidence of Rome's involvement in collecting taxes in Asia through *publicani* as early as 129.[40] Yet, subsequent scholars argued that the decree should actually be dated later.[41] This led P. Kay to propose that the inscription in fact refers to the implementation of a new tax system by C. Gracchus in 123.[42] According to this interpretation, the literary tradition attributing the introduction of a new taxation system in Asia to C. Gracchus would thus be confirmed by the *Senatus Consultum de Agro Pergameno* inscription.

In 66, in his speech in favour of the Manilian law conferring Pompey an extraordinary command against the 'Cilician' pirates, Cicero mentions that Asia was an exceptionally rich province. It yielded revenues to Rome in the form of *portoria* (custom taxes), *decumae* (agricultural tithes), and *scripturae* (grazing rights).[43] According to the writings of Velleius

[38] Cic. *Verr*. 2.3.12: "or it [taxation] is regulated by censors' contracts, as in Asia under the Sempronian Law." (*aut censoria locatio constituta est, ut Asiae lege Sempronia*). (Loeb translation slightly modified).

[39] Diod. Sic. 34/35.25.1: "by sacrificing the provinces to the reckless rapacity of the tax farmers he provoked the subject peoples to well-merited hatred of their rulers." (*καὶ τῇ μὲν τῶν δημοσιωνῶν τόλμῃ καὶ πλεονεξίᾳ τὰς ἐπαρχίας ὑπορρίψας ἐπεσπάσατο παρὰ τῶν ὑποτεταγμένων δίκαιον μῖσος κατὰ τῆς ἡγεμονίας*); Fronto, *Ep. ad Ver. Imp*. 2.1.13: "Gracchus already farming Asia and parcelling Carthage out among individual settlers." (*iam Gracchus locabat Asiam et Karthaginem viritim dividebat*). On tax farming, see Tan, *Power*, 40-67.

[40] On this inscription, see Robert Kenneth Sherk, *Roman Documents from the Greek East: Senatus Consulta and Epistulae to the Age of Augustus* (Baltimore: Johns Hopkins Press, 1969), no. 12.

[41] David Magie, *Roman Rule in Asia Minor* (Princeton: Princeton Legacy Library, 1950), 1005–6; Harold B. Mattingly, "The Date of the *Senatus Consultum de Agro Pergameno*," *AJPh* 93 (1972): 412–423.

[42] Kay, *Revolution*, 59-69; *idem*, "Attalids", 121-148.

[43] Cic. *Leg. Man*. 15: "Consequently neither from customs duties, tithes nor grazing dues can the revenues be maintained" *(Ita neque ex portu neque ex decumis neque*

Paterculus, C. Gracchus implemented new custom taxes.[44] P. Kay has argued that 'The Customs Law of Asia', an inscription dated to the reign of Nero but mentioning arrangements existing under the Attalids, actually referred to the establishment of these new custom taxes by C. Gracchus.[45] In terms of agricultural taxes, based on Appian and Cicero as well as 'The Customs Law of Asia', it is clear that under C. Gracchus, a system in which one-tenth of the agricultural production was taxed was introduced (*decuma*). The contracts for these taxes were given to *publicani* in Rome.[46]

What did Asia contribute to the Roman treasury? According to Plutarch, the Republic's annual revenues before Pompey's eastern campaigns amounted to some 50 million denarii.[47] However, it is difficult to say whether this

---

ex scriptura vectigal conservari potest). On early second-century *portoria* in Italy, see Taylor, *Silver*, 130.

[44] Vell. Pat. 2.6.3: "establishing new customs duties" (*nova constituebat portoria*).

[45] Kay, *Revolution*, 75: "The Customs Law is more likely to have been drafted as part of C. Gracchus' legislation, especially as literary sources credit the establishment of *portoria* to him. On this analysis, C. Gracchus set up a comprehensive customs system for taxing trade in an area extending from the Bosporus to Pamphylia, with the exception of a tax-free zone in Lycia. The Bosporus, the most lucrative source of tolls, was identified as the most important area for the collection of customs dues, generally at a rate of 2½ per cent. A system of *locatio* was used to let the taxes to *publicani*. All communities, including free cities, were liable to the *portorium* and no significant exemptions were allowed except for officials, soldiers, or *publicani* acting on Roman official business."

[46] Cic. *Leg. Man.* 15; *Att.* 5.13.1; App. *B. Ciu.*5.4.18-19: "But when the need arose, we did not tax you according to property valuations, so that we could collect a sum not subject to fluctuations, but we required you to pay a percentage of your yearly harvest in order that we would share with you the bad harvests too. And when those who bought their contract to collect these taxes from the senate were treating you violently and demanding much more, Gaius Caesar remitted a third of the sum of money you were paying them, and brought an end to their assaults." (*ἐπεὶ δὲ ἐδέησεν, οὐ πρὸς τὰ τιμήματα ὑμῖν ἐπεθήκαμεν, ὡς ἂν ἡμεῖς ἀκίνδυνον φόρον ἐκλέγοιμεν, ἀλλὰ μέρη φέρειν τῶν ἑκάστοτε καρπῶν ἐπετάξαμεν, ἵνα καὶ τῶν ἐναντίων κοινωνῶμεν ὑμῖν. τῶν δὲ ταῦτα παρὰ τῆς βουλῆς μισθουμένων ἐνυβριζόντων ὑμῖν καὶ πολὺ πλείονα αἰτούντων, Γάιος Καῖσαρ τῶν μὲν χρημάτων τὰ τρίτα ὑμῖν ἀνῆκεν ὧν ἐκείνοις ἐφέρετε, τὰς δ' ὕβρεις ἔπαυσεν*); Customs Law of Asia, line 72. On the *decuma*, system, see Andrew Lintott, *Imperium Romanum: politics and administration* (London and New York: Routledge, 1993), 76–7, 86–91. See also Kay, *Revolution*, 75-76.

[47] Plut. *Pomp*. 45. "In addition to all this the inscriptions set forth that whereas the public revenues from taxes had been fifty million drachmas, they were receiving from the additions which Pompey had made to the city's power eighty-five million" (*πρὸς δὲ τούτοις ἔφραζε διὰ τῶν γραμμάτων ὅτι πεντακισχίλιαι μὲν μυριάδες ἐκ τῶν*

refers to the state of affairs before or after Mithridates had overrun Asia and deprived Rome of its revenues. In 85/84 Sulla inflicted a very large fine of 20,000 talents (120 million denarii) on Asia to punish the cities of the province for their support of Mithridates.[48] Even if this sum meant several years of taxes, perhaps what had been lost since 89 as a result of the Pontic invasion, the annual income it provided to the Republic was considerable (120 million denarii spread over five years would amount to 24 million denarii annually).[49]

As historians have recently emphasized, the *lex Sempronia de provincia Asia* was an important step towards a more controlled and regular use of provincial resources for Rome's benefit.[50] Although, it was a gradual process, Rome was slowly beginning to firmly tap into the economies of its provinces to make up for the suspension of *tributum* in a more sustainable way. Since the majority of the Republic's spending was of a military nature, most of the money obtained from provincial resources went into the army.

τελῶν ὑπῆρχον, ἐκ δὲ ὧν αὐτὸς προσεκτήσατο τῇ πόλει μυριάδας ὀκτακισχιλίας πεντακοσίας λαμβάνουσιν).

[48] Plut. *Sull.* 25; *Luc.* 4; 20; App. *Mith.* 63. On the Mithridatic wars, see Toni Ñaco del Hoyo, Borja Antela Bernárdez, Isaias Arrayás Morales, and Salvador Busquets-Artigas, "The « Ultimate Frontier »: War, Terror and the Greek « Poleis » Between Mithridates and Rome," in *Frontiers in the Roman world: proceedings of the ninth Workshop of the International Network Impact of Empire (Durham, 16-19 April 2009)*, eds. Olivier Hekster and Ted Kaizer (Leiden and Boston: Brill, 2011), 291-304; Toni Ñaco del Hoyo and Isaias Arrayás Morales, "Rome, Pontus, Thrace and the Military Disintegration of the World Beyond the Hellenistic East," in *Rome and the Worlds Beyond Its Frontiers*, eds. Daniëlle Slootjes and Michael Peachin (Leiden and Boston: Brill, 2016), 3-19.

[49] The financial importance of Asia for Rome is also emphasized by Cicero (*Leg. Man.* 14): "For while the revenues of our other provinces, citizens, are barely sufficient to make it worth our while to defend them, Asia is so rich and fertile as easily to surpass all other countries in the productivity of her soil, the variety of her crops, the extent of her pastures, and the volume of her exports." (*Nam ceterarum provinciarum vectigalia, Quirites, tanta sunt ut eis ad ipsas provincias tuendas vix contenti esse possimus, Asia vero tam opima est ac fertilis ut et ubertate agrorum et varietate fructuum et magnitudine pastionis et multitudine earum rerum quae exportentur facile omnibus terris antecellat*); Val. Max. 4.8.4. Kay, *Revolution*, 72, proposed that the territory of the former Attalid kingdom perhaps yielded 700 talents.

[50] Ñaco del Hoyo, "Republican West," 223; Kay, *Revolution*, 14, 83.

2.4 Gaius Gracchus' *lex militaris*

Gaius Gracchus' more famous legislation sometimes overshadows other less well-known bills he passed. Among them is a *lex militaris* proposing that clothing should no longer be deducted from the soldiers' *stipendium*.[51] It was normal practice at the time for additional weapons and food to be deducted from the soldiers' *stipendium*. It may look surprising that the law mentions that clothes rather than weapons should be provided free of charge. Weapons are indeed the first items that come to mind as far as military service is concerned. They are what differentiated the citizen in civilian life from the citizen-soldier. Although no price survives, there can be no doubt that weapons were much more expensive than clothes. Providing free weapons would have removed the biggest deduction made on pay for soldiers who needed additional weapons on campaign. E. Gabba once argued that Plutarch, the author who recorded the law, abridged his text and that in this context the word *ἐσθῆτα* referred not only to clothes, but also to the weapons needed by a soldier.[52] From a modern perspective, Gabba's hypothesis seems quite plausible since it seems perhaps natural that a Mediterranean empire would stop relying on a citizen militia system in which soldiers provided their own weapons. Yet, modern interpretations on state funded armies should not lead us to over-interpret sources.

Gabba's argument rests on speculation since nothing indicates that *ἐσθῆτα* could be the equivalent of weapons (*ὅπλά*) or equipment (*παρασκευή*). More recently, other scholars have argued that since the law only mentions clothing, this would then suggest that weapons were already freely supplied to soldiers by the state.[53] To be sure, the revenues of the mines in Spain,

---

[51] Plut. *C. Gracch.* 5: "and ordained that clothing should be furnished to the soldiers at the public cost, that nothing should be deducted from their pay to meet this charge" (ἐσθῆτά τε κελεύων δημοσίᾳ χορηγεῖσθαι καὶ μηδὲν εἰς τοῦτο τῆς μισθοφορᾶς ὑφαιρεῖσθαι τῶν στρατευομένων).

[52] Emilio Gabba, *Republican Rome, the Army and the Allies* (Berkeley and Los Angeles: University of California Press, 1976), 7; followed by Paul Erdkamp, "The transformation of the Roman army in the second century BC," in *War and Territory in the Roman World. Guerra y territorio en el mundo romano*, eds. Toni Ñaco del Hoyo and Isaias Arrayás Morales (Oxford: BAR, 2006), 46.

[53] Nick Rosenstein, "Marriage and Manpower in the Hannibalic War: "Assidui', 'Proletarii'" and Livy 24.18.7-8," *Historia* 51 (2002): 176; Followed by John Serrati, "Warfare and the State," in *The Cambridge History of Greek and Roman Warfare*, vol. 1, eds. Philip Sabin, Hans van Wees, and Michael Whitby (Cambridge: Cambridge University Press, 2007) 489; Dominic Rathbone, "Military finance and supply," in *The Cambridge History of Greek and Roman Warfare,* vol 2, eds. Philip

Macedon, and the taxes of the province of Asia would have been more than adequate to cope with this. Yet, this suggestion remains problematic for two reasons. First, it is simply not attested in the sources, and such a measure would have required a significant overhaul in army financing, greatly increasing military expenditure. Second, a passage from Tacitus' *Annales* indicates that deductions for food, clothing, tents, and weapons were still in place in the early first century CE, which makes this hypothesis even more improbable.[54] In my view, Gaius' law should be interpreted along with his other legislation: that the revenues of the Republic should benefit the Roman people, even during their military service.[55] However, since Gaius and part of his political legacy fell victim to brutal violence and repression, I would argue that his *lex militaris* probably did not outlive him for long. Tacitus' passage demonstrates that there was a reversal to the old system of deductions, although this cannot be dated with certainty. The limited attempt at proposing state sponsored clothing is at best a timid move towards the idea of an army fully financed by the state, an idea that never became a reality in the Republic.

3 - Analysis: The Slow Beginnings of Imperial Military Fiscality

The exploitation of the Macedonian and Spanish mines brought considerable revenues to the Republic. If they indeed generated some 15 million denarii every year in the late second century, this money could have easily paid for the *stipendium* of the average legionary and naval deployment between 167 and 91. The total of the other costs associated with the army probably necessitated money from other sources of revenues, which were available through the contributions of other provinces and eventually the taxes of Asia.[56] Yet, the shift from an army sustained through

---

Sabin, Hans van Wees, and Michael Whitby (Cambridge: Cambridge University Press, 2007), 159.

[54] Tac. *Ann.* 1.17.6: "ten asses a day was the assessment of body and soul: with that they had to buy clothes, weapons and tents, bribe the bullying centurion and purchase a respite from duty" (*denis in diem assibus animam et corpus aestimari; hinc vestem, arma, tentoria; hinc saevitiam centurionum et vacationes munerum redimi*).

[55] A point emphasized by Kay, *Revolution*, 59-83.

[56] In Sicily, the existing fiscal structures were preserved by Rome, cf. Cic. *Verr.* 2.3.12-15. Pausanias (7.16.9-10) claims that Greece had to pay a *φόρος* after the Achaean War. His claim about the payment of *φόρος* is considered suspicious because he might have equated Roman hegemony over Greece with his second

tributum, indemnities and plunder in the first half of the second century is striking. According to M. Taylor, 77% of Roman income between 200 and 157 came from war indemnities, plunder, and *tributum*.[57] However, by the second half of the second century, most of these revenues were no longer available as indemnities had run out, plunder became much less abundant as the Mediterranean major powers had been either conquered or stripped of their wealth, and the *tributum* had been abolished. Yet, perhaps surprisingly, the senate was initially content with replacing these revenues by limited measures that were enough to fund the current military establishment.

Among the reasons to explain this was the lack of interest among high officials to increase public revenues as they would not personally benefit from this because of their short tenure. Besides, war indemnities and plunder from the wars fought in the early second century provided a certain financial cushion for a while. Only when this started to run dry in the mid-second century were new measures implemented. This did not immediately result in wide-ranging fiscal reforms, but rather in the reopening of the Macedonian mines and a more intensive exploitation of the Spanish mines. Only in the late 120s do we hear of a major effort to profit massively from the newly acquired province of Asia, and the Gracchan proposal proved to be quite controversial. Thus, the gradual exploitation of provincial resources in the late second century came to replace the money provided by the *tributum*, plunder, and war indemnities to finance the army, although it can hardly be said that this was the result of careful long-term planning. Thus, Rome was rather slow in extracting provincial revenues on a systematic and extensive scale, especially where there was not already a fiscal regime existing.

One of the reasons to explain the relative slowness to implement systematic provincial taxation was that Rome could count on tens of thousands of Italian allies who could be mobilized for cheap (only food was furnished by Rome). In addition to this, allied communities in the provinces regularly

century CE understanding of political realities. There is no other concrete evidence for tribute in Greece until the age of Sulla. After the fall of Carthage, a land and poll tax was imposed on communities who did not side with Rome, cf. App. *Pun.* 135.639–641. Cicero (*Verr.* 2.3.12) later claimed that Rome imposed a fixed tribute on most of the Carthaginians. On Roman fiscal policy in the Republic, see most recently Jérôme France, *Tribut. Une histoire fiscale de la conquête romaine* (Paris: Belles Lettres, 2021).

[57] Taylor, *Silver*, 131.

provided auxiliary troops for the Roman army that they financed themselves.[58] The proportion of these sometimes amounted to that of the Italian troops, that is about the same strength as the Roman contingent. Therefore, the availability of all this cheap or outright free military manpower surely went against any idea that a major reform of state revenues was necessary to fund the army. What limited measures were undertaken in the late second century regarding mines were therefore deemed adequate. Gaius Gracchus' new tax system in Asia was a novelty and, despite the violent opposition he faced that led to his death, the revenues it generated probably persuaded the senate that other measures were superfluous as far as military fiscality was concerned.

Conclusion

In 167, the *tributum*, the most stable source of funding for the army, ceased to be levied and would only be reintroduced in 43 when the funding of the huge armies raised during the civil wars of the late Republic required it. The influx of money obtained through war indemnities paid by wealthy defeated enemies such as Macedon and the Seleucid Kingdom likely motivated this decision. However, over the course of the second half of the second century, those payments eventually stopped. In the absence of *tributum*, provincial resources gradually replaced the money paid by *assidui* to finance the army. Yet, there was nothing resembling a wide-ranging effort to systematically implement taxation in the provinces. Rome was still learning how to manage the empire it had recently acquired. This changed with the *lex Sempronia de prouincia Asia*, which can be understood as the first attempt at maximizing profits from provincial resources in order to increase state revenues. Despite the fact that Gaius Gracchus did not implement his law with the strict intention of increasing revenues for the army, the money made available by the taxes of the provinces of Asia could naturally be used to pay for military expenditure. Furthermore, his *lex militaris* shows that he was not devoid of interest for the funding of the army. However, the outbreak of the Social War would see Rome lose its reservoir of cheap Italian military manpower. The enfranchisement of the hitherto-cheap-to-mobilize Italian allies would greatly increase military expenditure for Rome as the new citizens needed payment from the *aerarium* when conscripted.

---

[58] François Gauthier, "Auxiliaries and War Financing in the Republic," *Journal of Ancient History* 7 (2019): 251-268.

Bibliography

Arnold, W.T. *The Roman System of Provincial Administration to the Accession of Constantine the Great*. Oxford: Blackwell, 1914.

Brunt, P. A. *Italian Manpower, 225 BC – AD 14*. Oxford: OUP, 1971.

Cadiou, F. *Hibera in terra miles. Les armées romaines et la conquête de l'Hispanie sous la république* (218–45 av. J.-C.). Madrid: Casa de Velázquez, 2008.

Crawford, M. *Coinage and Money under the Roman Republic. Italy and the Mediterranean Economy*. Berkeley and Los Angeles: University of California Press, 1985.

Erdkamp, P. "The transformation of the Roman army in the second century BC." In *War and Territory in the Roman World. Guerra y territorio en el mundo romano*, edited by T. Ñaco del Hoyo and I. Arrayás, 41–51. Oxford: BAR, 2006.

France, J. *Tribut. Une histoire fiscale de la conquête romaine*. Paris: Belles Lettres, 2021.

Frank, T. *An Economic Survey of Ancient Rome*. Baltimore: Johns Hopkins U. P., vol. 1, 1933.

Gabba, E. *Republican Rome, the Army and the Allies*. Berkeley and Los Angeles: University of California Press, 1976.

Gauthier, F. "Auxiliaries and War Financing in the Republic." *Journal of Ancient History* 7 (2019): 251-268.

Kay, P. "What Did the Attalids Ever Do for Us? The View from the *Aerarium*." In *Attalid Asia Minor: Money, International Relations, and the State*, edited by P. Thonemann, 121-148. Oxford: OUP, 2013.

—. *Rome's Economic Revolution*. Oxford: OUP, 2014.

Kromayer, J. "Die Entwicklung der römischen Flotte vom Seeräuberkriege des Pompeius bis zur Schlacht von Actium." *Philologus*, 56 (1897): 426-491.

Hollstein, W. (ed.). *Metallanalytische Untersuchungen an Münzen der Römischen Republik*. Berlin: Gebr. Mann Verlag, 2000.

Lintott, A. *Imperium Romanum: Politics and Administration*. London and New York: Routledge, 1993.

Magie, D. *Roman Rule in Asia Minor*. Princeton: Princeton Legacy Library, 1950.

Mattingly, H. B. "The Date of the *Senatus Consultum de Agro Pergameno*." *AJPh* 93 (1972): 412–23.

Milan, A. "I socii navales di Roma." *Critica storica* 10 (1973): 193-221.

Ñaco del Hoyo, T. *Vectigal incertum. Economía de guerra y fiscalidad republicana en el occidente romano: su impacto histórico en el territorio, (218-133 a.C.)*. Oxford: BAR, 2003.

—. "The Late Republican West: Imperial Taxation in the Making?" In *Crises and the Roman Empire. VII Workshop Impact of Empire (Roman Empire, 200 BC – AD 476)*, edited by O. Hekster, G. de Kleijn, and D. Slootjes, 218-231. Leiden and Boston: Brill, 2007.

Ñaco del Hoyo, T., B. Antela Bernárdez, I. Arrayás Morales, and S. Busquets-Artigas, "The « Ultimate Frontier »: War, Terror and the Greek « Poleis » Between Mithridates and Rome." In *Frontiers in the Roman World: Proceedings of the Ninth Workshop of the International Network Impact of Empire (Durham, 16-19 April 2009)*, edited by O. Hekster, and T. Kaizer, 291-304. Leiden and Boston: Brill, 2011.

Ñaco del Hoyo, T., Arrayás Morales, I. "Rome, Pontus, Thrace and the Military Disintegration of the World Beyond the Hellenistic East." In *Rome and the Worlds Beyond Its Frontiers*, edited by D. Slootjes, and M. Peachin, 3-19. Leiden and Boston: Brill, 2016.

Nicolet, C. *Tributum. Recherches sur la fiscalité directe et indirecte sous la république romaine*. Bonn: Rudolf Habelt, 1976.

Rathbone, D. "Military Finance and Supply." In *The Cambridge History of Greek and Roman Warfare*, edited by P. Sabin, H. van Wees, and M. Whitby, vol 2, 158–75. Cambridge: CUP, 2007.

Rosenstein, N. "Marriage and Manpower in the Hannibalic War: "Assidui', 'Proletarii'" and Livy 24.18.7-8." *Historia* 51 (2002): 163–191.

—. "*Bellum Se Ipsum Alet?* Financing Mid-Republican Imperialism." In *Money and Power in the Roman Republic*, edited by H. Beck, M. Jehne, and J. Serrati, 114-130. Brussels: Latomus, 2016.

—. "*Tributum* in the Middle Republic." In *Circum Mare*: *Themes in Ancient Warfare*, edited by J. Armstrong, 80– 97. Leiden: Brill, 2016.

Serrati, J. "Warfare and the State." In *The Cambridge History of Greek and Roman Warfare*, edited by P. Sabin, H. van Wees, and M. Whitby, vol. 1, 461–497. Cambridge: CUP, 2007.

Sherk, R.K. *Roman Documents from the Greek East: Senatus Consulta and Epistulae to the Age of Augustus*. Baltimore: Johns Hopkins Press, 1969.

Steinby, C. *The Roman Republican Navy, from the sixth century to 167 B.C.* Helsinki: Societas Scientiarum Fennica, 2007.

Tan, J. *Power and Public Finance at Rome, 264–49 BCE*. Oxford: OUP, 2017.

—. "The Fiscality of Foreign Relations in the Roman Republic (241–146 BCE)." In *Ancient Taxation: The Mechanics of Extraction in*

Comparative Perspective, edited by J. Valk, and I. Soto Marín, 262-289. New York: NYU Press, 2021.

Taylor, M. J. "State Finance in the Middle Roman Republic: A Reevaluation." *AJPh* 138 (2017): 143–180.

—. *Soldiers and Silver. Mobilizing Resources in the Age of Roman Conquest*. Austin: University of Texas Press, 2020.

Thiel, J. H. *Studies on the History of Roman Sea Power in Republican Times*. Amsterdam: North-Holland Publishing, 1946.

FROM BATTLE TO TEXT: A REVIEW OF THE WORD *ACTIUM* IN LATIN LITERATURE AS A RESOURCE FOR IDENTITY AND GEOGRAPHICAL MONUMENTALITY

NELSON HENRIQUE DA SILVA FERREIRA[1]

Actium is a small promontory in the Epirus region, near the island of Leucate, and gave its name to a naval battle that would define the destinies of the Roman Empire. Actium implies more than a simple location and Latin literature expresses that as part of historical chronicles, as literary topic, as rhetorical theme and as toponyms (in a travel context).

Toponym	Marcus Manilius, *Astronomica* 1.906-16, 5.52; Ov., *Met.* 1.3; Ov. *Met*. 13.715; Petron. *Sat*. 121.1; Plin. *HN* 7.149, Plin. *HN* 14.148, Plin. *HN* 32.3; Suet. *Aug*. 18.2; Tac. *Ann*. 2.53;
Epic/Heroi c Theme (battle)	Hor. *Ep*. 1.18.61-64; Petron. *Sat*. 121.1; Prop. 2.16; Prop. 4.6; Verg. *Aen*. 3.278-88; Verg. *Aen.* 8.704; Verg. *Aen*. 8. 675-84
Propagand a (battle/ peace)	O. *Met*. 1.438-51; Prop. 2.15.34; Prop. 2.34; Prop. 4.6.55-68; Ov. *met*. 13.715; Ov. *Fast*. 1.717; Suet. *Aug*. 9.1; Vell. Pat. 2.84; Vell. Pat. 2.84.1.2
Historical Chronicals	Livy, *Ab urbe condita* 133.1; Vell. Pat. 2.84.1.1-10; Vell. Pat. 2.84.1.2; Tac. *Hist*. 1. 1.; Tac. *Ann*. 1.3; Tac. *Ann*. 4.5; Tac. *Ann*. 2. 53; Tac. A*nn.*15.23; Plin. *HN* 19.22; Suet. *Aug*. 9.1; Suet. *Aug*. 17.3; Suet. *Aug*. 18.2; Plin. *HN* 7.149, Plin. *HN* 14.148
Rhethorical Resource	Hor. *Ep*. 1.18.61-64; Plin. *HN* 19.22; Plin. *HN* 21.12 ; Tac. *Ann.*15.23

[1] Orcid ID: 0000-0003-2637-3211 ; Ciência ID: 0513-1A91-26B0.

Based on an extensive survey of Latin literature, the aim of this paper is to analyze references to the word *Actium* in texts produced between the late 1st century BCE and the 3rd century CE. The survey considered only explicit references to the lexeme, excluding implicit references to the battle. The aim is to attempt a description of what the word *Actium* might represent for the educated Roman.

1. Historical account and chronicles of a place[2]

In early September of 31 BCE, during the consulate of the future Caesar Augustus and Messala Corvinus, a naval battle defined the fate of Marc Antony and Cleopatra VII. It was not the final act of the war between Marc Antony and Caesar Augustus, but the victory led by general Agrippa (c. 64-12 BCE) was crucial in ending the civil struggle. The fighting that followed the battle of Actium was no more than skirmishes, as Marc Antony's (c. 83-31 BCE) resources and resilience were low, and his supporting party felt discredited.[3] The consequence was a dramatic end with the suicide of Marc Antony and Cleopatra (Livy, *Per.* 133.1).

Besides some tangible artifacts, the historical event was mainly preserved through chronicles, storytelling, propaganda, and heroic/romanticized poetry during the golden age of Latin literature in the first century BCE. But its preservation in collective memory concerns much more than propaganda, as it unquestionably marks the beginning of the end of the civil war through a naval battle of great proportions. For the latter reason alone, it can be justified as good material for monumental and epic commemorations in that sense.

Regarding historical accounts, Livy (c. 59 BCE – 17 CE) has one of the earliest known references to the battle of Actium. The ancient historian provides data on the event and states that the clash in the Ambracia region

---

[2] On the description of the landscape and the geographical context of the battle, see Bettina Reitz-Joosse, "Land at Peace and Sea at War: Landscape and the Memory of Actium in Greek Epigrams and Propertius'Elegies," in *Valuing Landscape in Classical Antiquity: Natural Environment and Cultural Imagination*, ed. by Jeremy McInerney, Ineke Sluiter, and Bob Corthals (Leiden: Brill, 2016), 276–96.

[3] For an historical overview on the civil war and the specific developments of this naval battle see Barry S. Strauss, *The War That Made the Roman Empire: Antony, Cleopatra, and Octavian at Actium.* Simon&Schuster hardcover edition. (New York: Simon & Schuster, 2021).

was the engine of the second period of peace in the empire.[4] According to the historian, the first period of peace comes after the Third Punic War (149-146 BCE). It is also the first time that an association has been made between this battle and a period of peace as a consequence of the war. Other authors claim the battle as the moment that decided the victor, but none defined it as a chronological marker for internal peace so explicitly before Livy.

Although the 'idea of *actium*' in Livy's text *Ab urbe condita* seems to imply 'an end', Livy would not have in mind the end of military actions, since he referred to the events that followed the battle. He noted with emphasis the flight of Cleopatra and Marc Antony to Egypt (133), which was an important passage not only for the narration of the course of the most remarkable naval confrontation in the history of Rome, but also for understanding the consequences it would have on a civil war still going.

Octavian's (63 BCE-19 CE) victory and Marc Antony's defeat found their expression in the term *Actium*, and so it would be from then on, regardless of the events that followed. It may be extrapolative to regard Livy as the one who first considers this battle as the final blow in the war. Although there are no known records that mention in detail these historical episodes, the very development of the civil war and the rapid end after the naval skirmish may have propitiated and popularized Actium as a definitive event among those closest to it, and it is possible the Roman historian was following this assumption.

For authors who lived through the troubled period of the civil war and were on the winning side, *Actium* was synonymous with relief and the tittle *actiacus* assumed by Roman soldiers seems to attest it.[5] In fact, one may say that there was an 'ideology of Actium'[6] and regarding historical narrative

[4] *M.Antonius ad Actium classe victus Alexandriam profugit, obsessusque a Caesare in ultima desperation rerum, praecipua occisae Cleopatrae falso rumore inpulsus se ipse interfecit. Caesar Alexandria in poteate<m> redacta, Cleopatra, ne in arbitrium victoris veniret, voluntaria morte defuncta, in urbem reversus tres triumphos egit, unum ex Illyrico, alterum ex Actiaca victoria, tertium de Cleopatra, inposito fine civilibus bellis altero et vicesimo anno. M. Lepidus Lepidi, qui triumvir fuerat, filius coniuratione adversus Caesarem facta bellum moliens oppressus et occisus est.* (Livy, 133.1)

[5] L. J. F. Keppie, "A Note on the Title Actiacus." *The Classical Review* 21, no. 3 (1971): 329–30, https://www.jstor.org/stable/708579.

[6], Walter Wilson, "Pious Soldiers, Gender Deviants, and the Ideology of Actium: Courage and Warfare in Philo's De Fortitudine". *The Studia Philonica Annual* 17 (2005): 1–32.

Velleius Paterculus (1st century CE) is probably the Roman writer who better represents it. In *Historia Romana* (Vell. Pat. 2.84),[7] Velleius devotes special attention to the Battle of Actium and its historical value, and it is noteworthy that there is great proximity between what Velleius narrates about the time of Augustus and the information conveyed in the *Res Gestae (Divi Augusti)*.[8]

In addition to accounting for the events that preceded and instigated the confrontation, he describes the circumstances of the characters involved. Velleius' approach seems rather biased, with a strong inclination towards the victors, probably due to his chronological proximity to the events. Loading his text with propagandistic potential that greatly elevated the emperor and Rome, Velleius legitimized Octavian's actions against those who opposed him: Octavian is presented subduing his opponents with weapons and words, for this is how surrender is achieved. A certain symbolism can be identified in the representation of the ruler as a military commander of excellence and, at the same time, as a great statesman. This image may be, together with the *Res Gestae (Divi Augusti)*, Tacitus' target when he criticizes the loss of historical truth. (*vide infra*)

However, by the time *Historia Romana* may have been written, few people would have known a regime other than Augustus', so pure political propaganda may not have been the major factor that powered Velleius' choices. His comment may imply that strong opposition from different political regimes and with popular support was non-existent and that the side of the victorious faction, was the only significant side standing.

Velleius emphasizes that particular period of history, not only to the battle, but also to the whole period of Octavian's entry into the Roman political scene. Now, precisely because of the chronological proximity, much of the information made available would obviously coincide with the trends of the regime and the heroics of Caesar Augustus. Although, that does not mean

---

[7] *Caesare deinde et Messala Coruino consulibus, debellatum apud Actium, ubi longe ante quam dimicaretur, exploratissima Iulianarum partium fuit uictoria. Vigebat in hac parte miles atque imperator, <in> illa marcebant omnia: hinc remiges firmissimi, illinc inopia adfectissimi; nauium haec magnitudo modica nec celeritati aduersa, illa specie [et] terribilior; hinc ad Antonium nemo, illinc ad Caesarem cotidie aliquis transfugiebat; denique in ore atque oculis Antonianae classis per M. Agrippam Leucas expugnata, Patrae captae, Corinthus occupata, bis ante ultimum discrimen classis hostium superata.* (Vell. Pat. 2.84.1.1-10)

[8] About this text, see Alison Cooley. '*Res Gestae Divi Augusti': Text, Translation, and Commentary*. (New York: Cambridge University Press, 2009).

Tacitus was wrong, as such propagandistic presentation of the period, was common among Latin authors of the time.

Thus, for Velleius it was a struggle pitting Romans against Romans, a civil war that made little sense to the author, so the heroic treatment of the adversary, who could be taken as a traitor, appears to unite the military factions and at the same time recall Caesar Augustus' greatest achievement: the end of the civil war.

Velleius frequently mentions the defections of Antony's supporters (e.g. Vell. Pat. 2.83.1), highlighting that of King Amyntas of Galatia (ruler c. 37-25 BCE). Antony's faction is presented as demoralized, under weak leadership, inclined to flee and supported by a general who had lost his mind due to the influence of Queen Cleopatra VII. Doubting Antony's autonomy, Velleius suggests that he acted under Cleopatra's will.[9]

In fact, Velleius suggests that the battle was lost from the start and that its outcome was merely a conclusion of Agrippa's previous military victories, the commander of Octavian's forces. In relation to the combat itself, Agrippa is a name inseparable from the Battle of Actium, because of the action he took before and during it. Velleius notes this very clearly in his narrative, distancing himself from other authors who, in praise of Augustus, relegate Agrippa to the background. To this end, one may recall the description of Aeneas' shield (Verg. *Aen.* 8.657ss), which we discuss later in this paper. (*vide infra*)

Velleius apologizes the soldiers opposing Octavian stressing that they only followed a general corrupted by a barbarian queen and were only doing their duty (Vell. Pat. 2.84.1). Velleius states that Antony's soldiers fought bravely and only accepted surrender after their own general had deserted them. The historian also alludes to what may have been a serious military error: the use of large ships at the expense of greater agility and speed (Vell. Pat. 2.84.1.). Given the geographical space and the large concentration of vessels, the disadvantage would have been enormous. Perhaps, Antony's likely intention was not to gain a victory over Octavian's army, but rather to break the blockade to which he was subjected. In this sense, he will not have

---

[9] *Rex Amyntas meliora et utiliora secutus; nam Deillius, exempli sui tenax ut a Dolabella <ad Cassium, a Cassio ad Antonium, ita ab Antonio transiit> ad Caesarem; uirque clarissimus Cn. Domitius, qui solus Antonianarum partium numquam reginam nisi nomine salutauit, maximo et praecipiti periculo transmisit ad Caesarem* (Vell. Pat. 2.84.1.2).

had time to prepare an adequate fleet and set the conditions of battle. This information is absent from Velleius' text, and it is not possible to distinguish with certainty a deliberate omission from a simple lack of knowledge. But it is plausible that the historian was unaware of all the circumstances of the fighting, namely a strategy by Marc Antony to postpone the fight to another time and under different circumstances and only disrupt the blockade.

Regardless of historical accuracy or lack of it, Velleius is the author who most recounts the action of the battle, so the image he leaves us would have been the most common in the thought of the authors of the late Republic and early Empire.

Cornelius Tacitus, who lived between the years 55 and 120 CE, comments on the radical change that the Battle of Actium caused in the Roman universe (Tac. *Hist*. 1. 1).[10] For Tacitus the battle culminated in a subsequent weakening of the political system, literary production and the preservation of historical truth.[11] Tacitus may have in mind the great accumulation of political powers in the emperor figure and all the consequences that come from it: little participation of citizens (in this case of political elites) and a consequent propaganda or adulation that would influence literary production and the accuracy of historical records, as can be inferred from examples by Velleius Paterculus.

Tacitus refers to Actium recalling that the oldest among the Romans had been born during the civil wars and the youngest after the Battle of Actium, so they did not know the republican regime from their own experience (Tac. *Ann*. 1.3).[12] Although there is no clear intention to load the term with a negative symbolism, the consequences the author associates it with reveal

---

[10] *postquam bellatum apud Actium atque omnem potentiam ad unum conferri pacis interfuit, magna illa ingenia cessere; simul veritas pluribus modis infracta, primum inscitia rei publicae ut alienae, mox libidine adsentandi aut rursus odio adversus dominantis: ita neutris cura posteritatis inter infensos vel obnoxios. sed ambitionem scriptoris facile averseris, obtrectatio et livor pronis auribus accipiuntur; quippe adulationi foedum crimen servitutis, malignitati falsa species libertatis inest* (Tac. *Hist*. 1. 1).

[11] John Marincola, "Tacitus' Prefaces and the Decline of Imperial Historiography," *Latomus* 58, no. 2 (1999): 391–404, http://www.jstor.org/stable/41538749.

[12] *bellum ea tempestate nullum nisi adversus Germanos supererat, abolendae magis infamiae ob amissum cum Quintilio Varo exercitum quam cupidine proferendi imperii aut dignum ob praemium. domi res tranquillae, eadem magistratuum vocabula; iuniores post Actiacam victoriam, etiam senes plerique inter bella civium nati: quotus quisque reliquus qui rem publicam vidisset* (Tac. *Ann*. 1.3)?

another possible dimension of *Actium* for the Romans, at least for those who would defend the return of the Republic. Tacitus also gives us a fact that was ignored in the works mentioned above: Octavian would have captured some of the boats and later manned them with an equipped crew (Tac. *Ann.* 4. 5).

After Tacitus, the battle is loaded with a symbolic value: the end of the Republic and the beginning of a totalitarian and dictatorial regime, an assumption reflected in other literary works. Potter,[13] following Dio Cassius (c. 155-235 CE), notes "(…) The battle itself is unlike any other, for Philippi marked the end of the Republic and Actium, the beginning of the monarchy. Pharsalus, though, was a contest between the two greatest Romans of their time. But the result would not bring about a change in government: The government Caesar established had a precedent in that of Sulla, for it was the Republic run by a dictator" (Cass. Dio 47.39.1–3).[14]

Actium also means a topographical monument and a religious and commercial site, probably after the great construction works done by the Emperor Augustus in the region, which made it an interesting spot in antiquity. Travelling to Greece was both a means of education and recreation, very common among Roman patricians. Because sailing in antiquity was mainly along the coast, many ports were traditional stops for visiting, whether for religious reasons, tourist recreation or scientific curiosity. The promontory and the nearby town of Nicopolis were on the map of the ancient traveler. One can take the example of Pliny the Elder as the most famous representative of travelling across the empire, made during his public offices. Actium met all the conditions to be included in a sea voyage through the Mediterranean by Pliny, or at least appear in his reports, considering the undeniable historical notoriety of *Actium*, the privileged geographical space regarding maritime traffic, the location in the vicinity of a city of relevant magnitude, and the fact that it had an important temple dedicated to Apollo.[15] Most of the time, the occurrence of the term *Actium*

---

[13] David Potter, "Dio and Pompey: Explaining the Failure of the Republic," in *The Intellectual Climate of Cassius Dio: Greek and Roman Pasts*, eds. Adam Kemezis, Colin Bailey, and Beatrice Poletti (Leiden: Brill, 2022), 53.

[14] See Cass. Dio 41.13.1. See also commentary on Velleius' context by Edward Bispham, "Time for Italy in Velleius Paterculus", in *Velleius Paterculus: Making History*, ed. Eleanor Cowan, Edward Bispham (Swansea: Classical Press of Wales, 2011), 17–58.

[15] On the temple of Apollo after Actium and the worship of Apollo, see John Miller, "Ovid and Augustan Apollo," *Hermathena*, 177/178 (2005): 165-180.

is merely geographical, with Pliny identifying a specific spatial spot, without any allusion to the meaning it might have beyond toponymic. Such occurrences (Plin. *HN* 7.149, *HN* 14.148, *HN* 32.3), without any information identifying *Actium* as the region of the famous battle, were not very common in authors of the 1st and 2nd centuries CE. In fact, most occurrences like those in *Naturalis Historia* predate the battle.[16] This is due to the connotation acquired by the site, whose historical and political value was only omitted in certain circumstances, such as writings of a personal or encyclopaedic nature.

With the end of the civil war and the political reforms came a wave of municipal restructuring and construction in the Italic peninsula and throughout the empire, promoted by Augustus.[17] The archaeological remains of Nicopolis, the city built by Augustus to celebrate his victory, indicate a high population density and a relevant commercial flow supported by important infrastructures. Marcus Manilius (c. 1st century CE) in *Astronomica* (5. 52) refers to *Actium* for reasons of geographical definition, which aligns with its regional importance in the itineraries. Such a note implies a general awareness of this region, probably enhanced by the relevance of Nicopolis. Tacitus, among other authors, corroborates this awareness by referring to *Actium* in this context (Tac. *Ann*. 2. 53), although he also mentions the memory of battle.[18] The historian mentions that Germanicus (15 BCE– 19 CE) of the Julio-Claudian family, Caesar Augustus' general, who enjoyed great fame and was appointed as Tiberius' successor, visited the promontory of Actium during one of his journeys. This detail becomes quite interesting, considering the importance of the character referred to and the presentation of this space as an archaeological site, something only accessible to places of great importance at a given moment in history. Bearing in mind Tacitus', Pliny's and Manilius'

---

[16] As an example, one could consider Cicero's letters *ad familares* (Cic. *fam*. 16.9).

[17] On the municipalization of Italy after *Actium,* see Edward Bispham, *From Asculum to Actium: The Municipalization of Italy from the Social War to Augustus* (Oxford: Oxford University Press, 2007), 405-446.

[18] *Sequens annus Tiberium tertio, Germanicum iterum consules habuit. sed eum honorem Germanicus iniit apud urbem Achaiae Nicopolim, quo venerat per Illyricam oram viso fratre Druso in Delmatia agente, Hadriatici ac mox Ionii maris adversam navigationem perpessus. igitur paucos dies insumpsit reficiendae classi; simul sinus Actiaca victoria inclutos et sacratas ab Augusto manubias castraque Antonii cum recordatione maiorum suorum adiit. namque ei, ut memoravi, avunculus Augustus, avus Antonius erant, magnaque illic imago tristium laetorumque* (Tac. *Ann*. 2. 53).

comments, it seems safe to say that in the last third of the first century, *Actium* already had an established value as a physical monument. Otherwise, Tacitus would have ignored it, as this location was not of great relevance to the traveler.

The knowledge of the Actium region by the Romans, motivated by the naval battle that took place there, is suggested by many colloquial parallels in which appears the occurrence of the lexeme Actium. In other words, the fame of that geographic space would be so rooted in popular culture that its use as a landmark, comparative adjective or chronological reference would be a rhetorical resource, attested in literature. For example, when Tacitus compares the festivities in honor of Claudia Augusta, daughter of the Emperor Nero (63 CE), to the *Actiacae religionis* (*Ann.* 15.23), he has in mind the popularity of that region and the consequent accessibility of such comment to a wide audience.[19] *Actiacae religionis* was a festival (quinquennial games) celebrated in Nicopolis, decreed by Caesar Augustus, as a commemoration of victory at the Battle of Actium. Indeed, this religious dimension was evoked by Ovid, who presents Apollo killing Python (*Met.* 1.438-51, cf. Prop. 4.6.55-68), which according to Miller, "metaphorically represents Octavian's victory at Actium. Ovid's Phoebus commemorates his victory by establishing the Pythian Games (…)".[20] The institution of these traditional games on Actium seems to manifest political motivations, since the process of Octavian's divinization was definitely crystallized in toponymic designations and is related to Apollo and his temple through the designations of Apollo Palatinus or Apollo Actius. Ovid (*Met.* 1.3) reflects Vergil's Actian Apollo, who participated in the battle as described in Aeneas' shield (Verg. *Aen.* 8.704). The same Aeneas who

[19] *iam senatus uterum Poppaeae commendaverat dis votaque publice susceperat, quae multiplicata exolutaque. et additae supplicationes templumque fecunditatis et certamen ad exemplar Actiacae religionis decretum, utque Fortunarum effigies aureae in solio Capitolini Iovis locarentur, ludicrum circense, ut Iuliae genti apud Bovillas, ita Claudiae Domitiaeque apud Antium ederetur* (Tac. *Ann.* 15.23).

[20] Miller 2005 (171). On the myth of Actium in Propertius and similes of Apollo after the Battle of Actium and its relation with the historical event see, Gottfried Mader, "The Apollo Similes at Propertius 4.6.31-36," *Hermes* 118, no.3 (1990): 325–34, http://www.jstor.org/stable/4476765.

supposedly visited the ancient shrine (Verg. *Aen.* 3.278-88),[21] while Ovid laces Aeneas in the gulf of Ambracia (Ov. *met.* 13.713-15)[22]

Pliny the Elder (c. 23 CE-79 CE) has left us some occurrences of the term *Actium* that are not intended to reflect events and the space associated with them or to provide any information that was not already known to the reader. In one example, Pliny records some aspects concerning the fight and the flight of Marc Antony (*HN* 19.22), but such a reference appears as a note to the exposition of another theme – the author briefly pauses the topic being discussed with complementary information. He considers linen as a fabric, and its possible uses, and as an example, he recalls the sails of Cleopatra's vessel, dyed purple and distinct from the rest of the fleet. With this brief pause, through a slight thematic change, Pliny makes the reading easier for the interlocutor. The same feature is used in *HN* 21.12 on Antony's relationship with Cleopatra. In these passages, *Actium* is almost a anedocte used for rhetorical proposes.

Returning to geographical references, one also finds the lexeme *Actium* in Petronius' S*atyricon* (Petron. *Sat.* 121.1).[23] Uncertainties about the authorship and the historical character behind *Satyricon* make it complex to decode some of the text's intentions regarding political contexts and a satirical dimension. But at least it can be said that the place is known for its importance in the Roman's historical and monumental imagery and for being a popular tourist destination. The author of *Satyricon* satirizes many of the customs considered refined by the Roman economic elite. Among some of these habits are recreational trips, which would have many mandatory stops so that the trip would be legitimized by a certain cultural refinement. However, this passage does not refer specifically to this aspect, but satirizes a high narrative genre, through a very tragic and epic lamentation: a prayer that evokes great Roman victories, naming the spaces

---

[21] *Actiaque Iliacis celebramus litora ludis* (Verg. *Aen.* 3.280). On the divine relation between Apollo and Octavian see Carsten Hjort Lange, *'Res Publica Constituta': Actium, Apollo, and the Accomplishment of the Triumviral Assignment. Impact of Empire* (Leiden: Brill, 2009), 39-46, 166-180.

[22] On the relation between those passages see Miller, "Ovid and Augustan Apollo," 168-169.

[23] *Cerno equidem gemina iam stratos morte Philippos / Thessaliaeque rogos et funera gentis Hiberae / et Libyae; cerno tua, Nile, gementia claustra / Actiacosque sinus et Apollinis arma timentes. / iam fragor armorum trepidantes personat aures.* (Petron. *Sat.* 121.1) Petronius, mentioned in Tacitus' *Annales* (passage reference), was considered the first Latin novelist and an influential personality at Nero's court.

where they took place, among which appears the Actium promontory, the last of the territories where these glorious battles were fought according to the poet. It is important to point out that the monumental dimension associated with this place is clearly present, and if we do not ignore that the participants in the action were not of high social status, it is possible to conjecture about a possible knowledge common to the social group represented by the narrator – who, despite not being rich, had access to education.

Regarding epic dimensions, Suetonius (c. 70 CE – 130 CE), in *Aug*. 17.3 comments on the Battle of Actium, saying that it lasted quite a long time into the night. This underlines its exceptional character, since confrontation on the battlefield in antiquity could not last long. Moreover, it was a naval fight, being dependent on oarsmen, and this type of locomotion has physical limitations inherent in the human condition.

> *nec multo post nauali proelio apud Actium uicit in serum dimicatione protracta, ut in naue uictor pernoctauerit* (Suet. *Aug.* 17.3).
> *(...)*
> *quoque Actiacae uictoriae memoria celebratior et in posterum esset, urbem Nicopolim apud Actium condidit ludosque illic quinquennales constituit et ampliato uetere Apollinis templo locum castrorum, quibus fuerat usus, exornatum naualibus spoliis Neptuno ac Marti consecrauit.* Suet. *Aug.*18.2).

Suetonius also recalls the foundation of the city of Nicopolis and the renovation of an ancient temple of Apollo to commemorate the Battle of Actium. This temple enjoyed great fame in antiquity and, apparently, great importance came after the battle and the building program promoted by Augustus in the region (Suet. *Aug*. 18.2). Its foundation (or refoundation) is also known from Pausanias 5.23.3 who noted that Ambracia's importance was replaced by Nicopolis', a new foundation by Octavian in honor of Apollo.

Also, and as already mentioned, Ovid (43 BCE – 17 CE) (Ov. *met*. 13.715), recalls the region of Ambracia, famous for the Temple of Apollo Actium, which, before the dispute between men, had been the object of dispute between the gods. The reference is apparently only a geographical one, so that the battle which took place there is implied. The fame as a memorable region of glorious times is undeniable – the place now assumes that value in itself, without the need to mention the battle. In the context of a historical or geographical record, this would not be noteworthy, but, when evoked in a high aesthetic genre, it already denotes a distance from physical and archaeological reality, presenting itself as an idyllic and almost mythical place.

The city, founded in 31 BCE north of the promontory, became itself an evocation of the military clash, so references to the Roman city tend not to ignore this fact. And the same happens with the Temple of Apollo Actium, which owes its fame and importance to the same historical event, since the temple is not named before the conflict, and from then on became an obligatory reference on sea voyages through the region.

Suetonius also mentioned Actium as part of the list of the five great civil battles: Modena, Philippi, Perusia, Sicily and Actium (Suet. *Aug*. 9.1). In relation to this, note that Marc Antony was involved in both the first and the last battles. It is therefore the last major battle of the last great internal conflict, and consequently the most valuable memory in Roman history. Although not being part of the scope of the literary survey for this paper, one shall note that Pausanias (c. 110-180 CE) refers to *Actium* as an already consecrated term loaded with symbolism in the military and historical context, as a mark of an end and change (Paus. 5.23.3) that one can assume obeys a well-established tradition.

2. *Actium* in verses

Contrary to what one might expect, the term *Actium* does not occur often in poetic production. Apart from a few occurrences among the elegiacs contemporary to Augustus, *Actium* appears only sporadically among poets. This commentary refers to the great authors of Latin verse of the 1st and 2nd centuries, for it is possible to find the term *Actium* in some fragments whose authors are not known and in the commentaries on the works of Horace and Vergil. Often, there is only a reference that functions as a stylistic resource, a quick reminder of a past action or simply a famous geographical space. However, in Vergil's *Aeneid* (70 BCE – 19 BCE), the term *Actium* appears with special importance in the description of the shield of Aeneas (*Aen*. 8. 675), treated in an epic style. In the center of the shield was the sea and a scene very similar to what Velleius Paterculus would later describe, since the conflict is represented in the center, as if this event were the most dramatic and glorious in the history of Romc. Thc battle seems to be placed on a mythical plane and, although the text evokes a distant future, the way it is presented refers to a past glory.

> *In medio classis aeratas, Actia bella,*
> *cernere erat, totumque instructo Marte uideres*

> *feruere Leucaten auroque effulgere fluctus.*
> *Hinc Augustus agens Italos in proelia Caesar*
> *cum patribus populoque, penatibus et magnis dis,*
> *stans celsa in puppi, geminas cui tempora flammas*
> *laeta uomunt patriumque aperitur uertice sidus.*
> *Parte alia uentis et dis Agrippa secundis*
> *arduus agmen agens, cui, belli insigne superbum,*
> *tempora nauali fulgent rostrata corona* (*Aen.* 8. 675-84)

Caesar Augustus led all the Italic peoples and senators, as well as the Penates and the great gods, who favored him. This scene, drawing a whole culture and its gods, finds a similarity in the Greek advance towards Troy. In addition to the epic motif, there is a complete separation of factions. Here one does not find a fratricidal struggle, but a battle between Romans against barbarians commanded by a general intoxicated by the love of Cleopatra. This is a different approach to that of Velleius Paterculus, and contradictory to some of Augustus' propagandistic ideas, such as the end of hostilities between Romans. Two dimensions of the ideology surrounding the Battle of Actium can then emerge: that of heroic triumph over an enemy of Rome, and that of the victory of the political faction immune to barbarian influence, which puts an end to civil wars and the republican system. Summing up, in Vergil's *Aeneid*, *Actium* means not the pacifying battle for Rome, but a glorious victory over a mighty coalition of barbarian peoples. Augustus is the unifying leader and the warrior hero, not so much the statesman. In fact, the emperor is portrayed as an almost divine entity, again a rather different approach from the writings of Velleius Paterculus. Simply put, one could say that the difference lies in the fact that one text is an epic poem aimed at glorifying Roman achievements and its heroes, and the other a historiographical work, which would pass the facts through the filter of his affection for the regime.

Octavian presents himself on such a high plane that he surpasses the level of heroes and Agrippa occupies the role of human hero, and in this dimension he can, in no way, cast a shadow on the very image of Augustus. Agrippa is one of the main actors of the action that took place in Actium and, in this respect, the coincidence with the information transmitted by Livy, Velleius and Tacitus is great. However, while Vergil places Agrippa on a lower level, Velleius seems to surpass him in relation to Octavian by his decisive role in the entire campaign and battle. Velleius reveals, in a way, what may have instigated some 'shy commentaries': the future emperor's little or no direct participation in the battle, given the total commitment of his general. But it is also certain that neither Livy nor Tacitus add more information than what Paterculus conveys.

As we have seen, in Vergil the term *Actium* takes on a value of its own, alien to the crystallization that the lexeme may have undergone in the cultural, historical, or even mythical sphere. Political issues are forgotten in favor of the glorification of the victors with a role in the conflict.

The time of Augustus was the "Golden Age" of Latin literature, for the artistic greatness of many of his creations, but also for the variety of genres that some authors managed to cultivate.[24] Besides Vergil, we can count with the already mentioned Ovid, who bequeathed us works such as *Metamorphoses*, *Fasti*, *Amores*, *Ars Amatoria*, Propertius, and Horace, author of some of the greatest poems in the Latin language and at the same time a great reformer of the satire. Horace's satire sought comedy through situations comparable to known realities but set against a context of disproportion and ridicule. *Actium* serves that propose, when Horace presents the dispute between two brothers over certain lands that have a lake (*Ep*. 1.18).

> *partitur lintres exercitus, Actia pugna*
> *te duce per pueros hostili more refertur;*
> *adversarius est frater, lacus Hadria, donec*
> *alterutrum velox Victoria fronde coronet* (*Ep*. 1.18.61-64).

Setting and environment are presented as if they were fighting the battle of Actium: the first of the brothers appears as an authentic general, intent on waging a tremendous war on a lake transfigured into the Adriatic Sea. This step not only exposes the disproportion of a conflict over such a small matter, but also reminds us of a serious aspect: two brothers entering into litigation. This is a vision that goes beyond the criticism of the confrontation between two men who, in addition to being fellow citizens, had family ties, triggering civil war and the confrontation between Roman armies. Octavian and Marc Antony promoted a series of marriages between their families, even becoming active participants in them – Marc Antony married Octavia, Octavian's sister, in 40 BCE. Considering the author's chronological proximity to these events, such an interpretation makes sense, since the end of the wars between Romans was one of Augustus' propaganda banners. The text implies a linear comic interpretation: the ridiculous exaggeration of a

---

[24] Mario Citroni, *Literatura de Roma Antiga*. (Lisboa: Fundação Calouste Gulbenkian, 2006), 433-446.

trivial dispute between brothers. But the simple fact that it involves two brothers can by no means be innocent, given the context in the background.

The Latin elegy had its highest moment in the time of Augustus, with a great thematic range that could sing from the most epic of battles to the most trivial love affair. In fact, even when it had a central theme, as in Ovid, the elegiac poem would not be exempt from containing other themes that might even clash with the first. Propertius (c. 50 BCE-16 CE), the poet of love par excellence, did not exclude himself from addressing themes of war in his poetry, although he pointed to his choice of the song of love over warlike acts.[25] The use of these themes varies according to the image intended by the poet, so it is possible to find occurrences of *Actium* in different registers, providing different "canvases" for the thematic framework. For example, Propertius describes the majesty of Caesar Augustus' victory parade in Rome by painting a picture in which the spurs of Actium's ships shine. (Prop. 2.15.34)[26] The Baroque canvas is a hymn to victory, however, without the active factor of the battle being expressed. The weapons and ships, although evoking the conflict of war, appear more as an aesthetic element. Although, concerning material memory, one must note that there is evidence that Octavian's monument at Nicopolis was studded with rams, probably from captured ships.[27]

Actium is a national motif and, therefore, the poet mentions it among other memorable deeds (Prop. 2.15), to be associated with glory, as a mark of

---

[25] For a more thorough discussion of Propertius' approach on the battle and its space, see Bettina Reitz-Joosse, "Land at Peace and Sea at War: Landscape and the Memory of Actium in Greek Epigrams and Propertius'Elegies," in *Valuing Landscape in Classical Antiquity: Natural Environment and Cultural Imagination*, ed. Jeremy McInerney, Ineke Sluiter, and Bob Corthals (Leiden Boston: Brill, 2016), 276–96.

[26] See Frances V. Hickson, "Augustus 'Triumphator': Manipulation of the Triumphal Theme in the Political Program of Augustus," *Latomus* 50, no. 1 (1991): 124–38. http://www.jstor.org/stable/41535965.

[27] William M. Murray, Photios M. Petsas, "The Spoils of Actium." *Archaeology* 41, no. 5 (1988): 28–35. http://www.jstor.org/stable/41740172. Besides, must note the construction of an arch in honor of victory at Actium (Dio 51.19.1), among other constructions, Carsten Hjort Lange, *Res Publica Constituta*: Actium, Apollo, and the Accomplishment of the Triumviral Assignment. (Leiden : Brill, 2009), 95-124 and coinage commemorating the battle. Hickson, Augustus 'Triumphator' , Zarrow 2007 99-101), Edward Martin Zarrow, *The Image and Memory of Julius Caesar between Triumvirate and Principate (44 BCE-CE 14)*. (Yale University (PhD Diss.), 2007).

identity. The choice could have fallen on other battles or deeds of great importance for Roman history, certainly not ignored by the author. But the value of *Actium* in the sociopolitical context of the text and its chronological proximity dictated the choice. Indeed, Propertius contradicts himself when he refuses to sing the war theme of Actium. While this is a pleasing theme for poets like Vergil, it is not so for Propertius (Prop. 2.34), who criticizes this battle because so many Romans had fallen there. One should note that Pliny the Elder also alludes to this volume of death (*HN* 7.149, *HN* 14.148). The negative effects of a victorious battle are not always remembered because conquest intoxicates the defeat of human life. And Propertius does not forget it. Similarly, it would be an idea present in the mindset of the Romans chronologically close to the battle. The way the verse appears in the poem, and the degree of lamentation contained in it, give the battle a meaning that is both epic and realistic: from the most important of battles, it makes the best argument for rejecting war.

The rejection of the epic song is a mere literary theme, for Propertius does indeed sing the deeds of the Divine Augustus at Actium and associates the events that took place there with the Homeric battles and heroes (Prop. 4.6). The poet also speaks of the opposing faction, more specifically of Cleopatra, for whom he seems to feel some fascination, especially for her iron will: she, the cause of the whole conflict, fled the fight, gaining only the possibility of choosing the day of her death herself. Propertius presents the Egyptian queen's escape, followed by Marc Antony, as the cause of the defeat (Prop. 2.16).

The poet connects the battle with the origin of the Temple of Apollo Actium in that region. In fact, Actium corresponds to the temple, the festival, the place, the battle and its actors, but, above all, to the shameful love of the Roman general for the barbarian Cleopatra. Love that the poet understands, had he not also been bewitched and fallen into a madness of love. (Prop. 2.16) In the elegiac context, Actium thus assumed two main motifs: the glorious *Bellico* and the tragic *Amoroso*.

Ovid, who also ignored war in favor of love, praises *Actium* and the peace that the past events brought to a Rome immersed in civil wars (*Fast.* 1.717). This praise of Actium may not be innocent. Considering the silence in his texts, Ovid would not have felt much inclination to sing such events; however, the sensitive situation in which he found himself involved during the writing period of *Fasti* may have been the main trigger of such a disposition: Ovid was exiled by Augustus (8 CE). The great merit he attributes to such a conflict is peace. Given that the literary rejection of weapons was common to the great elegiacs, the idea of peace after such a traumatic period would be a topic worthy of praise and heartfelt singing (in a literary context).

Conclusion

The period in which the battle of *Actium* took place, and its circumstances, made it memorable. Had the literary flourish not been so great, it is possible that its memory would not be very different from that of the campaigns of Scipio Africanus or Pompey the Great. It was not only the historical importance that preserved and nurtured memory, for time is relentless. It was the architectural, artistic, and written records that, after the generations involved in it were long gone, preserved, and nourished the memory and the idea of the Battle of Actium.

The purposes and means of its references are diverse: propaganda, scientific evidence, astronomical information, geographical location, parody, epic motifs, and love themes, among others. The diversity of its meaning and purpose was great, as the examples from Ovid show, the same author could use Actium as lamentation, rejection of military glory, and thus epic glorification.

For all that, it is not possible to reach a clear and irrefutable meaning of the term *Actium* in the Roman literary context of the first two centuries of our era. Since each author assigned it a specific value, according to his literary, political, or informative intentions, the common Roman, who was aware of it, could also do so, regardless of the accuracy of the information he had about the region and the event. At least in literature, the term seems to be adaptable and, regarding collective memory, although education or propaganda could influence its meaning, it may have been quite open to different approaches.

Bibliography

Ash, Harrison Boyd. *On Agriculture. Volumes. 1: I - IV*. The Loeb Classical Library 361. Cambridge: Harvard Univ. Press, 2010.

Bispham, E. 'Time for Italy in Velleius Paterculus.' In *Velleius Paterculus: Making history*, edited by E. Cowan & E. Bispham, 17–58. Swansea: Classical Press of Wales, 2011.

—. *From Asculum to Actium: The Municipalization of Italy from the Social War to Augustus*. New York: Oxford University Press, 2007.

Citroni, M. *Literatura de Roma Antiga*. Translated by Isaías Hipólito and Margarida Miranda. Lisboa: Fundação Calouste Gulbenkian, 2006.

Cooley, Alison. *Res Gestae Divi Augusti: Text, Translation, and Commentary*. New York: Cambridge University Press, 2009.

Cowan, Eleanor, and Edward Bispham (eds.). *Velleius Paterculus: Making History*. Swansea : Classical Press of Wales, 2011.

Hickson, Frances V., "Augustus 'Triumphator' : Manipulation of the Triumphal Theme in the Political Program of Augustus." *Latomus* 50 no. 1 (1991): 124–38.

Lange, Carsten Hjort. *Res Publica Constituta: Actium, Apollo, and the Accomplishment of the Triumviral Assignment*. Leiden : Brill, 2009.

Hübner, Wolfgang. *Manilius, Astronomica, Buch V*. Berlin : De Gruyter, 2010.

Marcovich, Miroslav. *Clementis Alexandrini Protrepticus. Supplements to Vigiliae Christianae*. Leiden : Brill, 1995.

Mader, Gottfried. "The Apollo Similes at Propertius 4.6.31-36." *Hermes* 118, no. 3 (1990): 325–34.

Marincola, John. "Tacitus' Prefaces and the Decline of Imperial Historiography." *Latomus* 58 no. 2 (1999): 391–404.

McInerney, Jeremy, and Ineke Sluiter. *Valuing Landscape in Classical Antiquity: Natural Environment and Cultural Imagination*. Leiden: Brill, 2016.

Miller, J. F. "The Shield of Argive Abas at *Aeneid* 3.286." *The Classical Quarterly* 43 no 2 (1993): 445–50.

Miller, John. "Ovid and Augustan Apollo." *Hermathena*, no. 177/178 (2005): 165–80.

Murray, William M., and Photios M. Petsas. "The Spoils of Actium." *Archaeology* 41 no.5 (1988): 28–35. http://www.jstor.org/stable/41740172.

van Niejenhuis, J. E. *Appiani Historia Romana. 3: Index nominum*. Leipzig 1939. Lipsiae: Teubner, 1992.

Potter, David S. "Dio and Pompey: Explaining the Failure of the Republic." In *The Intellectual Climate of Cassius Dio: Greek and Roman Pasts*, edited by Adam Kemezis, Colin Bailey, and Beatrice Poletti, 33–58. Leiden: Brill, 2022.

Reitz-Joosse, Bettina. "Land at Peace and Sea at War: Landscape and the Memory of Actium in Greek Epigrams and Propertius'Elegies'." In *Valuing Landscape in Classical Antiquity: Natural Environment and Cultural Imagination*, edited by Jeremy McInerney, Ineke Sluiter, and Bob Corthals, 276–96. Leiden Boston: Brill, 2016.

Strauss, B. S. *The war that made the Roman Empire: Antony, Cleopatra, and Octavian at Actium*. Simon & Schuster: New York, 2021.

Yardley, J. C and Anthony A Barrett. *The Roman History: From Romulus and the Foundation of Rome to the Reign of the Emperor Tiberius.* Indianapolis: Hackett Pub. Co., 2011.

Watt, William Smith. *Vellei Patercvli 'Historiarvm ad M. Vinicivm consvlem libri dvo'. Bibliotheca scriptorvm Graecorvm et Romanorvm Tevbneriana. Stvtgardiae* Lipsiae: B. G. Tevbner, 1998.

Zachos, Kōnstantinos L., and Deborah Brown Kazazis. *An Archaeological Guide to Nicopolis: Rambling through the Historical, Sacred, and Civic Landscape*. Monuments of Nicopolis 10. Athens: Ministry of Culture & Sports, DIPCA, 2015.

Zarrow, Edward Martin. *The Image and Memory of Julius Caesar between Triumvirate and Principate (44 BCE-CE 14)*. Yale University (PhD Diss.), 2007.

The Impact of the Tacfarinas Revolt beyond Africa

Jared Kreiner

1 - Introduction

By the end of 19 CE, Tacfarinas' Revolt had been raging for at least three years in the Roman province of Africa Proconsularis. During this time, Tacfarinas' *razzias* (or raids) on the coastal hinterlands of the region's most productive zones had disrupted farming (see Figure 1) as his forces pillaged, burned, and seized rural property, animals, and provincials. It is my contention that these actions played an indirect, yet vital role in the threat to imperial order in the city of Rome brought on by a grain crisis and the illness and subsequent death of Germanicus. Sections two and three of this paper explain how a revolt in one part of the empire could impact events elsewhere. Meanwhile, section four examines imperial-level decision-makers' (Tiberius' court's) multi-pronged responses to the revolt in Africa and to issues in Rome which threatened order in the city that they may or may not have known were related to it. In Rome itself, the state found temporary solutions in fixing grain prices, subsidizing grain merchants, and expelling Jews and followers of the cult of Isis from Rome – the latter in my opinion as scapegoats and to relieve pressure on the imperial court. Beyond Rome, Tiberius ordered reinforcements to Africa from Sardinia and Illyricum to shore-up Rome's grain supply. I contend that Tiberius only sent reinforcements to Africa because Tacfarinas had indirectly impacted imperial order and security in the city of Rome. This paper concludes with a brief hypothesis about why Tacitus does not connect these seemingly interconnected events.

2 - Outline of Revolt through 19/20 CE

There is a particular challenge in outlining the course of early events in this revolt because of Tacitus' sporadic treatment of Tacfarinas' activities between 17 and 21 CE, making a play-by-play reconstruction unobtainable.

What we can capture a glimpse of, however, is the types of rebel activities and their geographical extent, which reveals how greatly Tacfarinas threatened Rome's grain interests in the region. It may be inferred that, presumably prior to 17 CE, Tacfarinas' forces began as a roving band of auxiliary deserters and tribesmen raiding on the fringes of Roman power. These early successes led to him gathering a coalition of forces including some Mauri led by Mazippa, the Cinithii, and Musulamii.[1] Their activities brought the governor, M. Furius Camillus, to the field in 17 CE, who would then defeat Tacfarinas' forces in a pitched battle (Tac. *Ann.* 2.52). After the defeat, the Cinithii and Mauri abandoned Tacfarinas, leading him to transition towards traditional north African forms of unconventional warfare for the duration of the revolt, namely guerilla warfare and raids/*razzias*. For the next few years, Tacfarinas would operate with reduced numbers and experienced varying success. Coming out from the steppes around the Aures Mountains, rebel forces engaged in surprise attacks on Roman forces/strongholds (Tac. *Ann.* 3.20-1) and raids on villages, towns, and their hinterlands.[2] In these endeavors he plundered as much as possible

[1] For the most recent accounts of the revolt, see Gil Gambash, *Rome and Provincial Resistance* (New York: Routledge, 2015), 68-77; Wouter Vanacker, "Adhuc Tacfarinas: Causes of the Tiberian War in North Africa (AD Ca. 15-24) and the Impact of the Conflict on Roman Imperial Policy," *Historia: Zeitschrift für Alte Geschichte* 64, no. 3 (2015): 336-356; Frédéric Hurlet, "Rejeter le contrôle de Rome: les formes de résistance aux structures fiscales et administratives de l'Empire romain," in *Économie et inégalité: ressources, échanges et pouvoir dans l'Antiquité classique: huit exposés suivis de discussions*, ed. Pascale Derron (Vandoeuvres: Fondation Hardt pour l'étude de l'antiquité Classique, 2017), 213-214; Jared Kreiner, "Win, Lose, or Draw: Roman Provincial Administration and Resistance in the Early Principate" (PhD Diss., University of Chicago, 2020), 288-301. It surely would have taken some time to grow from a roving band into an army trained along Roman lines. Just from the viewpoint of size and training he would have had to garner success to attract members and form alliances with other groups.

[2] Stephen Dyson, "Native Revolt Patterns in the Roman Empire," in *Aufstieg und Niedergang der römischen Welt: Geschichte und Kultur Roms im Spiegel der neueren Forschung*, ed. Joseph Vogt et al. (Berlin: de Gruyter, 1975), 163; Wouter Vanacker, "Ties of Resistance and Cooperation. Aedemon, Lusius Quietus and the Baquates," *Mnemosyne* 66, no. 4–5 (2013): 712-3; Catherine Wolff, "La guerre de Tacfarinas," in *La guerre dans l'Afrique romaine sous le Haut-Empire: actes du CXXXVIe Congrès national des sociétés historiques et scientifiques "Faire la guerre, faire la paix", Perpignan, 2011*, eds. Michèle Coltelloni-Trannoy and Yann Le Bohec (Paris: Comité des travaux historiques et scientifiques, 2014), 56, 59; Gambash 2015, 64, 70, 73, 75-6. Thala and coastal areas: Tac. *Ann.* 3.21; Lepcis Magna: Tac. *Ann* 3.74; *IRT* 330; *AE* 1961, 108; Oea: *AE* 1961, 107. Tac. *Ann.* 3.21 surely is referring to the coastal areas in the east of the province, and we must

whenever he could but did experience setbacks (cf. Tac. *Ann*. 3.21, 73). As the map in figure 1 shows, Tacfarinas threatened Roman interests in a wide arc from the eastern portions of the Aures Mountains to Tripolitana in the east. The situation was serious enough, especially regarding the security of Rome's interests in the region, that by the end of the winter 19/20 CE a decision was made to send reinforcements to Africa (see below).

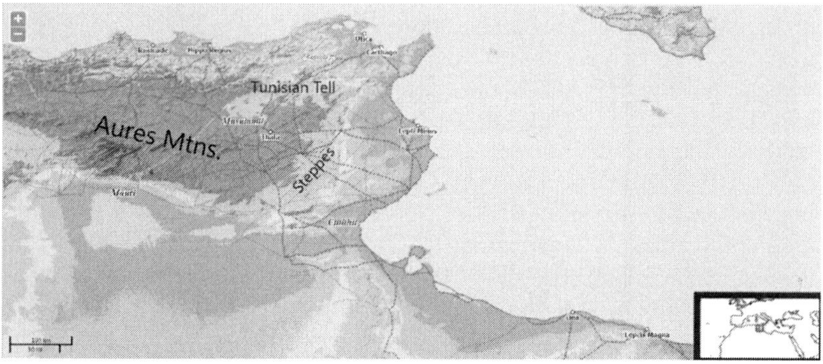

Fig. 10.1 Major Sites and Range of Tacfarinas' Actions c.19 CE

3 - Food Crisis in Rome

Back in Rome, the situation was tense. This was a city which had already experienced issues with floods of the Tiber, fires, and grain supply since the death of Augustus five years earlier.[3] In 19 CE, Tacitus records riotous conditions with grain price hikes, conditions that were compounded by reports of the illness of Germanicus, emperor Tiberius' adopted son, and his eventual death in October (Tac. *Ann*. 2.87; Suet. *Calig*. 5-6).[4] I contend that

---

presume that Tripolitana as a whole was threatened at one point by rebel actions given that Leptis Magna and Oea (or their hinterlands) were threatened (Vanacker 2015, 340-1). On raids: Dyson 1975, 163; Vanacker 2013, 712-713.

[3] Margaret Williams, "The Expulsion of the Jews from Rome in A.D. 19," *Latomus* 48 (1989): 781 (with references).

[4] Tac. *Ann*. 2.87: *Saevitiam annonae incusante plebe statuit frumento pretium quod emptor penderet, binosque nummos se additurum negotiatoribus in singulos modios*. With the plebs censuring the savagery of the supply rates, (Tiberius) established the price for grain which the buyer should pay, and, for the businessmen, he himself would add sesterces for each individual measure (trans. A.J. Woodman). Suet. *Calig*. 6: *Romae quidem, cum ad primam famam valitudinis attonita et maesta civitas sequentis nuntios opperiretur, et repente iam vesperi incertis auctoribus convaluisse*

the grain situation in Rome was intimately connected to both Tacfarinas' raiding in North Africa and more indirectly by Germanicus opening state reserves in Egypt to Egyptians earlier that year because of famine.[5] While it is not possible to be certain how much either of these events impacted the grain supply to the city, reports of the devastation in Africa and the opening of reserves in Egypt would have propelled fears of a future shortage, driving prices upwards.

It is widely recognized that the city of Rome relied heavily on grain from Africa Proconsularis, as much as it did upon Egypt at this time. Without wishing to enter on the debate of quantifying or verifying his claim,[6] Josephus demonstrates the prominent role of north African grain in supplying Rome (*BJ* 2.383, 386). It was a major grain growing region and Rome benefitted from the short distance required to transport grain from its coastal regions to the capital. The province contained several key ports, as shown on the map, and infrastructure existed to connect grainland to ports with artificially enhanced harbors and quays (though some of these developed after our period of concern).[7]

---

tandem percrebruisset, passim cum luminibus et victimis in Capitolium concursum est ac paene revolsae templi fores, ne quid gestientis vota reddere moraretur.... Et ut demum fato functum palam factum est, non solaciis ullis, non edictis inhiberi luctus publicus potuit duravitque etiam per festos Decembris mensis dies. At Rome when the community, in grief and consternation at the first report of his illness, was awaiting further news, and suddenly after nightfall a report at last spread abroad, on doubtful authority, that (Germanicus) had recovered, a general rush was made from every side to the Capitol with torches and victims, and the temple gates were all but torn off, that nothing might hinder them in their eagerness to pay their vows. ...But when it was at last made known that (Germanicus) was no more, the public grief could be checked neither by any consolation nor edict, and it continued even during the festal days of the month of December (trans. J.C. Rolfe).

[5] Tac. *Ann.* 2.59.1; Suet. *Tib.* 52.2. Germanicus was active in Egypt already by January 25th, 19 CE (Robert Sherk, ed. and. trans., *The Roman Empire: Augustus to Hadrian* (Cambridge University Press, 1988), 34C).

[6] Neville Morley, *The Roman Empire: Roots of Imperialism* (London: Pluto Press, 2010), 96. Emphasizing Egypt over Africa, G.E. Rickman, "The Grain Trade under the Roman Empire," *Memoirs of the American Academy in Rome* 36 (1980): 263-4; David Mattingly and Greg Aldrete, "The Feeding of Imperial Rome: The Mechanics of the Food Supply System," in *Ancient Rome: The Archaeology of the Eternal City*, eds. J.C. Coulston and Hazel Dodge (Oxford School of Archaeology, 2000), 145, 152. cf. Joseph. *BJ*, 2.383, 386.

[7] Key ports lie at sites such as Carthage, Hadrumentum, Hippo Regius, Leptis Magna, Leptis Minor, Oea, Rusicae, and Utica. On artificial harbors: David

Revolts, like foreign and civil wars, have the potential to strain grain supplies to the city of Rome.[8] Tacfarinas' wide-ranging raids and operations across the province disrupted the supply system from Africa to Rome. As Mattingly and Aldrete point out, major export regions, like Africa Proconsularis, had thousands of people involved in the growing of grain as well as "making amphorae, operating presses, transporting the agricultural products by land and river, overseeing production and transport, and shipping goods to Rome."[9] The raids would have resulted in crops, animals, and humans being carried off or destroyed by rebels and would have prevented or intimidated (some) affected locals from carrying out their routine tasks. While the province was not at risk of being overthrown by Tacfarinas, the security of Roman interests, especially grain, were at stake during the revolt. Furthermore, Rome's grain supply was possibly further hampered by Germanicus opening imperial storehouses in Egypt to resolve a local famine and maybe by continued brigandage in another grain producing province, Sardinia, at least intermittently since 6 CE.[10] Even though Germanicus' actions may not necessarily have hampered the grain supply to Rome in 19 CE, the limited availability of quick alternative sources of grain to meet the city's needs meant that prices in Rome soared as limited shipments of grain arrived, or even just fear that shipments would be reduced, from these two provinces after many years of disruptions (and possibly with increased costs associated with guaranteeing its arrival due to the raiding).[11]

Mattingly and Bruce Hitchner, "Roman Africa: An Archaeological Review," *Journal of Roman Studies* 85 (1995): 165-213; David Stone, "Africa in the Roman Empire: Connectivity, the Economy, and Artificial Port Structures," *American Journal of Archaeology* 118 no.4 (2014): 565-600; cf. Rickman 1980, 264.

[8] See esp. Peter Garnsey, *Famine and Food Supply in the Graeco-Roman World* (Cambridge University Press, 1988). Oftentimes in warfare, it is the concentration of troops in campaign areas that results in the diversion of grain and other supplies to the military originally destined for Rome (Garnsey 1988, 228-9). The impact of diverted grain to support the military in Africa, however, surely pales in comparison to the disruptions brought on by Tacfarinas' actions, since the additional forces sent to the region are not extensive.

[9] Mattingly and Aldrete 2000, 157.

[10] Tac. *Ann.* 2.59.1; Cass. Dio 55.28.1; cf. Str. 5.2.7. On Sardinia, see further below.

[11] cf. Morley 2010, 96.

4 - Tiberian Responses to Crisis

In response, Tiberius, and his court, took a multipronged approach to shore up the grain supply and to reduce tensions in the city. At the urban level, he set grain prices, subsidized grain merchants, and expelled Jews and followers of Isis from the city. To combat continued threats to the African grain supply, Tiberius sent units from Illyricum and Sardinia to Africa and conscripted expelled freedmen Jews of military age and sent them to Sardinia.[12] This reveals that the emperor sought to treat not just the symptoms, namely the ire at the grain price, but the underlying causes as well, specifically brigandage and disruptions in grain producing regions, which goes well beyond Tacitus' surface-level treatment of these episodes.

Let us examine these responses in turn. Since the state, *sic* emperor, was responsible for providing adequate quantity and reasonable prices of grain to the urban masses, Tiberius had to act in 19 CE. In one sense, Tiberius was responsible for providing monthly grain doles of five *modii* to roughly 200,000 male citizens in the city of Rome. In another sense, the emperor and appointed officials were meant to ensure that adequate supplies of foodstuffs reached Roman markets and at reasonable prices. Combined with Germanicus' illness and eventual death, the quickly rising prices in grain created tensions within the city that threatened order.[13]

It is possible that Tiberius' range of responses was limited by Germanicus opening the state reserves in Egypt in response to local famine, probably in January or February 19 CE (Tac. *Ann.* 2.59.1-2; cf. Suet. *Tib.* 52.2).[14] This raises some questions concerning Tiberius' ability to transfer grain from the state's stores in Egypt to alleviate problems in Rome later that year. It is entirely uncertain if the famine continued into the current grain season in Egypt or even if state reserves were depleted there by Germanicus. How much of an impact this situation had on Roman supply in 19 CE is uncertain, but imperfect information or rumor in the city would still have impacted prices. These limitations may explain Tiberius' ultimate effort to limit grain prices and subsidize merchants at some indeterminable date during the year. According to Rickman, Tiberius would have ordered the *praefectus annonae* to release local reserves and sell them for an artificially lower price, which would have forced the private market to follow suit. This alone

---

[12] Tac. *Ann.* 2.85, 2.87, 3.9; Suet. *Tib.* 36; Joseph. *AJ*, 18.3.5; *AE* 1929, 169; *CIL* 8.23252.

[13] Rickman 1980, 270; Mattingly and Aldrete 2000, 143, 146.

[14] Germanicus was in Egypt by January 25th, 19 CE (Sherk 1988, 34C).

would not have been enough to resolve the immediate issue because private sellers could either be ruined by this move or seek out alternative markets to gain a profit. So, to entice these private merchants to continue selling at Rome at this new lower price, Tiberius incentivized them with a special subsidy of two sesterces per modius (Tac. *Ann.* 2.87).[15] This move may also have further encouraged merchants to divert their shipments towards Rome and thus help secure Rome's immediate needs.

Next, presumably between the death of Germanicus in October and the end of the year, imperial decision makers sought to relieve tensions in the city by expelling Jews and followers of the cult of Isis from the city of Rome (Joseph. *AJ* 18.65-84; Tac. *Ann.* 2.85.4; Suet. *Tib.* 36.1).[16] In connection to this, Roman authorities conscripted 4,000 Jewish freedmen of military age, probably many of the same expelled Jews, and sent them to Sardinia (Tac. *Ann.* 2.85.4; Suet. *Tib.* 36.1; Joseph. *AJ* 18.84). There is little agreement on why the expulsions occurred as well as why these specific populations became targets.[17] This is not the place to delve into the intricacies of the debates, but I believe that this episode cannot be treated separately from the general tensions swelling in Rome brought on by the rising grain prices and Germanicus' illness and death. The news of his failing health, recovery, and then final decline in October of 19 CE would have reached a capital already distressed from grain issues and would have placed ever more stress on the imperial administration to relieve tensions.

Following Erich Gruen, I believe that the Jews and Isis followers were expelled as scapegoats to divert attention away from the pressure mounting on the Senate and Imperial Court. As Rutgers and Wendt have shown, Rome

---

[15] Rickman 1980, 270; Mattingly and Aldrete 2000, 145.

[16] On issue of chronology of this episode, see Erich Gruen, *Diaspora: Jews Amidst Greeks and Romans* (Cambridge, MA: Harvard University Press, 2002), 33.

[17] The literature is vast, and the following only represents a small sampling of viewpoints: Mary Smallwood, "Some Notes on the Jews under Tiberius," *Latomus* 15, no. 3 (1956): 314-329; Margaret Williams, "The Expulsion of the Jews from Rome in A.D. 19," *Latomus* 48 (1989): 765-784; Martin Goodman, *Mission and Conversion: Proselytizing in the Religious History of the Roman Empire* (Oxford: Clarendon, 1994); Leonard Victor Rutgers, "Roman Policy towards the Jews: Expulsions from the City of Rome during the First Century C.E.," *Classical Antiquity* 13, no. 1 (1994): 56-74; Gruen 2002, 29-34; David Woods, "Tiberius, Tacfarinas, and the Jews," *Arctos* 42 (2008): 267-284; Heidi Wendt, "Iudaica Romana: A Rereading of Judean Expulsions from Rome," *Journal of Ancient Judaism* 6 (2015): 97-126. See below on why the conscripts were probably sent to Sardinia.

frequently responded to urban crises by expelling various groups, and these two populations are of eastern origin and were known for their expert knowledge of divination and magic.[18] Given the rumors swirling of Germanicus' death being caused by magic, curses, and poisoning (Suet. *Calig*. 1.2, 3.3; Tac. *Ann*. 2.69.3) and the fact that Germanicus had been in Egypt and Syria just before his demise (Tac. *Ann*. 2.59-61, 69), these populations were perfect scapegoats for expulsion. The effectiveness of carrying out the expulsions does not matter in my opinion; it is the very act itself that does. The purpose of the act could simply be to buy time for the Roman administration to find solutions for the urban crises (short-term guaranteeing grain and long-term hitting the root cause of it, Tacfarinas) while simultaneously relieving tensions by giving the population targets for their frustrations.[19]

With the pressing matters inside the city now handled, Tiberius could turn his attentions towards Tacfarinas and Africa in the winter of 19/20. It is not certain how many units were transferred to Africa at this time, but it is possible to trace the movements of three units: *legio IX* and *cohors XV Voluntariorum* from Illyricum and a *cohors Lusitanorum* from Sardinia.[20] This would represent a minimum of roughly 6,000 soldiers transferred to the warfront. All three of these units had recent counterinsurgency experiences fighting rebels and brigands in the highlands of Sardinia and Illyricum. In 6 CE, Sardinia was so overrun by brigands that it required a change in governance (Cass. Dio 55.28.1; cf. Str. 5.2.7). From then until the reign of Nero, c.67 CE, Sardinia was an imperial province under the control of *praefecti Sardiniae* and even held a prefect delegated with command over the *civitatum Barbariae* in the mountainous interior of the island, accompanied by an auxiliary cohort c.14 CE (Cass. Dio 55.28.1; Paus. 7.17).[21] This is an obscure feature of imperial administration deployed in

[18] Rutgers 1994, 69, 74; Gruen 2002, 29-34; Wendt 2015.

[19] In this vein I follow Wendt (2015, 116-117) in disregarding Gruen's notion of the expulsion as an act to expiate the city out of concern for ancestral divinities. I do, however, generally favor his reconstruction.

[20] Paul Holder, *Studies in the Auxilia of the Roman Army from Augustus to Trajan* (Oxford: BAR Publishing, 1980), 64; Yann Le Bohec, *Le Sardaigne et l'armee romaine sous le Haut-Empire* (Sassari: C. Delfino 1990), 30; D.J. Knight, "The Movements of the Auxilia from Augustus to Hadrian," *Zeitschrift für Papyrologie und Epigraphik* 85 (1991): 190; Wolff 2014, 54-55. *Legio IX*: Tac. *Ann*. 3.9. *Cohors Lusitanorum*: *AE* 1929, 169. *Cohors XV Voluntariorum*: *CIL* 8.23252. On the chronology of the decisions for which units to transfer, see below.

[21] On these prefects, see Kreiner 2020, 47-48, 358-360; *CIL* 14.2954 = *ILS* 2684. Cf. presence of a *pro legato* in Sardinia who could have had command over legionaries

the first decades of the 1st century CE where *praefecti/legati provinciae* delegated command and administration of difficult regions (in terms of pacification and geography) to other prefects. Sardinia was an island with a long history of resistance, especially from the mountainous interior.[22] In the final decade of Augustus' reign, *cohors Lusitanorum* likely gained plenty of experience in irregular, highland, warfare, the sort of experiences beneficial to handling Tacfarinas' forces who would hit Africa's productive territories before disappearing back into the steppes and Tunisian Tell on the east and north sides of the Aures Mountains.

Since there were still perceived threats to Sardinia's stability, it makes sense that the 4,000 conscripted Jews were sent to that island in 19 or 20 CE. Even if, as some scholars assert, brigandage was only an intermittent problem on the island,[23] stripping the garrison so soon after bandits had posed issues for imperial control (and we cannot be sure that there were not more units transferred at this time) would have represented a risk and have presented an opportunity for brigands to raid coastal parts of the island. The movement of fresh recruits to replace units destined for a warfront is quite common in the Roman world, and I think this is the most logical explanation for why Tiberius conscripted some of the Jews exiled from the city of Rome. As Jonathan Roth and others have shown, Jews serving in the Roman military is a non-issue. Judaean Jews were explicitly exempt from military service, but nothing stopped Jews from volunteering for service nor for Jews outside of that province from being conscripted.[24] Furthermore, Sardinia could be utilized as another source for guaranteeing Rome's grain supply, and this is especially important given the issues Rome just experienced acquiring grain at reasonable prices in 19. The island had played an important role in the city's supply less than 100 years earlier, as such maintaining the garrison

---

or auxiliaries in 13/14 (*ILS* 105; Jaro Šašel, "Pro legato," *Chiron* 4 (1974): 467-478; Le Bohec 1990, 22.

[22] Raimondo Zucca, "Le *Civitates Barbariae* e l'occupazione militare della Sardegna: aspetti e confronti con l'Africa," *L'Africa Romana* 5 (1987): 349-373; Kreiner 2020, 31-57, 332-365 with literature.

[23] Such as Le Bohec 1990, 23; Woods 2008.

[24] On movement of fresh recruits: cf. D.B. Saddington, *The Development of the Roman Auxiliary Forces from Caesar to Vespasian: 49 B.C.-A.D. 79* (Harare: University of Zimbabwe Press, 1982), 167-168. Jews and Roman military service: Andrew Schoenfeld, "Sons of Israel in Caesar's Service: Jewish Soldiers in the Roman Military," *Shofar* 24, no. 3 (2006): 115-126; Jonathan Roth, "Jews and the Roman Army: Perceptions and Realities," in *The Impact of the Roman Army (200 BC – AD 476): Economic, Social, Political, Religious and Cultural Aspects*, eds. Lukas de Blois and Elio Lo Cascio (Leiden: Brill, 2007), 417-418.

size to stave off potential bandit lowland raids would have been pertinent to further secure grain supplies.[25]

Like the Sardinian force(s), *cohors XV Voluntariorum* and *legio IX* gained experience in counterinsurgency and taking rebel strongholds during the Great Illyrian Revolt of 6 to 9. This was a long and arduous campaign to end indigenous resistance quite like the Tacfarinian Revolt. After the first year of the revolt, during which the two rebel leaders, both named Bato, had unsuccessfully gone on the offensive to seize key strategic points and cut off Roman operational routes in the region, the remaining years of the revolt entailed rebel forces holding hillforts and strongpoints, leading raids against populations not in revolt, and to ambush Roman forces. This resulted in a difficult war for Tiberius in which he had to divide his forces to trap rebel cells and force them into pitched battles, to capture strongholds, and to wear down rebels by plundering their territories and causing famine.[26]

Even though units from the Danube region were ideal to handle the situation in Africa Proconsularis, recent geopolitical shifts along the Danube were the reason behind Tiberius' choice. Since 17 CE, the emperor's son Drusus had been on a special mission in Illyricum, in part to undermine Maroboduus' Marcomannic kingdom through diplomacy and statecraft. Concurrently, Maroboduus was at war with Arminius of the Cherusci for supremacy in Germania Magna, one which he would lose before the end of 18; Rome refused to come to the Marcomannian's aid as Maroboduus had done a decade earlier against Arminius (Tac. *Ann.* 2.44-46). The consequences of this war were massive for Roman security on the Rhine and Danube frontiers. Not only would the victor, Arminius, soon be assassinated by disgruntled tribesmen (Tac. *Ann.* 2.88), but Maroboduus, who had posed a major security concern to the Roman government for at least 15 years, would be overthrown by Catualda of the Gotones and would spend his remaining days in exile at Ravenna (Tac. *Ann.* 2.62-63).

[25] Rickman 1980, 263. Woods maintains that the conscripts were destined for Africa to deal with Tacfarinas. While he raises several interesting and salient points about the sources concerning the conscription of Jews and transferring them to Sardinia, the concerns outlined in the paragraph outweigh the likelihood of the fresh conscripts going to Africa and Wood's need to explain this idea away against the source material (Woods 2008, esp. 271, 277, 283-284).

[26] For recent narrative presentations of the revolt in English, see: Danijel Džino, *Illyricum in Roman Politics, 229 BC-AD 68* (Cambridge University Press, 2010), 149-153; Kreiner 2020, 259-263.

For the time being, in 19 CE, Drusus was able to ensure Catualda as a friendly king just north of Roman territory,[27] allowing for one of the Danubian legions to be temporarily transferred to Africa. While we do not know exactly where *Legio IX* was stationed at this time, it was most likely the legion closest to Italy, maybe in the region of Burnum. At any rate, we know that Drusus was in Italy in December of 19 around the time Tiberius ordered a cessation of public business on account of Germanicus' death (Tac. *Ann.* 3.2.3; *fasti ost.* = Ehrenberg and Jones 1955, 41), which had postponed his triumphal ovation. This would have given ample time for Tiberius and Drusus to discuss what forces to send to end the Tacfarinian Revolt.

Drusus then left Italy for Illyricum at some point between December 8th and April 4th to make arrangements for the transfers and to settle affairs in the region (Tac. *Ann.* 3.7.1).[28] He then must have returned to Italy before May 28th for his ovation (*CIL* 14.244, cf. Tac. *Ann.* 3.11.1). *Legio IX* also made its way through Italy sometime around late April to early May, 20 CE, where Piso ran into it near Ancona on his way to Rome (Tac. *Ann.* 3.9).[29] This provides us with a tight timeline for Tiberian responses to the crises of 19 CE, wherein most of his actions probably took place between October and the end of December (but perhaps extending into March, 20 CE). However, some of the responses, such as the setting of grain prices and subsidies to grain merchants, could have been enacted as early as March, 19 CE.

5 - Conclusion

To summarize, the year 19 CE offers an excellent opportunity to assess how the emperor Tiberius responded to crisis. Tiberius enacted swift and sweeping actions, some aimed at immediately relieving stress within the city and others aimed at ensuring future grain supplies. It is not entirely certain, however, that Tiberius would have sent reinforcements to Africa if Tacfarinas' raiding had not impacted grain prices at Rome. Africa was not

---

[27] Jenö Fitz, "La division de l'Illyricum," *Latomus* 48 (1988): 24-25; Frédéric Hurlet, *Les collègues du prince sous Auguste et Tibère. De la légalité républicaine à la légitimité dynastique* (Rome: École Française de Rome, 1997), 214-218.

[28] The emperor began the *iustitium* on December 8th according to the *fasti Ostienses* (*VI idus Dec. iustitium ob excessum G[er]manici*) and then ended it shortly before the Megalesian Games, which traditionally started at the latter date (Tac. *Ann.* 3.6.3).

[29] Christopher Mackay, ""Questiones Pisonianae": Procedural and Chronological Notes on the "S.C. de Cn. Pisone Patre"," *Harvard Studies in Classical Philology* 101 (2003): 358, 364, 368.

in danger of being lost and, in fact, *Legio IX* returned to Pannonia before Tacfarinas was killed in 24 CE (Tac. *Ann.* 4.23, 25), so reinforcements may not have been required in the first place.

Regardless, I wish to briefly conclude by hypothesizing why Tacitus does not connect these disparate events. It would be too easy to maintain that Tacitus may not have viewed these events as interrelated. Rather, I wonder if Tacitus had been at least partially aware of Tiberius' programmatic response to crisis in 19 CE and chose not to draw explicit attention to it because such a response was not consistent with his desired portrayal of the emperor. These events and the death of Germanicus in 19 CE are right at the cusp of ancient historians' divide in the presentation of Tiberius' rulership, namely that the emperor is portrayed in a very negative light in Tacitus' and Cassius Dio's accounts after the death of Germanicus.[30] Moreover, if Tacitus wants to present the emperor Tiberius in the worst possible light, he surely cannot present him as a responsive and capable administrator at that critical moment where he wants to emphasize the rivalry between the princeps and the prince. It is worth noting that Tacfarinas' Revolt (17-24 CE) also spans this critical chronological divide between Tiberius the capable administrator kept in check by his rival Germanicus (up to 19 CE) and the savagery of his reign following the rise of Sejanus (23 CE). Does Tacitus present the revolt sporadically just so he can make it appear like the emperor was neither taking it seriously nor had a clue on how to handle it properly? Just a thought.

Bibliography

Dyson, Stephen L. "Native Revolt Patterns in the Roman Empire." In *Aufstieg und Niedergang der römischen Welt: Geschichte und Kultur Roms im Spiegel der neueren Forschung*, edited by Joseph Vogt, Hildegard Temporini, and Wolfgang Haase, 2 pt.3, 137–175. Berlin: W. de Gruyter, 1975.

Džino, Danijel. *Illyricum in Roman Politics, 229 BC-AD 68*. Cambridge University Press, 2010.

Ehrenberg, Victor and A.H.M. Jones. *Documents Illustrating the Reigns of Augustus & Tiberius*. 2nd Edition. Oxford: Clarendon Press, 1955.

---

[30] cf. R.H. Martin and A.J. Woodman, *Tacitus: The Annals, Book IV* (Cambridge University Press, 1989), 80; C.T. Mallan, *Cassius Dio Roman History: Books 57 and 58 (The Reign of Tiberius)* (Oxford University Press, 2020), 245. In Tacitus, the negative presentation of Tiberius' reign truly begins in 23 CE with the rise of Sejanus, first introduced at *Ann.* 4.1.1.

Fitz, Jenö. “La division de l’Illyricum.” *Latomus* 48 (1988): 3-25.

Gambash, Gil. *Rome and Provincial Resistance*. New York: Routledge, 2015.

Garnsey, Peter. *Famine and Food Supply in the Graeco-Roman World*. Cambridge University Press, 1988.

Goodman, Martin. *Mission and Conversion: Proselytizing in the Religious History of the Roman Empire*. Oxford: Clarendon, 1994.

Gruen, Erich. *Diaspora: Jews Amidst Greeks and Romans*. Cambridge, MA: Harvard University Press, 2002.

Holder, Paul. *Studies in the Auxilia of the Roman Army from Augustus to Trajan*. BAR International Series 70. Oxford: B.A.R., 1980.

Hurlet, Frédéric. *Les collègues du prince sous Auguste et Tibère. De la légalité républicaine à la légitimité dynastique*. Rome: École Française de Rome, 1997.

—. “Rejeter le contrôle de Rome: les formes de résistance aux structures fiscales et administratives de l’Empire romain.” In *Économie et inégalité: ressources, échanges et pouvoir dans l’Antiquité classique: huit exposés suivis de discussions*, edited by Pascale Derron, 197–225. Entretiens sur l’Antiquité classique, LXIII. Vandoeuvres: Fondation Hardt pour l’étude de l’antiquité Classique, 2017.

Knight D.J. “The Movements of the Auxilia from Augustus to Hadrian.” *Zeitschrift für Papyrologie und Epigraphik* 85 (1991): 189–208.

Kreiner, Jared. “Win, Lose, or Draw: Roman Provincial Administration and Resistance in the Early Principate.” PhD diss., University of Chicago, 2020.

Le Bohec, Yann. *Les unités auxiliaires de l’armée romaine en Afrique proconsulaire et Numidie sous le Haut Empire*. Paris: Editions du Centre national de la recherche scientifique, 1989.

—. *Le Sardaigne et l’armee romaine sous le Haut-Empire*. Sassari: C. Delfino, 1990.

Mackay, Christopher S. ““Questiones Pisonianae”: Procedural and Chronological Notes on the “S.C. de Cn. Pisone Patre”.” *Harvard Studies in Classical Philology* 101 (2003): 311-370.

Mallan, C.T. *Cassius Dio Roman History: Books 57 and 58 (The Reign of Tiberius)*. Oxford University Press, 2020.

Marasco, G. "Tiberio e l'esilio degli Ebrei in Sardegna nel 19 d.C." *L'Africa Romana* 8 (1990): 649–659.

Martin, R.H. and A.J. Woodman. *Tacitus: The Annals, Book IV*. Cambridge University Press, 1989.

Mattingly, David J. and Greg Aldrete. “The Feeding of Imperial Rome: The Mechanics of the Food Supply System.” In *Ancient Rome: The*

Archaeology of the Eternal City, edited by J.C. Coulston and Hazel Dodge, 142-165. Oxford School of Archaeology, 2000.
Mattingly, David J. and Bruce Hitchner. "Roman Africa: An Archaeological Review." *Journal of Roman Studies* 85 (1995): 165-213.
Morley, Neville. *The Roman Empire: Roots of Imperialism*. London: Pluto Press, 2010.
Rickman, G.E. "The Grain Trade under the Roman Empire." *Memoirs of the American Academy in Rome* 36 (1980): 261-275.
Roth, Jonathon P. "Jews and the Roman Army: Perceptions and Realities." In *The Impact of the Roman Army (200 BC – AD 476): Economic, Social, Political, Religious and Cultural Aspects. Impact of Empire 6,* edited by Lukas de Blois and Elio Lo Cascio, 409-420. Leiden: Brill, 2007.
Rutgers, Leonard Victor. "Roman Policy towards the Jews: Expulsions from the City of Rome during the First Century C.E." *Classical Antiquity* 13, no. 1 (1994): 56-74.
Saddington, D.B. *The Development of the Roman Auxiliary Forces from Caesar to Vespasian: 49 B.C.-A.D. 79*. Harare: University of Zimbabwe Press, 1982.
Schoenfeld, Andrew J. "Sons of Israel in Caesar's Service: Jewish Soldiers in the Roman Military." *Shofar* 24, no. 3 (2006): 115-126.
Sherk, Robert K. (ed. and. trans.). *The Roman Empire: Augustus to Hadrian*. Cambridge University Press, 1988.
Smallwood, E. Mary. "Some Notes on the Jews under Tiberius." *Latomus* 15, no. 3 (1956): 314-329.
Stone, David L. "Africa in the Roman Empire: Connectivity, the Economy, and Artificial Port Structures." *American Journal of Archaeology* 118 no.4 (2014): 565-600.
Šašel, Jaro. "Pro legato." *Chiron* 4 (1974): 467–478.
Vanacker, Wouter. "Ties of Resistance and Cooperation. Aedemon, Lusius Quietus and the Baquates." *Mnemosyne* 66, no. 4–5 (2013): 708–733.
—. "Adhuc Tacfarinas: Causes of the Tiberian War in North Africa (AD Ca. 15-24) and the Impact of the Conflict on Roman Imperial Policy." *Historia: Zeitschrift für Alte Geschichte* 64, no. 3 (2015): 336–356.
Wendt, Heidi. "Iudaica Romana: A Rereading of Judean Expulsions from Rome." *Journal of Ancient Judaism* 6 (2015): 97-126.
Williams, Margaret H. "The Expulsion of the Jews from Rome in A.D. 19." *Latomus* 48 (1989): 765-784.
Wolff, Catherine. "La guerre de Tacfarinas." In *La guerre dans l'Afrique romaine sous le Haut-Empire: actes du CXXXVIe Congrès national des sociétés historiques et scientifiques "Faire la guerre, faire la paix",*

Perpignan, 2011, edited by Michèle Coltelloni-Trannoy and Yann Le Bohec, 53–68. Collection CTHS histoire, 54. Paris: Comité des travaux historiques et scientifiques, 2014.

Woods, David. "Tiberius, Tacfarinas, and the Jews." *Arctos* 42 (2008): 267-284.

Zucca, Raimondo. "Le *Civitates Barbariae* e l'occupazione militare della Sardegna: aspetti e confronti con l'Africa." *L'Africa Romana* 5 (1987): 349–373.

Neglected Iranian and Syriac Evidence on the Persian Campaigns of Gordian III and Galerius

Byron Waldron

In studying the Roman-Sasanian wars of the third century AD, one is left to work with precious few details. This is especially true of the campaigns of Gordian III and Galerius, which took place in the early 240s and late 290s respectively. However, in their reconstructions of both of these campaigns, scholars have neglected non-Roman historical traditions that help us to better understand the events that took place.

In this article, I aim to integrate these traditions into current reconstructions of the campaigns of Gordian and Galerius. I defend the historical value of these traditions and discuss how they change our understanding of the campaigns. Ultimately, this article proposes that future scholarship on the Roman-Sasanian wars fully employs the many non-Roman historical traditions in order to challenge existing Romano-centric narratives.[1]

1. Shapur I, Gordian III, and the Northern Rebellions, AD 241-242

The received narrative for Gordian's campaign is brief. The Sasanians renewed their attacks on Rome's eastern provinces during the reign of Maximinus I (r. 235-238). Through these attacks, the Sasanians took possession of Nisibis and Carrhae.[2] Whereas Nisibis was located among the

---

[1] I thank Graham Wrightson and Jared Kreiner for their feedback. I am grateful to Armin Mandehzadeh for providing sources and advice on the campaigns of Ardashir I and on the importance of Gundeshapur.

[2] *Hist. Aug., Gord.* 26.6, 27.6; Sync. 681 (*CSHB* 22); Zonar. 12.18; see also Hdn. 6.6.6; *Hist. Aug., Max. et Balb.* 13.5. Hauser prefers to date the capture of Nisibis and Carrhae to 240/1, the aftermath of the fall of Hatra: Stefan R. Hauser, "Where is the Man of Hadr, Who Once Built it and Taxed the Land by the Tigris and

easternmost territories of Roman-held northern Mesopotamia, Carrhae was in western Osrhoene.[3] This suggests that the Sasanians captured numerous other towns and fortresses between the two locations, and had therefore conquered much or most of Roman-held Upper Mesopotamia.[4] In 240 or 241, perhaps after a two-year siege, the Persians then captured the independent trade metropolis of Hatra, which was located in eastern Upper Mesopotamia and, having resisted Trajan and Septimius Severus in the past, may now have been receiving Roman protection.[5]

Gordian III (r. 238-244) had likely responded with emergency defensive measures, and in 239 he appears to have visited Antioch.[6] Relatedly, in 238/9 he reinstated the client kingdom of Osrhoene, which was centred on Edessa. Its new king, Aelius Septimius Abgar, belonged to the traditional Abgarid dynasty but ceased to rule by the middle of 242.[7] Regardless of any defensive efforts undertaken by Abgar, the loss of so much territory of strategic value demanded an imperial campaign.[8]

For what happened next, we rely heavily on Greek and Latin accounts, especially Gordian's biography in the *Historia Augusta* (early fifth century), which likely employs the third-century *Chronicle* of Dexippus for the

---

Chaboras? On the Significance of the Final Siege of Hatra," in *Hatra: Politics, Culture and Religion between Parthia and Rome*, Oriens et Occidens 21, ed. Lucinda Dirven (Stuttgart: Franz Steiner Verlag, 2013), 137-138.

[3] *Barrington Atlas* (henceforth *Barr.*) 89.

[4] David S. Potter, *Prophecy and History in the Crisis of the Roman Empire: A Historical Commentary on the Thirteenth Sibylline Oracle* (Oxford & New York: Oxford University Press, 1990), 192; Karin Mosig-Walburg, *Römer und Perser: vom 3. Jahrhunderts bis zum Jahr 363 n. Chr.* (Gutenberg: Computus, 2009), 29-30; Peter Edwell, *Rome and Persia at War: Imperial Competition and Contact, 193-363 CE* (London & New York: Routledge, 2021), 77.

[5] Cod. Manichaicus Coloniensis 18; al-Tabari, 827-830; Hauser, "On the Significance of the Final Siege of Hatra." Mosig-Walburg, *Römer und Perser*, 30-31, is sceptical of the hypothesis that a Roman garrison was present in Hatra.

[6] Steven K. Ross, *Roman Edessa: Politics and Culture on the Eastern Fringes of the Roman Empire, 114-242 CE* (London & New York: Routledge, 2001), 76-79; Lukas De Blois, *Image and Reality of Roman Imperial Power in the Third Century AD: The Impact of War* (London & New York: Routledge, 2019), 62.

[7] Ross, *Roman Edessa*, 69-82. The coins minted by Abgar X at Edessa are collated in *RPC* 7.2, 3404-3425. He was granted the status of a consular governor (Ross 79-81) and may have originally come to prominence defending his city against the Sasanians.

[8] Potter, *Prophecy and History*, 189-192; Mosig-Walburg, *Römer und Perser*, 31-34; De Blois, *Image and Reality*, 62; Edwell, *Rome and Persia*, 74-77.

relevant chronological passages.[9] Under the year 241, the *Historia Augusta* claims that war broke out with the Persians and that Gordian made the experienced officer Timesitheus his father-in-law and praetorian prefect (*Gord.* 23.5-6). The former appears to reflect an official decision to campaign against the Persians. The *Historia Augusta* then records that in 242 Gordian opened the gates of Janus and marched by land to the east, and that in the same year he won repeated battles, drove the Persian king Shapur I (r. 240-271) out of Roman territory, and afterwards retook Carrhae and Nisibis (26.3-27.3). This is supported by other sources,[10] and without providing a year, the fourth-century historian Ammianus notes that Gordian defeated Shapur in a battle near Resaina (23.5.17), which was located on the route between Carrhae and Nisibis.[11]

As David S. Potter argues, it is likely that Timesitheus conducted these initial campaigns, with Gordian only reaching Mesopotamia towards the end of the year. This is suggested by a) the testimony of an apocalypse of Elijah, b) the movements of the emperor by land (as attested by literary and numismatic evidence), and c) epigraphic evidence for the presence of Roman fleets, which presumably were ferrying other forces to the east independent of Gordian's movements.[12] However, in 243 Timesitheus died of illness,[13] and in 244 Gordian marched into Sasanian-held Lower Mesopotamia, where, as the Sasanians record, Shapur defeated him in the

[9] Timothy D. Barnes, *The Sources of the Historia Augusta*, Collection Latomus 155 (Brussels: Latomus Revue d'Études Latines, 1978), 109-110; Potter, *Prophecy and History*, 192-193.

[10] These are collected in Michael H. Dodgeon & Samuel N. C. Lieu (eds.), *The Roman Eastern Frontier and the Persian Wars AD 226-363: A Documentary History* (London & New York: Routledge, 1991), 36-45. Gordian's marriage: Zos. 1.17.2; Zonar. 12.18. The recapture of Carrhae and Nisibis: Sync. 681 (*CSHB* 22); Zonar. 12.18. The Latin epitomes, likely derived from Victor's lost *Historia*, claim that Gordian won great battles (Eutr. 9.2; Festus, *Brev.* 22; cf. Aur. Vict. *HAb* 27.8).

[11] *Barr.* 89. Zosimus similarly records that the Romans won a first engagement (1.18.2). *Hist. Aug.*, *Gord.* 26.5-6, 27.5 claims that Gordian recovered Antioch, which supposedly had been under Persian rule, but this is generally considered to be a fabrication (Edwell, *Rome and Persia*, 77), and no other source mentions this.

[12] Potter, *Prophecy and History*, 192-194. On Gordian's journey east, see also Erich Kettenhofen, *Die römisch-persischen Kriege des 3. Jahrhunderts n. Chr.: nach der Inschrift Šāhpuhrs I. an der Ka'be-ye Zartošt (ŠKZ)*, Beihefte zum Tubinger Atlas des Vorderen Orients, Reihe B, Geisteswissenschaften, no. 55 (Wiesbaden: Ludwig Reichert, 1982), 23-27; Katrin Herrmann, *Gordian III. Kaiser einer Umbruchszeit* (Speyer: Kartoffeldruck-Verlag, 2013), 145-161.

[13] Zosimus claims that this diminished Gordian's confidence (1.18.2).

battle of Misiche.[14] Although fourth-century Roman accounts claim that Gordian was slain by his own men, Sasanian official media and a Greek historical tradition preserved in Byzantine sources relate that Gordian was killed or mortally wounded in the battle.[15]

All modern accounts of the war fought between Gordian and Shapur follow the broad outline presented above,[16] and aside from Sasanian testimony on Misiche and the death of Gordian, the Sasanian perspective figures little in this reconstruction. However, a key event in Sasanian history should be taken into account when considering this war, namely, the death of Ardashir I, their first king. In 224 Ardashir overthrew the Parthian Arsacid Dynasty, thereby establishing the Sasanian Dynasty, but on April 17/18, 240, his son Shapur became king or joint king,[17] and Zosimus notes that Shapur had succeeded his father by the time Gordian decided to go to war (1.18.1). Ardashir appears to have died soon after his son's accession. He is last attested in 240 or early 241 in connection with the capture of Hatra, and only Shapur is attested during the Roman campaign. A badly damaged passage of the Codex Manichaicus Coloniensis (164), which employs third-century sources, appears to indicate that Shapur mourned Ardashir when the prophet Mani was aged 25 (April 14, 241 – April 13, 242).[18] Moreover, it

---

[14] *ŠKZ* 7-8 (Maricq). This is supported by the less specific testimony of the inscription of Abnon (D. N. Mackenzie, "The Fire Altar of Happy *Frayosh," *Bulletin of the Asia Institute*, New Series 7 [1993]: 105-109), and by a rock relief at Bishapur (Beate Dignas & Engelbert Winter, *Rome and Persia in Late Antiquity: Neighbours and Rivals* [Cambridge & New York: Cambridge University Press, 2007], 78 fig. 6).

[15] David MacDonald, "The Death of Gordian III: Another Tradition," *Historia* 30, no. 4 (1981); Potter, *Prophecy and History*, 201-212; Herrmann, *Gordian III.*, 164-169. MacDonald persuasively argues that Gordian was mortally wounded in battle. On the treaty that followed, which favored the Persians, see Potter, *Prophecy and History*, 221-225; Mosig-Walburg, *Römer und Perser*, 34-43.

[16] E.g. Kettenhofen, *römisch-persischen Kriege*, 19-37; Potter, *Prophecy and History*, 189-194; Ulrich Huttner, "Von Maximinus Thrax bis Aemilianus," in *Die Zeit der Soldatenkaiser: Krise und Transformation des Römischen Reiches im 3. Jahrhundert n. Chr. (235-284)*, ed. Klaus-Peter Johne in assoc. with Udo Hartmann & Thomas Gerhardt (Berlin: Akademie Verlag, 2008), 1.184-188; Touraj Daryaee, *Sasanian Persia: The Rise and Fall of an Empire* (London & New York: I.B. Tauris, 2009), 6-7; Mosig-Walburg, *Römer und Perser*, 31-34; Herrmann, *Gordian III.*, 132-170; De Blois, *Image and Reality*, 62-63; Edwell, *Rome and Persia*, 74-78.

[17] Cod. Manichaicus Coloniensis 18.

[18] A. Henrichs & L. Koenen, "Der Kölner Mani-Kodex (P.Colon. Inv.Nr. 4780) Περὶ τῆϲ γέννηϲ τοῦ ϲώματοϲ αὐτοῦ: Edition der Seiten 121-192," *ZPE* 48 (1982): 4-5, 44.

seems relevant that Iranian sources are divided on the length of Shapur's reign. The Arabic-Iranian historian al-Tabari notes that his sources either claim that he reigned for 30 years and 15 days, or for 31 years, 6 months and 19 days (831). While the number of days and months diverges from this dichotomy in the other surviving sources, most of the sources can be divided between those who claim a little more than 30 years and those who record 31 and a half years.[19] I am tempted to suggest that the latter represents Shapur's total reign, whereas the former is Shapur's sole reign, or his reign following the death of his father, in which case Ardashir died around October 241.[20]

It had been less than eighteen years since Ardashir had founded the dynasty, and one can expect such a young regime to have faced challenges during the subsequent change in leadership. In fact, an often-overlooked Syriac source, the *Chronicle of Arbela*, provides evidence that the challenges were quite serious. This chronicle records the history of Christianity within the region of Adiabene in north-eastern Mesopotamia. It completes its narrative in the sixth century, and internal evidence indicates that the original version of the *Chronicle* was written before c. 569.[21] It thus appears to reflect a historical tradition that was developed within the Sasanian Empire. Crucially, it records the following on Shapur's accession (9):

> At this time Ardashir, the king of the Persians, died. And Shapur followed him. This man was very hard in his nature. And in the first year he had a war with the Chorasmians and the Medes of the mountains, and in one hefty stroke he overwhelmed them. And from there he pushed and subjugated the Gilans and the Dailamites and the Gurganians, which live in the distant mountains in the proximity of the rearmost sea [the Caspian Sea].[22]

---

[19] The sources are collected in S. H. Taqizadeh, "The Early Sasanians: Some Chronological Points Which Possibly Call for Revision," *Bulletin of the School of Oriental and African Studies, University of London* 11, no. 1 (1943): 17-19.
[20] On the chronological questions surrounding the reigns of Ardashir and Shapur, see also Taqizadeh, "Early Sasanians"; Werner Sundermann, "Shapur's Coronation: The Evidence of the Cologne Mani Codex Reconsidered and Compared with Other Texts," *Bulletin of the Asia Institute*, New Series 4 (1990): 295-299; Dodgeon & Lieu, *Roman Eastern Frontier*, 354-355 n. 3; Lutz Richter-Bernburg, "Mani's Dodecads and Sasanian Chronology: Kephalaia, Shāpūragān, and Codex Manichaicus Coloniensis," *ZPE* 95 (1993): 71-80.
[21] Omert J. Schrier, "Syriac Evidence for the Roman-Persian War of 421-422," *Greek, Roman and Byzantine Studies* 33, no. 1 (1992): 83 n. 27.
[22] Peter Kawerau (ed.), *Die Chronik von Arbela*, *CSCO* 467-468, Scriptores Syri 199-200, 2 Vols. (Leuven: Peeters Publishers, 1985), 55-56. Translation by Timothy

Unfortunately, modern accounts of Gordian's war have consistently overlooked the *Chronicle of Arbela*, or they have sidelined its testimony. Admittedly, the *Chronicle* is a problematic source. First published in 1907 by the Orientalist Alphonse Mingana, discrepancies exist between the only surviving manuscript and Mingana's printed version, and it is apparent that the manuscript is a modern copy.[23] Indeed, in 1967 Jean Maurice Fiey argued that the *Chronicle* was a forgery by Mingana.[24] Fiey relied on insinuation rather than conclusive evidence,[25] but these issues have cast a shadow over perceptions of the *Chronicle*. In their sourcebook on the Roman-Sasanian wars of the third and fourth centuries, Michael H. Dodgeon and Samuel N. C. Lieu include a single passage of the *Chronicle*, which they confine to an endnote on Ardashir's overthrow of the Arsacids, noting that the text must be treated with extreme caution.[26] Similarly, in his discussion of Gordian's war, Erich Kettenhofen notes the existence of the *Chronicle* in a footnote and expresses concern over its credibility.[27]

However, already in 1968 Sebastian Brock demonstrated that key arguments for a modern forgery were invalid,[28] and while there are strange

---

Kroll, *The Chronicle of Arbela*, *CSCO* 468, Scriptores Syri 200 (Leuven: Peeters Publishers, 1985), 17-18, with alterations.

[23] Julius Assfalg, "Zur Textüberlieferung der Chronik von Arbela: Beobachtungen zu Ms.or.fol.3126," *Oriens Christianus* 50 (1966): 19-36.

[24] Jean Maurice Fiey, "Auteur et date de la Chronique d'Arbèles," *L'orient syrien* 12 (1967): 265-302; see also Joel Thomas Walker, *The Legend of Mar Qardagh: Narrative and Christian Heroism in Late Antique Iraq*, The Transformation of Classical Heritage 40 (Berkeley, Los Angeles & London: University of California Press, 2006), 287-290.

[25] Sebastian Brock, "Syriac Historical Writing: A Survey of the Main Sources," *Journal of the Iraqi Academy Syriac Corporation* 5 (1979/80): 24.

[26] Dodgeon & Lieu, *Roman Eastern Frontier*, 349-350 n. 3.

[27] Kettenhofen, *römisch-persischen Kriege*, 24 n. 18. In narrating Gordian's campaign, Frye mentions the testimony of the *Chronicle*, but he does not explore the possible connection, simply noting that "In the meantime, Shapur had been busy, according to the *Chronicle of Arbela*, in subduing the Khwarazmians and the people of Gilan on the Caspian Sea coast.": Richard N. Frye, "The Political History of Iran under the Sasanians," in *The Cambridge Iranian History 3(1): The Seleucid, Parthian and Sasanian Periods*, ed. Ehsan Yarshater (Cambridge: Cambridge University Press, 1983), 125.

[28] Sebastian Brock, "Review of G. Wiessner, *Untersuchungen zur syrischen Literaturgeschichte 1: Zur Märtyrerüberlieferung aus der Christenverfolgung Schapurs II.* Abhandlungen der Akademie der Wissenschaften in Göttingen, philologisch-historische Klasse, Dritte Folge 67 (Göttingen, 1967)," *JThS* 19 (1968): 308 n. 1.

omissions in the manuscript, it might be that the text was abbreviated during the Middle Ages.[29] The *Chronicle* also contains a considerable amount of historically reliable data,[30] and most notably, the date ascribed by the *Chronicle* to the battle of Hormozdgan (Wednesday, April 27, 224)[31] is very close to the correct date supplied by an inscription (April 28, 224, a Wednesday), which was found in 1935 after the *Chronicle*'s publication. Indeed, although Theodor Nöldeke had already argued for April 28 in 1879 on the basis of al-Tabari (818),[32] the *Chronicle* did not adopt Nöldeke's date, as one would expect from a forgery, but instead supplied a date that was close but not correct. As Omert J. Schrier notes, this led Mingana to attempt to re-date the battle from 224 to 225 so that it could still fall on a Wednesday,[33] which in no way accords with the work of a forger.[34] Ultimately, the manuscript cannot be a forgery.

Let us return to the *Chronicle*'s account of Shapur's accession. It notes that Ardashir died and that Shapur succeeded him, and it then relates that in the first year Shapur had to campaign against Chorasmians, Gilans, Dailamites and Gurganians. The Chorasmians resided north of the empire in the Oxus River Delta, and the other three peoples were subjects of the Sasanians and located south and east of the Caspian. The *Chronicle* implies a Chorasmian incursion and a major set of rebellions, with the Chorasmians assisting the rebels. Shapur defeated a combined force of Chorasmians and 'Medes of the mountains' (perhaps a catch-all term for the rebels), and then afterwards subjugated the Gilans, Dailamites and Gurganians.

It is not immediately clear to when these campaigns should be dated, since, as we have seen, from 240 to possibly 241 Shapur was king during his

---

[29] Schrier, "Syriac Evidence," 83 n. 26.

[30] See the commentaries of Kawerau, *Chronik von Arbela*, and Ilaria Ramelli, *Il Chronicon di Arbela: presentazione, traduzione e note essenziali*, Ilu. Anejo 8 (Madrid: Universidad Complutense de Madrid, 2002), as well as Christelle Jullien & Florence Jullien, "La *Chronique d'Arbèles*: Propositions pour la fin d'une controverse," *Oriens Christianus* 85 (2001): 41-83; cf. Erich Kettenhofen, "Die Chronik von Arbela in der Sicht der Althistorie," in *Simblos: Scritti di storia antica*, ed. Lucia Criscuolo (Bologna: CLUEB, 1995), 287-319.

[31] Kawerau, *Chronik von Arbela*, 50.

[32] Theodor Nöldeke, *Geschichte der Perser und Araber zur Zeit der Sasaniden: Aus der arabischen Chronik des Tabari übersetzt und mit ausführlichen Erläuterungen und Ergänzungen versehn* (Leiden: Brill, 1879), 409-411.

[33] Alphonse Mingana, *Sources syriaques* 1 (Mosul: Otto Harrassowitz, 1907), 109 n. 1.

[34] Schrier, "Syriac Evidence," 82 n. 26.

father's lifetime and was potentially joint ruler. However, it is most likely that the *Chronicle* refers to the first year of the period following Ardashir's death, at which time Shapur probably received the title of King of Kings.[35] Not only would this have presented the clearest opportunity for enemies and rebels to challenge Sasanian rule, but the *Chronicle* implies the same sequence, with the 'first year' and its events following the death of Ardashir.

In terms of historicity, the *Chronicle*'s testimony fits well with the claims of other sources, especially Iranian sources.[36] Specifically, when we consider the Sasanians of the third century and their relations with northern vassals and neighbours, we find considerable support for the *Chronicle*'s account. The Chorasmians were indeed a problem for the early Sasanians. Al-Tabari records that Ardashir, in a grand campaign across eastern Iran, fought against the Chorasmians and other enemies, perhaps in the late 220s or early 230s (819). Although writing in the ninth century, al-Tabari employed Sasanian historical traditions.[37] Relatedly, the Greek historian Herodian, a contemporary of Ardashir, narrates that the king subdued and imposed tribute on neighbouring barbarian nations (6.2.1, 7). The Abbasid-era *Kitab al-Taj* includes an anecdote where Khosrow I narrates a battle between Ardashir and the king of the 'Khazars' (Pellat 84). Although anachronistic for the third century, the story may preserve a muddled memory of war with the Chorasmians, considering the similar geographic location of the Khazars and the phonetic points of connection.

Moreover, despite having defeated and killed the Parthian king Artabanus IV in 224, Ardashir had had to spend much of the later 220s fighting

---

[35] On the question of when Shapur received the title King of Kings, see Sundermann, "Shapur's Coronation," 296.

[36] On the historicity of this passage, see also Jullien & Jullien, "*Chronique d'Arbèles*," 62; Ilaria Ramelli, "Il Chronicon di Arbela: Una Messa a Punto Storiografica," *Aevum* 80, no. 1 (2006): 156-157; cf. Kettenhofen, "Chronik von Arbela," 303-307. Kettenhofen (306) suggests that the passage is contradicted by the fact that Ardashir, as described in other sources, had already conquered these peoples. However, it would be surprising if the young Sasanian regime had not fought rebellions and reconquered territories in its early history, and as we will see below, the sources describe Ardashir's campaigns differently, with the Dailamites and Gilans, for example, conquered through diplomacy.

[37] On al-Tabari's *Tarikh*, see Ehsan Yarshater (gen. ed.), *The History of Al-Tabari*, 40 Vols. (Albany, NY: SUNY Press, 1985-2007); Hakan Rydving (ed.), *Al-Tabari's History: Interpretations and Challenges* (Uppsala: Uppsala University Publishing, 2007); Hugh N. Kennedy (ed.), *Al-Tabari: A Medieval Muslim Historian and His Work* (Princeton, NJ: Gerlach Press, 2008).

campaigns against Parthian vassals and Arsacid loyalists. Notably, he led a campaign into Media, where he conquered Hamadan and 'the rest of the mountain-land' (al-Tabari, 819). In what appears to be a reference to this Median campaign, the late Sasanian *Book of the Deeds of Ardashir* (5) claims that Ardashir even suffered a defeat to 'Madig, king of the Kurds', but that he then launched a night attack on Madig's camp and captured the king. In the tenth-century *Tarikh-i Qum* of Hasan ibn Muhammad Qumi (Tehrani 70), it is narrated that Ardashir crushed the kings of Hamadan and Ray, implicitly in two separate battles. It is plausible that the *Chronicle*'s Gilans, located on the Caspian's southern coast, and the Dailamites of the Alborz,[38] were involved in the Median resistance to Ardashir. Indeed, according to the *Book of the Deeds of Ardashir* (4.14), in the lead-up to his clash with Ardashir, Artabanus had acquired soldiers and provisions from 'Ray, Damavand, Dailaman and Patashkhvargar'. The late Sasanian *Letter of Tansar* claims that Gilan, Dailaman, Ruyan and Damavand belonged to King Goshnasp of Tabaristan and Pareshvar (5), who submitted to Ardashir only after careful consideration, and retained his realm through Ardashir's personal guarantee. Shapur eventually made one of his sons, Bahram, the king of Gilan (*ŠKZ* 47 [Maricq]), thus strengthening Sasanian control over the area. The region again posed a threat to Sasanian control in the late third century, for a panegyric delivered to Maximian in 291 claims that the Sasanian king Bahram II was currently facing a rebel coalition of Sakas, Kushano-Sasanians, and notably, Gilans, under the leadership of the king's brother Hormizd of Sakastan (*Pan. Lat.* 11[3].17.2).

The Dailamites were formidable in war. Sixth-century Roman authors note that they were ultimately able to preserve their autonomy within the Sasanian Empire, exploiting the mountainous terrain of Dailaman.[39] The historian Agathias praises Dailamite heavy infantry for their discipline, coordination, and flexibility in combat, and he and his contemporary Procopius note that they thrived in hilly terrain, moving with speed and seizing points of vantage (Procop. *Bell.* 8.14.7-9; Agath. 3.17.7-9).[40]

As for the *Chronicle*'s Gurganians, located on the south-eastern shore of the Caspian,[41] Ardashir marched through Sakastan, Gurgan, Abarshahr,

---

[38] *Barr.* 90.
[39] Procop. *Bell.* 8.14.6-7; Agath. 3.17.9.
[40] On the Dailamites, see also Wolfgang Felix & Wilferd Madelung, "Deylamites," *Encyclopaedia Iranica* 7, no. 4 (1995/2011): 342-347. They remained virtually unsubdued until the ninth century.
[41] *Barr.* 96.

Margiana and Kushan-held Bactria during the aforementioned eastern campaign. He reportedly killed many and sent their heads to the fire-temple of Anahita (al-Tabari, 819).[42]

The *Chronicle* thus helps to build a picture of the first Sasanians and their insecure hold over the northern territories. Its testimony indicates that in 241 and 242, the peoples to the south and east of the Caspian Sea used the death of Ardashir to rebel against his successor, Shapur. The Sasanian regime was still relatively fragile. Ardashir had been a highly successful military campaigner, but it had yet to be proven whether the Sasanian Empire could be more than a one-man operation centred around the dynasty's founder.[43] The fact that the Arsacids in Armenia had resisted Ardashir and still reigned over the Armenian kingdom, not far from these northern rebels, would have made the need to crush these rebellions all the more urgent.[44] As the *Chronicle* implies, Shapur had to deal with these rebellions and the Chorasmian incursion before he could campaign elsewhere.

Shapur's northern campaigns contextualize the Roman campaign of 242. Undoubtedly, a campaign against the Sasanians had been on Rome's to-do list since Sasanian attacks began anew during the reign of Maximinus. Indeed, by the beginning of 241, Gordian had installed a *dux* in Egypt, Domitius Philippus, who retained his existing post as *praefectus uigilum* but may have also taken over the functions of the Prefect of Egypt. Notably, Philippus remained in office only until the middle of 242, when the

[42] Cass. Dio 80.3.2-3 claims that Ardashir subjugated much of Media and Parthia through force and intimidation. The satrapy of Parthia, the homeland of the Arsacids, was likely invaded during the same eastern campaign. Relatedly, Al-Ya'qubi, 179 notes that Ardashir conquered some of the districts of Khorasan.

[43] Ardashir's campaigns against the Arsacids, their allies, and the Romans: Dodgeon & Lieu, *Roman Eastern Frontier*, 10-33, 276-280. Successful campaigns in East Arabia: V. Fiorani Piacentini, "Ardashir I Papakan and the Wars against the Arabs: Working Hypothesis on the Sasanian Hold of the Gulf," *Proceedings of the Seminar for Arabian Studies* 15 (1985), 57-77. A raid on the Punjab: Vincent A. Smith, "Invasion of the Panjab by Ardashir Papakan (Babakan), the First Sasanian King of Persia, A.D. 226-41," *Journal of the Royal Asiatic Society of Great Britain and Ireland* 2 (1920): 221-226. The conquest of the Kushan Empire by Ardashir and Shapur: Martha L. Carter, "A Numismatic Reconstruction of Kushano-Sasanian History," *Museum Notes (American Numismatic Society)* 30 (1985): 266-271.

[44] The resistance of Armenia: Cass. Dio 80.3.3; Agathangelos, *Hist. Arm.* 19-21; Moses Khorenats'i, *Hist. Arm.* 71-73.

campaign was underway.[45] Similarly, in 231 Severus Alexander appointed a *dux* in Egypt in the lead-up to his own campaign against the Sasanians.[46] In the case of these ad hoc appointments, the *dux* appears to have been charged with making military preparations.[47]

However, in 242 the praetorian prefect Timesitheus began the campaign without waiting for the emperor's arrival and proceeded to win engagements and recapture cities. One wonders whether Timesitheus had become aware of the rebellions and incursions, and tailored operations to exploit the opportunity. After all, this strategy would be utilized later in the same century. In 283 the emperor Carus sacked Ctesiphon while Bahram II was dealing with rebellions further east, and in the late 280s Diocletian installed Tiridates III as king likely over the western part of Armenia and secured a favorable treaty with Bahram.[48] The aforementioned panegyrical reference to rebellions in 291 suggests that between 283 and 291 Bahram's position was still insecure, and that Diocletian knew to exploit this. Relatedly, Odaenathus campaigned with success against the Sasanians during the mid-260s. Shapur does not appear to have personally contended against Odaenathus in Upper or Lower Mesopotamia, and indeed, around the same time, he was likely expanding his dominion into Kushan-held Gandhara.[49]

Tellingly, Shapur's inscription on the Ka'ba-i Zardost, which focuses on Shapur's wars with Rome, implies a connection between his accession and Gordian's campaign (6-7 [Maricq]): 'And when I was first established over the dominion of the nations, the Caesar Gordian… raised an army and marched against Assyria, against the nation of the Aryans (i.e. Iranians) and

---

[45] References listed in Caillan Davenport, *A History of the Roman Equestrian Order* (Cambridge & New York: Cambridge University Press, 2019), 529 n. 233.
[46] References listed in Davenport, *Equestrian Order*, 528 n. 232.
[47] Potter, *Prophecy and History*, 193; Davenport, *Equestrian Order*, 528-529.
[48] Internal strife in 283: Eutr. 9.18; *Hist. Aug., Car.* 8.1; see also Festus, *Brev.* 24. Armenia: *P.Sakaon* 59.2–4; Agathangelos, *Hist. Arm.* 46–7; Moses Khorenats'i, *Hist. Arm.* 82; Daryaee, *Sasanian Persia*, 12; Edwell, *Rome and Persia*, 134–135. Treaty: *Pan. Lat.* 10(2).7.5–6, 9.1–3, 10.6–7; Wolfgang Felix, *Antike literarische Quellen zur Aussenpolitik des Sasanidenstates 1: 224-309* (Vienna: Verlag der Österreichischen Akademie der Wissenschaften, 1985), 105–107; Dignas & Winter, *Rome and Persia*, 26–27; Bill Leadbetter, *Galerius and the Will of Diocletian* (London & New York: Routledge, 2009), 88–89.
[49] On the numismatic and epigraphic evidence for the conquest of Gandhara, see Carter, "Numismatic Reconstruction," 231-237, 271. On Odaenathus' campaigns, see the sources collected in Dodgeon & Lieu, *Roman Eastern Frontier*, 68-83.

against us.'[50] In reality, Shapur had been king since April 240, more than a year and a half before Timesitheus' offensive. But Ardashir's death came later, perhaps in 241, at which point Shapur became the one man at the helm of the Sasanian Empire. This may be the reason why Gordian chose 241 to begin preparing for war. Rebellions and incursions followed, testing the new king, and Timesitheus may have launched the attack ahead of schedule to put pressure on the Sasanian regime when it was at its most vulnerable.

In any case, almost certainly Shapur's efforts against Timesitheus were adversely affected by the crises around the Caspian. Shapur is now remembered for his later successes against the Romans, best exemplified by the battles of Misiche, Barbalissos and Edessa, the siege of Dura Europos, and the two sackings of Antioch.[51] However, at Resaina in 242 Shapur suffered a rare defeat to the Romans. Shapur's northern campaigns were probably hard fought. The king had to campaign through the mountains of the Alborz and battle formidable peoples such as the Dailamites. Shapur's army may have suffered appreciable casualties against his northern enemies, and yet soon after, the king had to divert his attention to Mesopotamia to protect his father's conquests from the Romans. By the time he faced Timesitheus in battle, he had to fight the Romans on the prefect's terms and with an exhausted army.

2. Galerius and the Siege of Gundeshapur, AD 298

Let us now turn to the campaign of Galerius (r. 293-311), to consider another example where a non-Roman historical tradition may expand our understanding of the fighting that took place. In c. 296 the Sasanian king Narseh (r. 293-302) invaded Roman-held Upper Mesopotamia and Arsacid-held western Armenia. However, in late 297 the Caesar Galerius launched a surprise attack on Narseh's camp, at the time located in western Armenia, and captured the wives, sisters and daughters of the king as well as the royal treasury and many of the nobility.[52]

---

[50] Κ[α]ὶ ὅ[τ]ε πρώτως ἐπὶ τὴν βασιλίαν τῶν ἐθνῶν ἔστημεν, Γορδιανὸς Καῖσαρ... [δύναμιν συνέλεξ]εν καὶ εἰ]ς [τὴ]ν Ἀ[σσ]υρίαν ἐ[πὶ τὸ] τῶν Ἀριανῶν ἔθνος καὶ ἡμᾶς ἐπῆλθεν· Trans. Dodgeon & Lieu, *Roman Eastern Frontier*, 35. Kettenhofen, *römisch-persischen Kriege*, 23-24 n. 18, notes the implied chronological connection.

[51] The sources for his campaigns are collected in Dodgeon & Lieu, *Roman Eastern Frontier*, 34-67.

[52] The sources for this campaign are collected in Dodgeon & Lieu, *Roman Eastern Frontier*, 124-131. For the date of Galerius' victory, see *Des. Cons.* 34 (Becker);

Galerius then used his victory to invade Sasanian lands. The fourth-century epitomator Eutropius claims that Galerius forced Narseh to take refuge in the remotest deserts of his dominions (9.25), and the Byzantine historian Zonaras narrates that Galerius pursued the wounded Narseh as far as the interior of Persia (12.31).[53] No surviving source provides a narrative of this invasion, but Galerius' *cognomina* included Persicus, Armenicus, Medicus and Adiabenicus. Scholars have demonstrated that Tetrarchic *cognomina deuictarum gentium* consistently reflected actual campaigns led by the Roman emperors themselves, and they were listed in the order that they were adopted.[54] Therefore, we can deduce that Galerius campaigned in Media Atropatene and Adiabene.[55] Since Adiabenicus was consistently listed after Medicus, this implies that the Romans campaigned in Atropatene first.

Chron. Pasch. 512, 513; Timothy D. Barnes, "Imperial Campaigns, A.D. 285-311," *Phoenix* 30 (1976): 180-186; Constantin Zuckerman, "Les campagnes des tétrarques, 296-298: Notes de chronologie," *AnTard* 2 (1994): 65-66, 68-70; Timothy D. Barnes, "Emperors, Panegyrics, Prefects, Provinces and Palaces (284-317)," *JRA* 9 (1996): 543-544; cf. Frank Kolb, "Zu chronologischen Problemen der ersten Tetrarchie," *Eos* 76 (1988): 115-124; Frank Kolb, "Chronologie und Ideologie der Tetrarchie," *AnTard* 3 (1995): 23-27. The fifth-century Armenian historian P'awstos Buzand locates the battle near the village of Oskha in the district of Basean (*Hist. Arm.* 72).

[53] There is also Peter the Patrician's statement that the scattered forces of Narseh regrouped 'around the Asprudis River of the Median' (fragm. 14 [*FHG* 4, p. 189]); but note that this is a problematic statement (Thomas M. Banchich, *The Lost History of Peter the Patrician: An Account of Rome's Imperial Past from the Age of Justinian* [London & New York: Routledge, 2015], 136-137).

[54] Almost every *cognomen* can be linked to independently attested campaigns in the field: Barnes, "Imperial Campaigns"; Timothy D. Barnes, *The New Empire of Diocletian and Constantine* (Cambridge, MA & London: Harvard University Press, 1982), 27; Simon Corcoran, "Galerius, Maximinus and the Titulature of the Third Tetrarchy," *BICS* 49 (2006): 233-239; Kevin W. Wilkinson, "The Sarmatian and the Indians: A New Satirical Epigram on the Victory Titles of Galerius," *ZPE* 183 (2012): 39-52. On the ideological significance of Tetrarchic *cognomina deuictarum gentium*, see Mark Hebblewhite, "Tetrarchic Messaging, the Army, and an Ideology of Collective Victory," in *The Tetrarchy as Ideology: Reconfigurations and Representations of an Imperial Power*, HABES 64, eds. Filippo Carlà-Uhink & Christian Rollinger (Stuttgart: Franz Steiner Verlag, 2023), 167-168.

[55] Considering the fact that Armenicus, Medicus and Adiabenicus were adopted alongside the second of three Tetrarchic adoptions of Persicus (in 297/8) and were not associated with the third adoption of Persicus (in 308/9), we can deduce that these *cognomina* reflect specific events in 297/8 and were not associated with victories over the Sasanians in a general sense.

Furthermore, in his *Oratio Ad Coetum Sanctorum*, Constantine claims that he had seen the ruins of idolatrous Babylon (16). The only situation in which Constantine could have visited Babylon, which was about 40 miles south of Ctesiphon,[56] was if he had been with Galerius' army during this campaign. Indeed, he had spent years of the Tetrarchy serving Diocletian and Galerius as a tribune and hostage, and the *Origo Constantini Imperatoris*, highly regarded for its accuracy, relates that he had served bravely in Asia under the eastern emperors (2.2).[57] Constantine's claim thus constitutes eyewitness evidence that Galerius had campaigned in the vicinity of Ctesiphon in Lower Mesopotamia.

Relatedly, the *Historia Augusta* notes the superstition that a Roman emperor cannot advance beyond Ctesiphon (*Car.* 9.1), which he states in reference to Carus, who in 283 sacked Ctesiphon but then died while marching further east. The author responds that cowardice should keep its devices for itself, for, as Galerius has shown, clearly it is granted to the Romans 'to conquer the Persians and advance beyond them' (*Persas uincere atque ultra eos progredi*; 9.3). The author's rhetoric may be responding to superstitions following the campaign of Julian, whose march beyond Ctesiphon in 363 had ended in disaster. The author seems to imply that Galerius, like Carus and Julian, had campaigned east of the city, but had avoided their misfortunes.[58] The author's knowledge may derive from Aurelius Victor's lost *Historia*, likely the *Historia Augusta*'s principal source.[59]

Other facts should be noted. Ammianus records encountering an old veteran of Galerius, who had settled in the Persian fortress of Anat on the Middle Euphrates after having been left behind due to illness (24.1.10). Joshua the Stylite (*Chron.* 7) records that the Romans recaptured Nisibis in year 609 of the Seleucid era (Oct. 297 – Sept. 298). Peter the Patrician relates that Galerius and Diocletian reconvened at Nisibis and from there conducted the

---

[56] *Barr.* 91.

[57] Barnes, *New Empire*, 41-42; Timothy D. Barnes, *Constantine: Dynasty, Religion and Power in the Later Roman Empire* (Oxford: Wiley Blackwell, 2011), 51-52. On Constantine's life at the courts of Diocletian and Galerius, see Byron Waldron, *Dynastic Politics in the Age of Diocletian, AD 284-311* (Edinburgh: Edinburgh University Press, 2022), 167-180.

[58] Barnes, "Imperial Campaigns," 183-184. *Contra* Barnes, "Emperors, Panegyrics, Prefects, Provinces and Palaces," 544, there is no evidence that Galerius captured Ctesiphon.

[59] On Victor and the *Historia Augusta*, see now Justin A. Stover & George Woudhuysen (eds.), *The Lost History of Sextus Aurelius Victor* (Edinburgh: Edinburgh University Press, 2023), 265-334.

negotiations with Narseh, who wanted his wives and children returned to him.[60]

It would appear then that Galerius' invasion happened as follows. After defeating Narseh in Armenia, Galerius pursued him south-east into Atropatene. He then marched south-west into Adiabene and followed the Tigris further south, reaching the vicinity of Ctesiphon on the Lower Tigris. After advancing further east, and then south to the Euphrates, he passed by Babylon and travelled north-west along the Euphrates back towards Roman territory. He then marched north along the Khabur before recapturing Nisibis.[61] The resulting treaty reextended Roman control to the Tigris, asserted Roman suzerainty over Armenia and Iberia, established Nisibis as the place of Roman-Sasanian transactions, and further extended Roman

---

[60] Petr. Patr. fragm. 13-14 (*FHG* 4, pp. 188-189). Some scholars have doubted that Galerius campaigned near Ctesiphon, or even that there was an invasion of the Sasanian Empire: Wilhelm Enßlin, *Zur Ostpolitik des Kaisers Diokletian*, Sitzungsberichte der Bayerische Akademie der Wissenschaft (Munich: Bayerische Akademie der Wissenschaft, 1942), 44-45; Felix, *Antike literarische Quellen*, 120; Bruno Bleckmann, *Die Reichskrise des III. Jahrhunderts in der spätantiken und byzantinischen Geschichtsschreibung: Untersuchungen zu den nachdionischen Quellen der Chronik des Johannes Zonaras* (Munich: Tuduv, 1992), 141-142 n. 317; Mosig-Walburg, *Römer und Perser*, 118-119. However, the arguments are not strong. For example, Mosig-Walburg suggests that the adoption of Medicus and Adiabenicus was a pleonasm, but this is undermined by existing scholarship on Tetrarchic *cognomina* (see nn. 54-55). *Contra* Mosig-Walburg, we also need not accept that Galerius would have needed to take his high-profile prisoners with him as he campaigned through Sasanian territory. Having captured Narseh's family in western Armenia, he could have charged a military unit with escorting them back to his base at nearby Satala in the province of Armenia Minor (Satala: P'awstos Buzand, *Hist. Arm.* 72). While Mosig-Walburg may be correct that Galerius' capture of the family was enough to secure a favorable treaty, we should remember that Roman emperors also fought campaigns for reasons of self-representation (see below). None of the sources is explicit concerning what Galerius did between the victory in Armenia and the treaty negotiations, but the cumulation of evidence points clearly to an invasion, likely narrated in the lost *Historia* of Aurelius Victor (on the *Historia*, see Stover & Woudhuysen, *Sextus Aurelius Victor*).

[61] *Barr.* 89-91. For similar reconstructions, see Barnes, "Imperial Campaigns," 183-184; Kolb, "Zu chronologischen Problemen," 124; Zuckerman, "campagnes des tétrarques," 70; Wolfgang Kuhoff, *Diokletian und die Epoche der Tetrarchie: Das römische Reich zwischen Krisenbewältigung und Neuaufbau (284-313 n. Chr.)* (Frankfurt: Peter Lang, 2001), 175-176; Leadbetter, *Galerius*, 94-95; Umberto Roberto, *Diocleziano* (Rome: Salerno Editrice, 2014), 113; Filippo Carlà-Uhink, *Diocleziano* (Bologna: Il Mulino, 2019), 88.

suzerainty beyond the Tigris, establishing their dominion over the strategically valuable satrapies of Ingilene, Sophene, Arzanene, Corduene and Zabdicene.[62]

However, the *Tarikh* of al-Tabari records a curious story that could potentially add more to this reconstruction. Al-Tabari collates various Arabic and Iranian literary traditions,[63] and one such story relates that Shapur II (r. 309-379) entered Syria disguised as a beggar to spy on the Romans. He was discovered at the banquet of an unnamed emperor, who imprisoned Shapur in the skin of a bull. Subsequently, the emperor invaded the Sasanian Empire, taking Shapur with him, and he devastated towns and villages, and chopped down fruit-trees, until he reached the city of Gundeshapur in Khuzistan. The emperor's catapults destroyed part of the city, but with the aid of oil applied by fellow prisoners, Shapur slipped out of his chains and entered the city by night. Revealing his identity to the city's defenders, he assembled and armed the inhabitants, and that same night, he led a surprise attack on the Romans, capturing the emperor and taking his women and treasures as booty. Shapur forced the emperor to restore the lands around Gundeshapur and Ctesiphon that he had destroyed. He then cut off the emperor's heels, shod him, and sent him on a donkey back to the Romans. After spending a long time in his own empire, Shapur then invaded the Roman Empire, killing many and taking many prisoners (844-845).[64]

Elements of this story bear the characteristics of mythology, but this muddled tale also draws upon history. It is ostensibly about Shapur II, but the Roman emperor's capture and the subsequent Sasanian invasion of the Roman Empire are clearly influenced by the fact that in 260 Shapur's namesake Shapur I captured the emperor Valerian and then raided Rome's eastern provinces, an episode well known to both Sasanians and Romans.[65]

---

[62] Petr. Patr. fragm. 14 (*FHG* 4, p. 189); Engelbert Winter, *Die sassanidisch-römische Friedensverträge des 3. Jahrhunderts n. Chr.: ein Beitrag zum Verständis der aussenpolitischen Beziehungen zwischen den beiden Grossmachten* (Frankfurt: Peter Lang, 1988), 152-215; Engelbert Winter, "On the Regulation of the Eastern Frontier of the Roman Empire in 298," in *The Eastern Frontier of the Roman Empire: Proceedings of a Colloquium Held at Ankara in September 1988*, British Institute of Archaeology at Ankara Monograph 11, eds. David H. French & Chris S. Lightfoot (Oxford: BAR, 1989), 2.555–571.

[63] See n. 37.

[64] A translation is provided in Dodgeon & Lieu, *Roman Eastern Frontier*, 294-295.

[65] Sasanian and Roman testimony on the capture of Valerian: Dodgeon & Lieu, *Roman Eastern Frontier*, 57-67, 283, 297; Dignas & Winter, *Rome and Persia*, 80-

Indeed, the story's representation of an emperor being made to restore the lands around Gundeshapur accords with what Iranian historical traditions reported about the Romans captured in 260. Shapur, on the Ka'ba-i Zardost inscription, claims that he settled Romans in Persia, Parthia, Susiana (Khuzistan), Assuristan and other nations (34-36 [Maricq]). Historical traditions relate that Roman prisoners were settled in Gundeshapur, and that Valerian was made to rebuild the dam-bridge of Shushtar, which was very near to Gundeshapur.[66] In fact, Gundeshapur, Bishapur and several bridges display evidence of Roman artisanship and engineering.[67] Like the emperor in this story, al-Tabari notes that, according to some, Valerian was released (827). Moreover, the tale of Shapur's surprise attack on the Roman camp, with the resulting capture of family members and treasures, bears a resemblance to Galerius' attack on the camp of Narseh. It is thus evident that there are genuine details preserved in this chimera of a story.

But let us now consider a different version of the same story, found in the *Tarikh* of Eutychius (186), a tenth-century Patriarch of Alexandria who likewise wrote in Arabic.[68] Eutychius' version is briefer, but whereas al-Tabari does not specify why Shapur was recognized at the emperor's banquet, Eutychius includes the detail that a sage recognized the king because his image was on a vase. Shapur was then threatened until he confessed his identity. The difference in detail suggests that al-Tabari and Eutychius drew upon a common tradition, parts of which had been condensed by the authors or their sources. This is significant, for Eutychius

84; Craig H. Caldwell III, "The Roman Emperor as Persian Prisoner of War: Remembering Shapur's Capture of Valerian," in *Brill's Companion to Military Defeat in Ancient Mediterranean Society*, Warfare in the Ancient Mediterranean World 2, eds. Jessica H. Clark & Brian Turner (Leiden & Boston: Brill, 2017), 335-358.

[66] Prisoners: *Chronicle of Se'ert* 2.220-221; al-Ya'qubi, 180. Shushtar: Al-Ya'qubi, 180; al-Tabari, 826-827; Ferdowsi, *Shahnamah* 25.

[67] David S. Potter, *The Roman Empire at Bay AD 180-395*, 2nd ed. (Oxford & New York: Routledge, 2014), 221; Edwell, *Rome and Persia*, 104-105.

[68] On Eutychius' *Tarikh*, see Michael Breydy, *Études sur Sa'id ibm Batriq et ses sources*, *CSCO* 450 (Leuven: Peeters Publishers, 1983); Michael Breydy (ed. & tr.), *Das Annalenwerk des Eutychios von Alexandrien; ausgewählte Geschichten und Legenden kompiliert von Sa'id ibn Baṭriq um 935 AD*, *CSCO* 471-472, 2 Vols. (Leuven: Peeters Publishers, 1985); Bartolomeo Pirone (tr.), *Eutichio, Patriarcha de Alessandria: Gli Annali*, Studia Orientalia Christiana Monographiae 1 (Cairo & Jerusalem: Franciscan Centre of Christian Oriental Studies, 1987); Sidney H. Griffith, "Eutychius of Alexandria," *Encyclopaedia Iranica* 9, no. 1 (1998/2012): 77-78.

also names the emperor: Maximianus, the *cognomen* of Galerius. Al-Tabari or his source may have dropped the name because of the chronological incongruence between Shapur II and Galerius. However, according to Eutychius, Galerius was the emperor. In his version of the story, Galerius marches into Sasanian lands, massacres inhabitants, burns down trees, and spends days besieging Gundeshapur. Shapur then enters the city by a ruse (the nature of the ruse is now unspecified), and he leads the inhabitants in an attack on the Romans, defeating them, killing many, and taking their money and possessions. Subsequently, Shapur invades Roman territory and destroys many towns.[69]

Could Galerius' unsuccessful siege of Gundeshapur be another genuine detail that has helped to form this confused story? Modern scholarship has neglected the possibility that Galerius campaigned against Gundeshapur,[70] but if any Roman emperor had attacked the city, he would be the most suitable candidate. Defeated in Armenia, Narseh fled into the Sasanian interior after having lost his treasury, much of his army and many of his generals and advisors. Galerius' victory proved so decisive that it ultimately forced Narseh to make major geopolitical, territorial and economic concessions that would rankle the Sasanians for decades.[71] The Sasanian Empire would have appeared especially vulnerable at this time, and a campaign into Khuzistan fits well with Galerius' march down the Tigris.

I propose that Galerius advanced into Khuzistan after reaching Ctesiphon and before his return march along the Euphrates. Bill Leadbetter hypothesizes that Galerius crossed from the Tigris to the Euphrates at its narrowest point, which was in the vicinity of Ctesiphon,[72] but it may rather be the case that Galerius continued his march down the Tigris to the south-east and then advanced into Khuzistan. Perhaps he travelled due east to Khuzistan from the curve of the Tigris near Kashkar,[73] or perhaps his route down the Tigris brought him as far as Mesene, which had only previously been reached by Trajan, from where he followed the Karun and Dez rivers

[69] A translation is provided in Dodgeon & Lieu, *Roman Eastern Frontier*, 296. Hamza, 52 briefly mentions the story, but like al-Tabari he does not name the emperor.
[70] Although they do so without comment, usefully Dodgeon & Lieu, *Roman Eastern Frontier*, 131, list this passage in their collection of sources on Galerius' campaign.
[71] Sasanian reactions to the treaty: Noel Lenski, *Failure of Empire: Valens and the Roman State in the Fourth Century A. D.* (Berkeley, Los Angeles & London: University of California Press, 2002), 153-167.
[72] Leadbetter, *Galerius*, 94-95; *Barr*. 91.
[73] *Barr*. 92.

north to Gundeshapur.[74] He then returned from Khuzistan to the Tigris before crossing the Mesopotamian Plain to the Euphrates. This would finally explain the *Historia Augusta*'s implication that Galerius had campaigned beyond Ctesiphon. Moreover, the Arch of Galerius in Thessalonica, which celebrates Galerius' campaign, tentatively supports the possibility that further engagements followed the victory in Armenia. Most of the artistic program no longer survives, but Margret S. Pond Rothman identifies friezes that depict a sequence of engagements, which she names as follows: Roman Cavalry Charge; Capture of the Persian Harem; Pursuit of the Persians over the River Tigris; and The Last Battle.[75] This supports the idea that fighting continued as Galerius advanced into the Sasanian Empire.

One should note that, in the lead-up to Galerius' Persian War, there was still a perception among Romans that they needed to avenge themselves on the Sasanians after decades of humiliating defeats, especially after Shapur's capture of Valerian.[76] Like Galerius' capture of Narseh's family, a far-reaching punitive invasion would have reinforced the sense that Rome was finally inflicting its revenge. Gundeshapur was an ideologically potent objective, a tempting conquest for a junior-ranking Caesar eager to bring further glory to his name. It was founded by Shapur I as Weh-Andiok-Shapur, 'Shapur's Better Antioch', probably in reference to the Antiochenes whom he had deported there following his two sackings of the Syrian capital.[77] Moreover, as previously discussed, the captivity of Valerian and his fellow prisoners-of-war was closely associated with Gundeshapur and its surrounds. For Romanists, Ctesiphon tends to occupy more attention than Gundeshapur. Formerly the Parthian capital, it ultimately became the capital of the Sasanians, and due to its relative proximity to the Roman Empire it was sacked by the Romans multiple times during the Parthian Era. However, Iranian, Syriac and Arabic sources indicate that Gundeshapur was a similarly important city, perhaps even more important than Ctesiphon to Sasanians in the late third and early fourth centuries. Dinawari narrates that

---

[74] *Barr*. 93.

[75] Margret S. Pond Rothman, "The Thematic Organization of the Panel Reliefs on the Arch of Galerius," *AJA* 81, no. 4 (1977): 432-436.

[76] Marc Tipold, "In the Shadow of Valerian: Galerius' Persian Campaigns and the Communication Strategies of Tetrarchic Eastern Policy," in *The Tetrarchy as Ideology: Reconfigurations and Representations of an Imperial Power*, HABES 64, eds. Filippo Carlà-Uhink & Christian Rollinger (Stuttgart: Franz Steiner Verlag, 2023), 267-286.

[77] Al-Tabari, 831; Hamza, 49; cf. *Chronicle of Se'ert* 2.221.

Bahram I (r. 273-276) was based in Gundeshapur when he executed Mani (*Al-Akhbar al-Tiwal*, ʿAmir & Shayyal 47), and Shapur II executed the Primate Shahdost in Gundeshapur in c. 343 (*Acta Martyrum et Sanctorum*, Bedjan 2.281). ʿOmar Kesra (according to al-Masʿudi, *Muruj* 1.295) claimed that the Sasanian kings resided there until the reign of Hormizd II (r. 302-309), and Hamza (52) relates that it was the residence of Shapur II for the first thirty years of his reign (from 309 to c. 339).[78]

Furthermore, Galerius had grand ambitions. In 303 he persuaded Diocletian to undertake the empire-wide persecution of the Christians,[79] and despite being the most junior member of the Tetrarchy, Galerius' influence likely played a role in the fact that in 305 his friend Severus and nephew Maximinus were co-opted into the imperial college at the expense of the sons of the emperors.[80] Lactantius claims that Galerius became arrogant following his victory over Narseh and had it suggested that Mars was his father (*De mort. pers*. 9.7-10). Although Lactantius was writing invective, as a contemporary author his view was almost certainly shared by others during the Tetrarchic period.

Fittingly, Galerius may have developed inflated ambitions in the immediate aftermath of his victory in Armenia. The *Epitome de Caesaribus* claims that the story of Galerius' divine birth was made to fit that of Alexander the Great (40.17), and the Arch of Galerius includes a visually prominent frieze where the Caesar, in Alexandrian pose, duels Narseh.[81] As highlighted by Peter Edwell, the Tetrarchs were the first of the post-Severan emperors not to use Parthicus and Persicus interchangeably, instead focusing on Persicus as well as the above-listed geographical *cognomina* to celebrate Galerius' victory. Parthicus encouraged comparisons with Trajan, Lucius Verus and Septimius Severus, who had all sacked the Parthian capital Ctesiphon. In contrast, Persicus evoked Alexander.[82] Perhaps the victory in Armenia had

---

[78] These testimonies are collated in A. Shapur Shahbazi & Lutz Richter-Bernburg, "Gondešapur," *Encyclopaedia Iranica* 11, no. 2 (2002/2012): 131-135.
[79] Lactant. *De mort. pers*. 10-14; Euseb. *Hist. eccl*. 8.16.2-3, Append. 1, 4.
[80] That said, Galerius' role is exaggerated in the account of Lactantius (*De mort. pers*. 18-19), as Diocletian also played an active role in determining the succession (Waldron, *Dynastic Politics*, 115-165).
[81] Panel B.II.20. Pond Rothman, "Thematic Organization," 432, notes the Alexandrian inspiration behind Eutropius' statement that Galerius chased Narseh into the remotest deserts (9.25), and Malalas' claim that Narseh fled to India (12.39).
[82] Peter Edwell, "Persicus (Maximus) and Parthicus (Maximus): *Cognomina devictarum gentium* and Rome's Wars with Persia in the Third Century AD," *AnTard* 30 (2023): 57-66.

inspired Galerius to imitate Alexander by penetrating deep into Persian territory. Perhaps he aimed to secure conquests that would transcend rather than imitate those of the second-century emperors, bypassing Ctesiphon to march further east.[83]

Most tellingly, the fourth-century *Historiae Abbreviatae*, derived from Victor's lost account, relates that Diocletian, for uncertain reasons, forbade Galerius from carving out a new province in Sasanian territory (39.36).[84] The abbreviated text does not retain the lost details, but perhaps Diocletian's administration genuinely feared that Galerius had penetrated too deeply into enemy territory and was at risk of compromising his victory in Armenia. Although a very successful commander, there are suggestions that Galerius could be over-bold. In c. 296, when Narseh invaded Roman Mesopotamia, Galerius fought a series of holding actions with a smaller army and likely fewer cavalry, encountering the Sasanians in the open countryside between Carrhae and Callinicum. Although he achieved his likely objective of preventing a Sasanian advance into Syria, his army was mauled in the third encounter.[85] Latin, Greek and Armenian sources provide different versions of the story that Galerius personally reconnoitred Narseh's camp in Armenia,[86] and a Constantinian tradition claims that Galerius, while campaigning against the Sarmatians, had Constantine lead a successful attack into a swamp, an obviously risky approach to giving battle if there is any truth to it.[87]

We need not believe that Galerius suffered a major defeat outside the walls of Gundeshapur. The details of the narration of Galerius' defeat owe more to Shapur's capture of Valerian and, ironically, Galerius' attack on Narseh's camp. Based on all other information, Galerius' broader campaign was successful, decisive, and devastating for the Sasanians. Nevertheless, it is plausible that in 298 Galerius raided Khuzistan and made an unsuccessful

---

[83] It is relevant to point out that, immediately after becoming Augustus, Galerius led three campaigns against the Sarmatians between the summer of 305 and the spring of 308, perhaps so that he could gain the *cognomen* Sarmaticus V, thereby surpassing Diocletian's Sarmaticus IV (Wilkinson, "The Sarmatian and the Indians," 49).

[84] On this text, see Stover & Woudhuysen, *Sextus Aurelius Victor*.

[85] Leadbetter, *Galerius*, 90-91.

[86] Eutr. 9.25; Festus, *Brev*. 25; Syn. *De Regno* 17; P'awstos Buzand, *Hist. Arm*. 72-73. Leadbetter, *Galerius*, 92, suggests that Galerius really did undertake this task, noting that there was little possibility for interdependence between the three historical traditions.

[87] *Origo* 2.3. In 251 Decius suffered a disastrous defeat after Cniva's Goths lured the Roman army into a swamp near Abritus (see e.g. Zos. 1.23).

attack on Gundeshapur. The Sasanian army had been overthrown, the wounded Narseh had fled, and Galerius may have believed that he could exploit the fear and disorder of the enemy to capture the residence of the Sasanian kings, a great city founded by the notorious Shapur himself and populated with the descendants of Roman prisoners and deportees. However, as the Iranian tradition tells us, the inhabitants offered stubborn resistance. Rather than commit to a long siege deep in enemy territory, Galerius may have chosen to withdraw, much as he would later do when he marched against Maxentius in Rome.[88] This reconstruction explains both the Iranian memory of a successful defence of the city, which they celebrated and mythicized, as well as the silence of Tetrarchic media, which had no interest in preserving the memory of a military setback.

Conclusion

There are significant gaps in our knowledge of the history of the third century, but in researching the Roman-Sasanian wars, we are privileged to have access to a multitude of historical traditions that transcend Greek and Roman narratives. Persian, Arabic, Armenian and Syriac authors all wrote on the climactic wars fought between the likes of Gordian and Galerius on the one hand, and Shapur and Narseh on the other. We should be seeking to fully employ this array of traditions where appropriate, to challenge existing Romano-centric narratives that have solidified over time. By integrating two non-Roman historical traditions into current reconstructions of Roman campaigns, the case studies in this article argue that we should be reassessing the wars of Gordian and Galerius. Moreover, they suggest that important pieces of information in non-Roman sources might still be awaiting scholarly attention.

Bibliography

Assfalg, Julius. "Zur Textüberlieferung der Chronik von Arbela: Beobachtungen zu Ms.or.fol.3126." *Oriens Christianus* 50 (1966): 19-36.

Banchich, Thomas M. *The Lost History of Peter the Patrician: An Account of Rome's Imperial Past from the Age of Justinian*. London & New York: Routledge, 2015.

[88] Galerius' march on Rome: Lactant. *De mort. pers.* 27.2-8; Aur. Vict. *HAb* 40.8-9; Zos. 2.10.3; *Origo* 3.6-7.

Barnes, Timothy D. "Imperial Campaigns, A.D. 285-311." *Phoenix* 30 (1976): 174-193.

—. *The Sources of the Historia Augusta*. Collection Latomus 155. Brussels: Latomus Revue d'Études Latines, 1978.

—. *The New Empire of Diocletian and Constantine*. Cambridge, MA & London: Harvard University Press, 1982.

—. "Emperors, Panegyrics, Prefects, Provinces and Palaces (284-317)." *JRA* 9 (1996): 532-552.

—. *Constantine: Dynasty, Religion and Power in the Later Roman Empire*. Oxford: Wiley Blackwell, 2011.

Bleckmann, Bruno. *Die Reichskrise des III. Jahrhunderts in der spätantiken und byzantinischen Geschichtsschreibung: Untersuchungen zu den nachdionischen Quellen der Chronik des Johannes Zonaras*. Munich: Tuduv, 1992.

Breydy, Michael. *Études sur Sa'id ibm Batriq et ses sources. CSCO* 450. Leuven: Peeters Publishers, 1983.

—. (ed. & tr.). *Das Annalenwerk des Eutychios von Alexandrien; ausgewählte Geschichten und Legenden kompiliert von Sa'id ibn Baṭriq um 935 AD. CSCO* 471-472. 2 Vols. Leuven: Peeters Publishers, 1985.

Brock, Sebastian. "Review of G. Wiessner, *Untersuchungen zur syrischen Literaturgeschichte 1: Zur Märtyrerüberlieferung aus der Christenverfolgung Schapurs II.* Abhandlungen der Akademie der Wissenschaften in Göttingen, philologisch-historische Klasse, Dritte Folge 67 (Göttingen, 1967)." *JThS* 19 (1968): 300-309.

—. "Syriac Historical Writing: A Survey of the Main Sources." *Journal of the Iraqi Academy Syriac Corporation* 5 (1979/80): 1-30.

Caldwell III, Craig H. "The Roman Emperor as Persian Prisoner of War: Remembering Shapur's Capture of Valerian." In *Brill's Companion to Military Defeat in Ancient Mediterranean Society*, Warfare in the Ancient Mediterranean World 2, edited by Jessica H. Clark & Brian Turner, 335-358. Leiden & Boston: Brill, 2017.

Carlà-Uhink, Filippo. *Diocleziano*, Bologna: Il Mulino, 2019.

Carter, Martha L. "A Numismatic Reconstruction of Kushano-Sasanian History." *Museum Notes (American Numismatic Society)* 30 (1985): 215-281.

Corcoran, Simon. "Galerius, Maximinus and the Titulature of the Third Tetrarchy." *BICS* 49 (2006): 231-240.

Daryaee, Touraj. *Sasanian Persia: The Rise and Fall of an Empire*. London & New York: I.B. Tauris, 2009.

Davenport, Caillan. *A History of the Roman Equestrian Order*. Cambridge & New York: Cambridge University Press, 2019.

De Blois, Lukas. *Image and Reality of Roman Imperial Power in the Third Century AD: The Impact of War*. London & New York: Routledge, 2019.

Dignas, Beate & Engelbert Winter. *Rome and Persia in Late Antiquity: Neighbours and Rivals*. Cambridge & New York: Cambridge University Press, 2007.

Dodgeon, Michael H. & Samuel N. C. Lieu. (eds.). *The Roman Eastern Frontier and the Persian Wars AD 226-363: A Documentary History*. London & New York: Routledge, 1991.

Edwell, Peter. *Rome and Persia at War: Imperial Competition and Contact, 193-363 CE*. London & New York: Routledge, 2021.

—. "Persicus (Maximus) and Parthicus (Maximus): *Cognomina devictarum gentium* and Rome's Wars with Persia in the Third Century AD." *AnTard* 30 (2023): 57-66.

Enßlin, Wilhelm. *Zur Ostpolitik des Kaisers Diokletian*. Sitzungsberichte der Bayerische Akademie der Wissenschaft. Munich: Bayerische Akademie der Wissenschaft, 1942.

Felix, Wolfgang. *Antike literarische Quellen zur Aussenpolitik des Sasanidenstates 1: 224-309*. Vienna: Verlag der Österreichischen Akademie der Wissenschaften, 1985.

Felix, Wolfgang & Wilferd Madelung. "Deylamites." *Encyclopaedia Iranica* 7, no. 4 (1995/2011): 342-347, http://www.iranicaonline.org/articles/deylamites, accessed 12/28/2022.

Fiey, Jean Maurice. "Auteur et date de la Chronique d'Arbèles." *L'orient syrien* 12 (1967): 265-302.

Frye, Richard N. "The Political History of Iran under the Sasanians." In *The Cambridge Iranian History 3(1): The Seleucid, Parthian and Sasanian Periods*, edited by Ehsan Yarshater, 116-180. Cambridge: Cambridge University Press, 1983.

Griffith, Sidney H. "Eutychius of Alexandria." *Encyclopaedia Iranica* 9, no. 1 (1998/2012): 77-78, https://www.iranicaonline.org/articles/eutychius, accessed 12/29/2022.

Hauser, Stefan R. "Where is the man of Hadr, who once built it and taxed the land by the Tigris and Chaboras? On the significance of the final siege of Hatra." In *Hatra: Politics, Culture and Religion between Parthia and Rome*, Oriens et Occidens 21, edited by Lucinda Dirven, 119-139. Stuttgart: Franz Steiner Verlag, 2013.

Hebblewhite, Mark. "Tetrarchic Messaging, the Army, and an Ideology of Collective Victory." In *The Tetrarchy as Ideology: Reconfigurations and Representations of an Imperial Power*, HABES 64, edited by Filippo Carlà-Uhink & Christian Rollinger, 161-178. Stuttgart: Franz Steiner Verlag, 2023.

Henrichs, A. & L. Koenen. "Der Kölner Mani-Kodex (P.Colon. Inv.Nr. 4780) Περὶ τῆc γέννηc τοῦ cώματοc αὐτοῦ: Edition der Seiten 121-192." *ZPE* 48 (1982): 1-59.

Herrmann, Katrin. *Gordian III. Kaiser einer Umbruchszeit*. Speyer: Kartoffeldruck-Verlag, 2013.

Huttner, Ulrich. "Von Maximinus Thrax bis Aemilianus." In *Die Zeit der Soldatenkaiser: Krise und Transformation des Römischen Reiches im 3. Jahrhundert n. Chr. (235-284)*, edited by Klaus-Peter Johne in association with Udo Hartmann & Thomas Gerhardt, 1.161-221. Berlin: Akademie Verlag, 2008.

Jullien, Christelle & Florence Jullien. "La *Chronique d'Arbèles*: Propositions pour la fin d'une controverse." *Oriens Christianus* 85 (2001): 41-83.

Kawerau, Peter (ed. & tr.). *Die Chronik von Arbela. CSCO* 467-468, Scriptores Syri 199-200. 2 Vols. Leuven: Peeters Publishers, 1985.

Kennedy, Hugh N. (ed.). *Al-Tabari: A Medieval Muslim Historian and His Work*. Princeton, NJ: Gerlach Press, 2008.

Kettenhofen, Erich. *Die römisch-persischen Kriege des 3. Jahrhunderts n. Chr.: nach der Inschrift Šāhpuhrs I. an der Ka'be-ye Zartošt (ŠKZ)*. Beihefte zum Tubinger Atlas des Vorderen Orients. Reihe B, Geisteswissenschaften, no. 55. Wiesbaden: Ludwig Reichert, 1982.

—. "Die Chronik von Arbela in der Sicht der Althistorie." In *Simblos: Scritti di storia antica*, edited by Lucia Criscuolo, 287-319. Bologna: CLUEB, 1995.

Kolb, Frank. "Zu chronologischen Problemen der ersten Tetrarchie." *Eos* 76 (1988): 105-125.

—. "Chronologie und Ideologie der Tetrarchie." *AnTard* 3 (1995): 21-31.

Kroll, Timothy (tr.). *The Chronicle of Arbela. CSCO* 468, Scriptores Syri 200. Leuven: Peeters Publishers, 1985.

Kuhoff, Wolfgang. *Diokletian und die Epoche der Tetrarchie: Das römische Reich zwischen Krisenbewältigung und Neuaufbau (284-313 n. Chr.)*. Frankfurt: Peter Lang, 2001.

Leadbetter, Bill. *Galerius and the Will of Diocletian*. London & New York: Routledge, 2009.

Lenski, Noel. *Failure of Empire: Valens and the Roman State in the Fourth Century A.D.* Berkeley, Los Angeles & London: University of California Press, 2002.

MacDonald, David. "The Death of Gordian III: Another Tradition." *Historia* 30, no. 4 (1981): 502-508.

Mackenzie, D. N. "The Fire Altar of Happy *Frayosh." *Bulletin of the Asia Institute*, New Series 7 (1993): 105-109.

Mingana, Alphonse. *Sources syriaques* 1. Mosul: Otto Harrassowitz, 1907.

Mosig-Walburg, Karin. *Römer und Perser: vom 3. Jahrhunderts bis zum Jahr 363 n. Chr.* Gutenberg: Computus, 2009.

Nöldeke, Theodor. *Geschichte der Perser und Araber zur Zeit der Sasaniden: Aus der arabischen Chronik des Tabari übersetzt und mit ausführlichen Erläuterungen und Ergänzungen versehn.* Leiden: Brill, 1879.

Piacentini, V. Fiorani. "Ardashir I Papakan and the Wars against the Arabs: Working Hypothesis on the Sasanian Hold of the Gulf." *Proceedings of the Seminar for Arabian Studies* 15 (1985): 57-77.

Pirone, Bartolomeo (tr.). *Eutichio, Patriarcha de Alessandria: Gli Annali.* Studia Orientalia Christiana Monographiae 1. Cairo & Jerusalem: Franciscan Centre of Christian Oriental Studies, 1987.

Pond Rothman, Margret S. "The Thematic Organization of the Panel Reliefs on the Arch of Galerius." *AJA* 81, no. 4 (1977): 427-454.

Potter, David S. *Prophecy and History in the Crisis of the Roman Empire: A Historical Commentary on the Thirteenth Sibylline Oracle.* Oxford & New York: Oxford University Press, 1990.

—. *The Roman Empire at Bay AD 180-395.* 2nd ed. Oxford & New York: Routledge, 2014.

Ramelli, Ilaria. *Il Chronicon di Arbela: presentazione, traduzione e note essenziali.* Ilu. Anejo 8. Madrid: Universidad Complutense de Madrid, 2002.

—. "Il Chronicon di Arbela: Una Messa a Punto Storiografica." *Aevum* 80, no. 1 (2006): 145-164.

Richter-Bernburg, Lutz. "Mani's Dodecads and Sasanian Chronology: Kephalaia, Shāpūragān, and Codex Manichaicus Coloniensis." *ZPE* 95 (1993): 71-80.

Roberto, Umberto. *Diocleziano*, Rome: Salerno Editrice, 2014.

Ross, Steven K. *Roman Edessa: Politics and Culture on the Eastern Fringes of the Roman Empire, 114-242 CE.* London & New York: Routledge, 2001.

Rydving, Hakan (ed.). *Al-Tabari's History: Interpretations and Challenges.* Uppsala: Uppsala University Publishing, 2007.

Schrier, Omert J. "Syriac Evidence for the Roman-Persian War of 421-422." *Greek, Roman and Byzantine Studies* 33, no. 1 (1992): 75-87.

Shahbazi, A. Shapur & Lutz Richter-Bernburg. "Gondešapur." *Encyclopaedia Iranica* 11, no. 2 (2002/2012): 131-135, https://www.iranicaonline.org/articles/gondesapur, accessed 12/29/2022.

Smith, Vincent A. "Invasion of the Panjab by Ardashir Papakan (Babagan), the First Sasanian King of Persia, A.D. 226-41." *Journal of the Royal Asiatic Society of Great Britain and Ireland* 2 (1920): 221-226.

Stover, Justin A. & George Woudhuysen. (eds.). *The Lost History of Sextus Aurelius Victor*. Edinburgh: Edinburgh University Press, 2023.

Sundermann, Werner. "Shapur's Coronation: The Evidence of the Cologne Mani Codex Reconsidered and Compared with Other Texts." *Bulletin of the Asia Institute*, New Series 4 (1990): 295-299.

Taqizadeh, S. H. "The Early Sasanians: Some Chronological Points Which Possibly Call for Revision." *Bulletin of the School of Oriental and African Studies, University of London* 11, no. 1 (1943): 6-51.

Tipold, Marc. "In the Shadow of Valerian: Galerius' Persian Campaigns and the Communication Strategies of Tetrarchic Eastern Policy." In *The Tetrarchy as Ideology: Reconfigurations and Representations of an Imperial Power*, HABES 64, edited by Filippo Carlà-Uhink & Christian Rollinger, 267-286. Stuttgart: Franz Steiner Verlag, 2023.

Waldron, Byron. *Dynastic Politics in the Age of Diocletian, AD 284-311*. Edinburgh: Edinburgh University Press, 2022.

Walker, Joel Thomas. *The Legend of Mar Qardagh: Narrative and Christian Heroism in Late Antique Iraq*. The Transformation of Classical Heritage 40. Berkeley, Los Angeles & London: University of California Press, 2006.

Wilkinson, Kevin W. "The Sarmatian and the Indians: A New Satirical Epigram on the Victory Titles of Galerius." *ZPE* 183 (2012): 39-52.

Winter, Engelbert. *Die sassanidisch-römische Friedensverträge des 3. Jahrhunderts n. Chr.: ein Beitrag zum Verständis der aussenpolitischen Beziehungen zwischen den beiden Grossmachten*. Frankfurt: Peter Lang, 1988.

—."On the Regulation of the Eastern Frontier of the Roman Empire in 298." In *The Eastern Frontier of the Roman Empire: Proceedings of a Colloquium Held at Ankara in September 1988*, British Institute of Archaeology at Ankara Monograph 11, edited by David H. French & Chris S. Lightfoot, 2.555-571. Oxford: BAR, 1989.

Yarshater, Ehsan. (gen. ed.). *The History of Al-Tabari*. 40 Vols. Albany, NY: SUNY Press, 1985-2007.

Zuckerman, Constantin. "Les campagnes des tétrarques, 296-298: Notes de chronologie." *AnTard* 2, 65-70.

TRAUMATIC BRAIN INJURY IN ROMAN WAR

BRENDAN MCCARTHY

In 1901, the Museum of Fine Arts in Boston purchased a terracotta portrait bust found near Pozzuoli.[1] The portrait depicts an older man with a sagging neck, drooping cheeks, and furrows in his brow. Most intriguing, however, is the large wound depicted on his forehead above his left eye. The wound has healed with scar tissue forming a large starburst, indicating that this man received a skull fracture which then became inflamed but eventually healed. Recent analysis has found evidence that suggests that this bust was made from a wax cast.[2] Thus, this is a real man who survived a severe head injury which is accurately depicted.

This type of injury can be received in many ways, but war is as likely a factor as anything else. This particular statue has been dated to the first century BCE based on his hairstyle, which conforms to depictions of soldiers. This man very well may have been a soldier in the wars of the late Republic. Perhaps one of the sling stones looking for Fulvia in the Perusine War found him or perhaps he fought against Sulla in the Social War. Granted, he may have been kicked by a mule or hit his head on a jagged rock, or received the wound in a bar brawl like Plautus's Curculio.[3] Regardless of the cause, this man's scar stands as a stark reminder of the types of injury Roman men could survive whether in war or in peace and provides context for discussing brain trauma in the ancient world.

Traumatic brain injury (TBI) is a major topic of study in medicine because it affects many different populations, especially athletes and soldiers. Fans of contact sports have likely heard about "concussion protocols" in recent

---

[1] *Bust of an Older Man*, 1st c. BCE, Terracotta, 37.5 cm x 18 cm, 1st c. BCE, Museum of Fine Arts, Boston, https://collections.mfa.org/objects/155696/bust-of-an-older-man?ctx=2fabc98f-9361-4aab-9542-15a84655ab78&idx=5.

[2] Peter Schertz et al., "Made from Life: A Roman Terracotta Portrait in the Museum of Fine Arts, Boston," *Memoirs of the American Academy in Rome* 61 (2016): 22–39.

[3] Plaut. *Curc*. 392-400.

years after studies revealed that athletes who received multiple concussions, a mild form of TBI, were likely to develop Chronic Traumatic Encephalopathy (CTE), a permanent swelling of the brain and an increase in fluid in the skull, which can cause a host of psychological and behavioral problems.[4] Likewise, the high prevalence of combat stress disorders like Post Traumatic Stress Disorder in veterans of recent wars has been strongly linked to traumatic brain injuries. Thus, research focusing on the cause, pathology (the specific way injuries cause harm to the body), and prevention of TBI has blossomed in the last decade. In fact, the National Collegiate Athletic Association, the United States' governing body for university athletic competitions, and the United States Department of Defense have collaborated on researching concussions since 2014, demonstrating the link between athletic and military injuries.[5] These new studies can be used to evaluate the prevalence and harm of TBI in the ancient world.

Emerging research on TBI shows that brain injury is incredibly common and that even mild TBI, like concussions, can cause permanent changes in the brain. Re-investigating ancient accounts of battle and injury with this new information reveals a high risk of TBI in ancient battles. TBI had a large impact on ancient warfare affecting ancient soldiers for their entire lives. This paper will demonstrate the physical and social impact of TBI received in battle. Understanding the risks and results of traumatic brain injury is necessary for a complete understanding of the experience of Roman soldiers in battle and after retirement.

Understanding Traumatic Brain Injury

TBI is a broad category that includes any injury that affects the brain and can be divided into three major categories—open TBI, blast TBI, and penetrating TBI—based on how the injury is received. Open TBI is caused by blunt force jarring the brain inside the skull, blast TBI is caused by blast waves or shock waves (usually from explosives), and penetrating TBI

---

[4] NFL Head, Neck, and Spine Committee, "NFL Concussion Diagnosis and Management Protocol," 2022, https://static.www.nfl.com/image/upload/v1665264248/league/moexjmdzy2kvjtqsdpbx.pdf; "Concussion Safety Protocol Management," NCAA.org, accessed October 17, 2022, https://www.ncaa.org/sports/2016/7/20/concussion-safety-protocol-management.aspx.

[5] "NCAA, DoD Launch Concussion Study," NCAA.org, accessed October 17, 2022, https://www.ncaa.org/news/2014/5/29/ncaa-dod-launch-concussion-study.aspx.

covers any head injury that breaks the skin and results in brain injury.[6] Ancient soldiers would rarely face blast TBI (a collapsing building might cause a shock wave) but the other two, as I will argue, were common. Indeed, the man from southern Italy is a clear case of penetrating TBI.

TBI is very common in the United States. One study estimated that 1.7 million TBIs, including 52,000 fatalities, occur annually in the US.[7] Another estimated that between 1.6 and 3.8 million concussions occur every year due to athletic activity alone (the variance is due to the fact that the numbers are taken from self-reported data).[8] Approximately 450,000 US service members have been diagnosed with TBI from the start of the Afghan War, accounting for 15-20% of all soldiers.[9] Many of those injuries would be blast TBI and therefore not applicable to ancient warfare.[10] The results of blast TBI and non-blast TBI are, however, indistinguishable.[11] This means that evidence from the military is as useful as evidence from athletes when discussing the long-term impacts of brain injury on ancient soldiers. Indeed, ancient soldiers faced many of the same risks that modern athletes do: direct or glancing hits to the head, falls (especially for cavalry), and hits to the body that cause rotational force on the neck and skull.

Traumatic brain injury, even in its mildest forms, can have long-lasting impacts including physical effects like paralysis and mental effects like depression, and behavioral disorders. These issues can be caused by

---

[6] Je Yeong Sone et al., "Helmet Efficacy against Concussion and Traumatic Brain Injury: A Review," *Journal of Neurosurgery* 126, no. 3 (March 1, 2017): 770–71, https://doi.org/10.3171/2016.2.JNS151972.

[7] Sone et al., 770.

[8] Daniel H. Daneshvar et al., "Helmets and Mouth Guards: The Role of Personal Equipment in Preventing Sport-Related Concussions," *Clinics in Sports Medicine* 30, no. 1 (January 1, 2011): 146, https://doi.org/10.1016/j.csm.2010.09.006.

[9] "DOD TBI Worldwide Numbers," Military Health System, accessed June 15, 2022, https://armymedicine.health.mil/MHSHome/Military-Health-Topics/Centers-of-Excellence/Traumatic-Brain-Injury-Center-of-Excellence/DOD-TBI-Worldwide-Numbers; Sone et al., "Helmet Efficacy," 771.

[10] Andrew J. MacGregor et al., "Repeated Concussion among U.S. Military Personnel during Operation Iraqi Freedom," *Journal of Rehabilitation Research & Development* 48, no. 10 (December 15, 2011): 1272, https://doi.org/10.1682/JRRD.2011.01.0013 found that over 90% of concussions among US military members were blast-related.

[11] Christine L. Mac Donald et al., "Prospectively Assessed Clinical Outcomes in Concussive Blast vs Nonblast Traumatic Brain Injury Among Evacuated US Military Personnel," *JAMA Neurology* 71, no. 8 (August 1, 2014): 1000, https://doi.org/10.1001/jamaneurol.2014.1114.

physical damage to specific regions of the brain but are also present in people who receive multiple concussions or even "sub-concussive" injuries.[12] These effects can last for weeks or even years depending on the severity of concussion and how frequently concussions were received. Some studies have shown that people in a state of fear or anxiety at the time of concussion are more likely to need more time to recover from their injury and are more prone to chronic issues like depression resulting from the injury.[13]

All forms of TBI influence the brain in similar ways. Research has shown that TBI causes damage to the white matter (the tissue that connects the different parts of the brain), affects blood flow in the brain, and releases a protein called tau.[14] These various effects can change the brain in different ways. The brain is separated into sections with specific functions and the white matter joins them together and carries impulses from one part of the

[12] Kausar Abbas et al., "Effects of Repetitive Sub-Concussive Brain Injury on the Functional Connectivity of Default Mode Network in High School Football Athletes," *Developmental Neuropsychology* 40, no. 1 (January 2, 2015): 51–56, https://doi.org/10.1080/87565641.2014.990455; Victoria N. Poole et al., "Sub-Concussive Hit Characteristics Predict Deviant Brain Metabolism in Football Athletes," *Developmental Neuropsychology* 40, no. 1 (January 2, 2015): 12–17, https://doi.org/10.1080/87565641.2014.984810; Mallika Marar et al., "Epidemiology of Concussions Among United States High School Athletes in 20 Sports," *The American Journal of Sports Medicine* 40, no. 4 (April 1, 2012): 747–55, https://doi.org/10.1177/0363546511435626.

[13] Donna K. Broshek, Anthony P. De Marco, and Jason R. Freeman, "A Review of Post-Concussion Syndrome and Psychological Factors Associated with Concussion," *Brain Injury* 29, no. 2 (January 28, 2015): 228–37, https://doi.org/10.3109/02699052.2014.974674.

[14] Benoit Mouzon et al., "Chronic White Matter Degeneration, but No Tau Pathology at One-Year Post-Repetitive Mild Traumatic Brain Injury in a Tau Transgenic Model," *Journal of Neurotrauma* 36, no. 4 (February 15, 2019): 576–88, https://doi.org/10.1089/neu.2018.5720; Francesca Pischiutta et al., "Single Severe Traumatic Brain Injury Produces Progressive Pathology with Ongoing Contralateral White Matter Damage One Year after Injury," *Experimental Neurology* 300 (February 1, 2018): 167–78, https://doi.org/10.1016/j.expneurol.2017.11.003; Joseph O. Ojo et al., "Chronic Repetitive Mild Traumatic Brain Injury Results in Reduced Cerebral Blood Flow, Axonal Injury, Gliosis, and Increased T-Tau and Tau Oligomers," *Journal of Neuropathology & Experimental Neurology* 75, no. 7 (July 1, 2016): 636–55, https://doi.org/10.1093/jnen/nlw035; Chad A. Tagge et al., "Concussion, Microvascular Injury, and Early Tauopathy in Young Athletes after Impact Head Injury and an Impact Concussion Mouse Model," *Brain* 141, no. 2 (February 1, 2018): 422–58, https://doi.org/10.1093/brain/awx350.

brain to another. When white matter is damaged, those impulses become slower or may be blocked altogether. Similarly, blood plays an important role in keeping brain cells healthy, so blocking or changing blood flow can cause lasting damage to cells. Tau is a protein that helps brain cells interact with other cells in the skull and with each other.[15] As the brain develops, tau changes its structure, and specific tau molecules can become hyperphosphorylated (called "ptau"), at which point instead of aiding communication between cells, they create tangles that block impulses from traveling through the brain. These tangles are associated with major illnesses like Alzheimer's disease, CTE, and other neurodegenerative diseases.[16] In persons who receive multiple concussions even after long periods, the amount of tau grows and remains in the body for longer amounts of time.[17]

Research on brain injuries and how they affect the body is still controversial. Even similar experiments in mice have come to different conclusions as to whether TBI increases the amount of tau in the body. Edwards et al. found that mice exposed to TBI had higher rates of tau-based diseases while Mouzon et al. observed no tau-based illnesses in their mice but a clear pattern of white matter deterioration.[18] Reviews of research on human subjects are likewise inconclusive. Researchers targeting athletes and soldiers have not been able to confirm exactly how brain injury affects the production of tau or whether it increases the amount of tau or the creation of ptau and tau-based disease.[19] As researchers produce new ways to measure tau in living persons, these results may rapidly change.[20]

---

[15] Rudy J. Castellani and George Perry, "Tau Biology, Tauopathy, Traumatic Brain Injury, and Diagnostic Challenges," *Journal of Alzheimer's Disease* 67, no. 2 (January 1, 2019): 449, https://doi.org/10.3233/JAD-180721.

[16] George Edwards et al., "Traumatic Brain Injury Induces Tau Aggregation and Spreading," *Journal of Neurotrauma* 37, no. 1 (January 2020): 80, https://doi.org/10.1089/neu.2018.6348.

[17] Tagge et al., "Concussion, Microvascular Injury, and Early Tauopathy in Young Athletes"; Ojo et al., "Chronic R-MTBI Injury Results."

[18] Edwards et al., "TBI Induces Tau Aggregation and Spreading"; Mouzon et al., "Chronic White Matter Degenerationat One-Year Post-r-MTBI in a Tau Transgenic Model."

[19] Castellani and Perry, "Tau Biology, Tauopathy, Traumatic Brain Injury, and Diagnostic Challenges."

[20] Nikos Gorgoraptis et al., "In Vivo Detection of Cerebral Tau Pathology in Long-Term Survivors of Traumatic Brain Injury," *Science Translational Medicine* 11, no. 508 (September 4, 2019): eaaw1993, https://doi.org/10.1126/scitranslmed.aaw1993.

CTE is the tau-based disease most commonly linked to TBI and gives insight into the way TBI affects behavior. Some patients with suspected CTE experience memory loss, mood swings, suicidal ideation, disorientation, and other difficulties in thinking and expressing themselves. These symptoms tend to progress over time if someone receives more brain injuries and might include loss of motor function. Some scholars have estimated that up to 17% of athletes who received multiple concussions have developed the disorder.[21] Recent research has confirmed that CTE is caused by multiple head traumas, though the specific role of tau is still debated.[22] Since that study was published in July 2022, neuroscientists advocated for the United States National Institute of Health to list multiple head traumas as the cause of CTE which they did in October 2022, thus adding TBI through sports as a public health concern.[23] Even with a clear consensus on the cause of CTE and the biology behind it, the behavioral effects vary unpredictably from person to person.

Combat Trauma in Ancient Sources

The infamous Sophist Gorgias of Leontini argued that merely seeing an opposing army can have grave effects on the mind of a soldier:

> For immediately, whenever warring bodies arm themselves against warriors…if sight observes them, it is disturbed and it disturbs the soul, so that often they are dumbstruck and flee a coming danger <as if> it were a present one…And it has already happened that some people, seeing frightening things, have been driven out from their present mind in the present moment: in this way fear has extinguished and expelled thought. And many have fallen victim to groundless sufferings, terrible diseases, and incurable madnesses.[24]

---

[21] Michael W. Collins et al., "Cumulative Effects of Concussion in High School Athletes," *Neurosurgery* 51, no. 5 (November 2002): 1175–79; discussion 1180-1181, https://doi.org/10.1097/00006123-200211000-00011.

[22] Christopher J. Nowinski et al., "Applying the Bradford Hill Criteria for Causation to Repetitive Head Impacts and Chronic Traumatic Encephalopathy," *Frontiers in Neurology* 13 (2022), https://doi.org/10.3389/fneur.2022.938163.

[23] "NIH Chronic Traumatic Encephalopathy Diagnosis Conference | National Institute of Neurological Disorders and Stroke," accessed November 9, 2022, https://www.ninds.nih.gov/current-research/focus-disorders/focus-traumatic-brain-injury-research/nih-chronic-traumatic-encephalopathy-diagnosis-conference.

[24] Gorg. *Hel*. 16-17. Translated in André Laks and Glenn W. Most, eds., *Early Greek Philosophy, Volume VIII*, trans. André Laks and Glenn W. Most (Cambridge, Mass.: Harvard University Press, 2016),
https://www.loebclassics.com/view/LCL531/2016/volume.xml.

Indeed, ancient literature reveals multiple men who seem to have suffered from the mental burden of war. The Spartan general Clearchus was unable to live a life as anything but a soldier.[25] After receiving a wound in the Battle of Ticinus, Publius Cornelius Scipio refused to lead the army into the eventual battle of Trebia because, according to Livy, he was "more sick in mind than in body."[26] Marius was so anxious at night that he could not sleep and thus drank until he passed out to avoid the distress.[27]

The popularity of *Achilles in Vietnam* by Jonathan Shay, the work of Lawrence A. Tritle, and the use of ancient material like the *Iliad* and ancient tragedies in group therapy for veterans of the Vietnam War (and later wars) have inspired research into combat trauma in the ancient world.[28] Nonetheless, that research is controversial. As Owen Rees stated, "social parameters, medical and philosophical beliefs, and the religious rubric by which [ancient people] lived" affected the way brain injury and other trauma manifested in behavior and how those behaviors were perceived within their community.[29] By centering the focus of the mental effects of war on TBI, researchers can avoid one aspect of this debate: Romans could not avoid the physiological harms of TBI, but they did decide what they meant.

Rees uses the Athenian soldier Epizelus to contrast the way ancient observers and modern researchers view mental trauma from combat. Epizelus fought in the Battle of Marathon. He saw an apparition who killed the man next to him and was struck blind even though he was never hit.[30] Epizelus was a respected man, people believed his story, and he does not fit

---

[25] Lawrence A. Tritle, "Clearchus' Story: The Heroic Ideal Transformed," in *From Melos to My Lai* (New York: Routledge, 2000), 55–78; Lawrence A. Tritle, "Xenophon's Portrait of Clearchus: A Study in Post-Traumatic Stress Disorder," in *Xenophon and His World: Papers from a Conference Held in Liverpool in July 1999*, ed. Christopher Tuplin (Stuttgart: Franz Steiner Verlag, 2004), 325–39, https://public.ebookcentral.proquest.com/choice/publicfullrecord.aspx?p=5806270.

[26] Livy 21.53.2. Korneel Van Lommel, "The Recognition of Roman Soldiers' Mental Impairment," *Acta Classica* 56 (January 1, 2013): 155–85.

[27] Plut. *Mar*. 45.3.

[28] Jonathan Shay, *Achilles in Vietnam: Combat Trauma and the Undoing of Character* (New York: Scribner, 2003); Tritle, "Portrait of Clearchus"; Lawrence A. Tritle, *From Melos to My Lai: War and Survival* (London: Routledge, 2000), http://www.crcnetbase.com/isbn/9780203462485.

[29] Owen Rees, "We Need to Talk about Epizelus: 'PTSD' and the Ancient World," *Medical Humanities* 46, no. 1 (March 1, 2020): 46, https://doi.org/10.1136/medhum-2018-011557.

[30] Hdt. 6.117.1-3.

the way other Greeks described the men who succumbed to fear and psychological damage on the battlefield described by Gorgias.[31]

Just as Greeks did not recognize Epizelus as the victim of combat trauma, the Romans did not connect Marius's behaviors to his career. Plutarch instead claimed that Marius's anxiety and alcoholism were because of his coming war with Sulla.[32] The differences in the modern and ancient approaches to combat trauma are duplicated in ancient accounts of brain injury. Likewise, Livy claims that P. Cornelius Scipio was "sick in the mind" and avoided battle at Trebia because of the injury he received at the earlier battle at Ticinus, not because of trauma in war.[33]

Ancient medical texts often discussed the skull and surgical repairs for skull injuries, showing that head trauma was fairly common. Indeed, there was serious debate about the importance of the brain in ancient medical texts and philosophers. Hippocrates' *Diseases* II claims that consciousness was linked to the heart but also shows how head injuries could affect sensations like vision.[34] *On Wounds to the Head* never discusses symptoms of brain injury but focuses on repairing skull injuries. The brain is mentioned mostly to warn surgeons about the risks of penetrating the dura mater (the outer membrane) in the course of removing bone fragments or placing plates in the skull to repair injuries. In contrast, *Diseases* II discusses diseases of the head and brain in almost half of its chapters but does not always consider them important enough to name.[35]

The Latin author Celsus gives a much more rigorous account of brain and skull injuries. Unlike *On Wounds to the Head*, Celsus discusses the many symptoms which accompany brain injury: loss of vision, daze, bleeding from the nose or ears, loss of speech, spasming, and loss of consciousness.[36] Despite this greater appreciation for the symptoms of injury, Celsus claims that these symptoms never appear unless there is also a skull fracture. These symptoms are only useful for a surgeon to know to look for a fracture, precluding other forms of treating and understanding brain injuries. Celsus

---

[31] Rees, "We Need to Talk about Epizelus," 50.
[32] Plut. *Mar*. 45.2-3.
[33] Livy 21.53.
[34] Hippoc. *Morb*. 2.6.20.
[35] Julius Rocca, *Galen on the Brain: Anatomical Knowledge and Physiological Speculation in the Second Century AD*, Studies in Ancient Medicine (Leiden: Brill, 2003), 25.
[36] Celsus, *Med*. 8.4.1.

does add that on very rare occasions these symptoms (blindness in particular) are caused by aneurysm and not by skull fractures.[37]

Hippocrates and Celsus (and authors like Galen who reference both) focus on major head trauma while overlooking more common mild TBI, like concussions. Even though ancient doctors linked brain injury to the same short-term symptoms known to modern medicine, there was little concept of how common such injuries were or of how brain injury could cause long-term issues. Indeed, long-term effects that might have stemmed from brain injury were often attributed to something else. In addition to his claims about Marius, Plutarch described a man who fell unconscious after a brain injury. After three days, he awoke and changed his personality completely. Modern scientists would argue that the brain injury caused the change directly, but Plutarch claims that the man decided to change his life in a conscious decision after reflecting on the injury.[38] Often, Romans did not consider causes at all when discussing mental disorders among soldiers. Korneel van Lommel, in a survey of mental impairment among Roman soldiers, found that no source directly links combat experiences to the mental disorders identified in soldiers under court martial or requesting a discharge.[39]

Even the premier study of Roman combat trauma, "Caesar in Vietnam" by Aislinn Melchior, falls victim to underestimating the effects of minor TBI. Melchior's study is, like Rees, a critique of the notion that modern military experience is analogous to the ancient experience.[40] Melchior rightly focuses on the importance of TBI in the emergence of combat trauma. She argues that the prevalence of combat trauma in the ancient world must be much lower than in the modern world because the leading form of head trauma in modern combat is blast trauma. She further asserts that Romans were well-protected from the other forms of TBI by their helmets and thus likely did not have a great risk of combat trauma due to injury. Unfortunately, Melchior, writing in 2011, did not have access to recent studies about the causes, commonness, and effects of concussions. These studies, when compared with Roman accounts of battle, demand a re-evaluation of Melchior's conclusions. TBI would be common among

[37] Celsus, *Med.* 8.4.7.
[38] Plut. *De sera*, 563c ff.
[39] Van Lommel, "The Recognition of Roman Soldiers' Mental Impairment."
[40] Aislinn Melchior, "Caesar in Vietnam: Did Roman Soldiers Suffer from Post-Traumatic Stress Disorder?" *Greece and Rome: Journal of the Classical Association* Ser. 2, 58, no. 2 (2011): 209–23, https://doi.org/10.1017/S0017383511000052.

Roman soldiers, though that does not necessarily mean that PTSD and other combat traumas and disorders were common too.

Evidence from medical texts is born out in ancient accounts of injuries. Injuries to the head were common. Head injuries are the most common form of injury described in Homer and are often described in other fictional and many historical accounts of warfare. There are several surveys of injuries in the Homeric epics that show that head injury is the most common form of injury in the texts. One scholar calculates that 31 of the 190 injuries mentioned in Homer are head injuries and the vast majority were fatal.[41] Cyrus the Younger and Pyrrhus were both killed after receiving head injuries and falling from their horses.[42] Livy discusses one Roman leader whose skull was fractured in a battle against the Volsci.[43]

Analysis of human remains in the Roman world also reveals a surprising prevalence of cranial injuries. A survey of injuries from 17 subjects discovered in Herculaneum showed 7 individuals with broken bones in their skull or jaw which had healed (thus they were not received as a result of the eruption).[44] A survey of 160 skulls from Egypt dated to 332 BCE – 395 CE revealed 31 examples of cranial trauma with men and women having similar rates of injury at about 20%.[45] A study of 67 gladiators buried in Ephesus showed 11 had survived head injuries and 10 had died from a head injury.[46]

Specific examples from literature and archaeology are illustrative but they are not directly related to battle. The number of soldiers and veterans among the populations studied by archeologists is unknown, and the fighting performed by gladiators was likely quite different than a typical

---

[41] Nikitas Nomikos, "Injuries in the Greek Epics of Homer," *Chinese Journal of Traumatology* 21, no. 2 (April 1, 2018): 109, https://doi.org/10.1016/j.cjtee.2017.09.005.

[42] Eleni Konsolaki et al., "Cranial Trauma in Ancient Greece: From Homer to Classical Authors," *Journal of Cranio-Maxillo-Facial Surgery* 38, no. 8 (December 2010): 551–52, https://doi.org/10.1016/j.jcms.2010.02.009.

[43] Livy 4.28.7.

[44] Luigi Capasso et al., "Lesions Linked to Athletic Activities in the Ancient Roman Population from Herculaneum (Italy, First Century Ad)," *Anthropologie* 42, no. 2 (2004): 181–88.

[45] Moushira Erfan et al., "Cranial Trauma in Ancient Egyptians from the Bahriyah Oasis, Greco-Roman Period," *Research Journal of Medicine and Medical Sciences* 4 (January 1, 2009): 78–84.

[46] Fabian Kanz and Karl Grossschmidt, "Head Injuries of Roman Gladiators," *Forensic Science International* 160, no. 2 (2006): 208.

infantryman. Therefore, accounts of battle must be analyzed more generally in light of modern research in TBI to determine how great a risk Roman soldiers faced when they entered battle.

Roman Battle

According to Caesar, the Battle of Pharsalus began with his soldiers charging Pompey's forces while Pompey's remained in their place. After Caesar's soldiers threw their javelins, Pompey's cavalry forced Caesar's cavalry to retreat then attacked Caesar's soldiers from the flank. Caesar's fourth line forced Pompey's cavalry into flight, neutralized the archers, and attacked Pompey's left flank which allowed for Caesar's eventual victory.[47]

Pharsalus was not a textbook Roman battle, but it does show the various risks a soldier faced.[48] There was an atmosphere of anxiety, an exchange of missiles, a headlong charge (though usually, both armies would charge at each other), cavalry and light troops flanking the enemy, and the eventual breaking of the formation. Each phase of that battle introduced Romans to different risks of head trauma.[49] Enemies could hurl javelins, stones, and arrows as the legionaries themselves discharged their *pila*. Romans' long shields and bronze helmets protected them from penetrating injuries while their forward charge meant that the missiles that did hit them would hit them directly without rotational force. Cavalry, however, could launch missiles while maneuvering, or from the flank as at Pharsalus. When these found their mark, they caused rotational force, which has been linked to a higher risk of concussion.[50] Sling bolts were common and were likely to cause

---

[47] Caes. *BCiv*. 3.93-94.

[48] For general discussions of Roman battle, see Philip Sabin, "The Face of Roman Battle," *The Journal of Roman Studies* 90 (2000): 1–17, https://doi.org/10.2307/300198; Catherine Gilliver, "Battle," in *The Cambridge History of Greek and Roman Warfare*, ed. Philip Sabin, Hans Van Wees, and Michael Whitby, vol. II: Rome from the Late Republic to the Late Empire. (Cambridge: Cambridge University Press, 2007), 122–57; Adam O. Anders, "The 'Face of Roman Skirmishing,'" *Historia: Zeitschrift Für Alte Geschichte* 64, no. 3 (2015): 263–300.

[49] Caes. *BCiv*. 3.92, *BHisp*. 31.

[50] Svein Kleiven, "Why Most Traumatic Brain Injuries Are Not Caused by Linear Acceleration but Skull Fractures Are," *Frontiers in Bioengineering and Biotechnology* 1, no. 15 (November 7, 2013), https://doi.org/10.3389/fbioe.2013.00015.

concussion.[51] In a series of skirmishes against Caesar's camps leading to the Battle of Pharsalus, not a single member of Caesar's defenders was unwounded, and four lost an eye.[52] While this is a siege scenario and not an open battle, these injuries are clear examples of the effectiveness of missiles against Roman armor, and the blindness is clear evidence of major head trauma all the more impressive since the injured were the defenders.

Romans often used their shield as a primary weapon. The front line would bash with their shields leaving downed enemies for the second line to incapacitate.[53] This action would create exactly the type of rotational force that could cause minor concussions. Modern studies have shown that the positions most at risk for concussion in American football are linemen who do not make spectacular body-hurling hits, but constantly push and jar against each other.[54] Each push they give and receive creates force on their heads and necks even if they do not make direct contact between their head and another player, leading to dozens of opportunities for minor injury. Roman soldiers would have faced greater force because they had the added mass of their shields and the added force of soldiers around them. In addition, Romans cycled soldiers between the front lines of battle, as Caesar did in Pharsalus with his third line, thus exposing more soldiers to these hits but also potentially diminishing the number of hits any one soldier would take.[55]

Melchior argued that helmets would protect ancient soldiers from TBI in these circumstances.[56] Indeed, well-made bronze helmets would have been effective at preventing penetrating brain injury. In fact, a bronze helmet is capable of withstanding the force of any ancient weapon except for a large

---

[51] Celsus, *Med.* 5.26, 7.55. Adrian Keith Goldsworthy, *The Roman Army at War, 100 BC-AD 200* (Oxford: Clarendon Press, 2009), 185.
[52] Caes. *BCiv.* 3.53.
[53] Plut. *Caes.* 16.1; Tac. *Agr.* 36; Tac. *Ann.* 2.14, 14.36-37; Livy 9.41.19; Goldsworthy, *Roman Army*, 208.
[54] Julian E. Bailes et al., "Role of Subconcussion in Repetitive Mild Traumatic Brain Injury: A Review," *Journal of Neurosurgery* 119, no. 5 (November 1, 2013): 1235–45, https://doi.org/10.3171/2013.7.JNS121822; Giacomo Querzola et al., "A Semi-Quantitative Sport-Specific Assessment of Recurrent Traumatic Brain Injury: The TraQ Questionnaire and Its Application in American Football," *Neurological Sciences* 40, no. 9 (September 1, 2019): 1909–15, https://doi.org/10.1007/s10072-019-03853-z.
[55] Caes. *BCiv.* 3.94; Gilliver, "Battle," 133.
[56] Melchior, "Caesar in Vietnam," 219.

ax.[57] Unfortunately, helmets that have been excavated are often notable for poor craftsmanship and thus were less protective than the ideal modern analog.[58] In addition, helmets do not protect from open TBI caused by blunt force. Even the best football helmets only prevented concussions less than 20% better than no helmet at all.[59] This is because concussion is the result of the brain making contact with the skull. Hits that cause the head to move from side to side or cause the neck to twist cause more movement in the brain. That means that in the most common circumstances, helmets would not be able to prevent the most common source of brain injury. This is true of modern athletics and ancient warfare.

While battles carried a high risk of injury, a Roman soldier was not likely to face battle or endure a siege or rush a fortification more than a few times in his career. Therefore, it would be hard to claim that Romans were exposed to the same risks of head trauma that modern athletes face purely from battle. Romans did, however, engage in training exercises and maintenance work that may also expose them to harm.

Polybius and Josephus both claim that Roman soldiers engaged in grueling training. Josephus goes so far as to claim that Roman soldiers trained at the same ferocity as they battled, calling their training "bloodless battles" with wooden weapons.[60] Polybius reveals that the practice was ancient: during the Second Punic War, Scipio Africanus drilled his troops with wooden weapons every two out of five days with other physical exercises on the off days.[61] These programs would make the risks of battle a regular occurrence. Evidence from the Roman Empire reveals that soldiers faced great hardship. Tacitus described one soldier complaining about corporal punishment, marches, and labor.[62] A Roman soldier's body excavated at Viminacium revealed clear

---

[57] Richard A. Gabriel, *Man and Wound in the Ancient World: A History of Military Medicine from Sumer to the Fall of Constantinople*, vol. 1st ed (Washington, D.C.: Potomac Books, 2012), 6–7.
[58] Raffaele D'Amato and Graham Sumner, *Arms and Armour of the Imperial Roman Soldier from Marius to Commodus, 112 BC - AD 192* (London: Frontline Books, 2010), 34.
[59] Sone et al., "Helmet Efficacy."
[60] Joseph. *BJ*, 3.5.1
[61] Polyb. 10.20.1-7.
[62] Tac. *Ann.* 1.17.4.

stress on the shoulder blades, pelvis, knees, and ankles from hard work.[63]

In the Republic, commanders were often replaced every year or two. Under the empire, a soldier's life depended on the commander on base. Some, like Scipio Aemilianus (adopted grandson of Africanus) and Marius, were known for drilling their men hard and focusing on training.[64] These men gained their reputation because many of their peers did not focus on drilling their soldiers. Indeed, the success of the army was inconsistent and very reliant on the commander which indicates that the soldiers themselves were not always well-trained and relied on leadership, not training, to know what to do.[65] Appian describes the soldiers in Spain as lazy and pampered before Aemilianus arrived, strongly implying criticism of Aemilianus's peers.[66] Other ancient authors describe training as less extreme. Vegetius claims that Roman soldiers drilled by aiming at posts hammered into the ground.[67] While this would expose the soldier to forces that could cause minor concussions, it was much safer than practicing the same actions against another soldier as Josephus implies.

Life after Injury

Roman jurists discuss mental disorders of soldiers in several places. Soldiers could be discharged for both physical and mental reasons, and those discharges could be honorable or dishonorable depending on the situations in which they received their injury.[68] Ulpian specifically mentions deaf and mute soldiers could make a will and be discharged.[69] Both injuries are linked to brain trauma. Soldiers who attempted suicide were court-martialed, and jurists list several extenuating circumstances that could grant soldiers leniency. Among them were sadness, weariness of life, and

---

[63] Snezana Golubovic, Nemanja Mrdic, and C. Scott Speal, "Killed by the Arrow: Grave No. 152 from Viminacium," in *Waffen in Aktion: Xantener Berichte 16. Akten Der 16. Internationalen Roman Military Equipment Conference (ROMEC) Xanten, 13.-16. Juni 2007*, ed. Alexandra W. Busch and Hans-Joachim Schalles (Mainz: Verlag Philipp Von Zabern, 2009), 55–64.

[64] App. *Hisp*. 85-86; Plut. *Mar*. 13. Plutarch claims that Marius was imitating his mentor, Aemilianus.

[65] Goldsworthy, *Roman Army*, 174.

[66] App. *Hisp*. 84.

[67] Veg. *Mil*. 1.11, 3.4.

[68] *Dig*. 49.16.13.3, 3.2.2.2; Korneel Van Lommel, "Terminology of the Medical Discharge," *Ancient History Bulletin* 27 (2013): 67.

[69] *Dig*. 29.1.4.

madness.[70] The same passage also discussed soldiers who harmed themselves (or perhaps another soldier) while drunk or in a manic state.[71] Suicidal ideation, depression, anger, alcoholism, and the inability to control impulses are commonly linked to TBI as discussed above. Curiously, Romans did not seem to link these behavioral issues to the soldiers' experiences in the military, even though Romans enlisted men based on physical health and behavior.[72]

Blinded soldiers, accounts of head injuries, and the focus on skull fractures in Roman medical texts reveal that TBI was a clear risk for soldiers even if the behavioral issues outlined above are impossible to link. Beyond the jurists, Roman soldiers and veterans had a reputation for being brutish, violent, and temperamental. Juvenal describes soldiers who can beat people on the street with impunity.[73] Soldiers were commonly described as "*impius*" in Roman poetry.[74] In the late Republic, Rome began offering veterans land after their service ended, but those soldiers seem to have had issues making a life there. Sallust claims that Sulla's veterans joined Catiline after they "spent their money too extravagantly" but his judgment may be masking a simple inability to adapt to farming after over a decade at war.[75] While some historians have argued that these were urban poor who were thrust into a lifestyle unfamiliar to them, it is equally likely that these were sons of tenant farmers or from families of farmers who had recently left the land.[76] In either case, they received land and often spoils which should have enabled them to succeed in their family's business even if they were not farmers.

While this is often discussed as an economic or social issue, there could be a pathological explanation. Combat trauma and brain injury often manifest in behavioral issues that make it difficult for a person to follow social norms. A man like Catiline described as violent, unable to keep hold of his money, and unable to regulate his physical comfort, fits the pathology.[77] The men

[70] *Dig*. 48.19.38.12, 49.16.6.7; Van Lommel, "The Recognition of Roman Soldiers' Mental Impairment."
[71] *Per vinum aut lasciviam*.
[72] Veg. *Mil*. 1.6-7.
[73] Juv. *Sat*. 16.
[74] Verg. *Ecl*. 1.70; Mart. *Ep*. 5.69.
[75] Sall. *Cat*. 16.
[76] L. de Ligt, "Poverty and Demography: The Case of the Gracchan Land Reforms," *Mnemosyne* 57, no. 6 (2004): 725–57 offers a detailed summary of scholarship surrounding Roman agrarian laws.
[77] Sall. *Cat*. 5.

who followed him in Etruria who defaulted on loans or became addicted to gambling also follow that pathology. Interestingly, these issues arose around the decades-long service under Sulla, Pompey, and Caesar whose soldiers tried to settle back in Italy but are left to the jurists in periods when Rome's army was less demanding or mostly filled with non-Italians.[78] Indeed, even men with subtle injuries (mental and physical) from war would face difficulty forming a strong reputation, and the more visible the veteran was in the cultural centers of the empire, the more likely surviving texts would discuss his disorders.[79]

Framing ancient descriptions of soldiers with a strong understanding of TBI based on modern research allows us to better understand the risks of ancient war. While it is irresponsible to diagnose Romans according to modern definitions, TBI gives a clear mechanical and biological explanation for certain behaviors commonly linked to veterans.

Bibliography

Abbas, Kausar, Trey E. Shenk, Victoria N. Poole, Meghan E. Robinson, Larry J. Leverenz, Eric A. Nauman, and Thomas M. Talavage. "Effects of Repetitive Sub-Concussive Brain Injury on the Functional Connectivity of Default Mode Network in High School Football Athletes." *Developmental Neuropsychology* 40, no. 1 (January 2, 2015): 51–56. https://doi.org/10.1080/87565641.2014.990455.

Alston, Richard. *Soldier and Society in Roman Egypt : A Social History*. London ; New York : Routledge, 1995. http://archive.org/details/soldiersocietyin0000alst.

Anders, Adam O. "The 'Face of Roman Skirmishing.'" *Historia: Zeitschrift Für Alte Geschichte* 64, no. 3 (2015): 263–300.

Bailes, Julian E., Anthony L. Petraglia, Bennet I. Omalu, Eric Nauman, and Thomas Talavage. "Role of Subconcussion in Repetitive Mild

---

[78] In the Republic, Romans served between 6 and 16 years, Augustus raised that to 20: Polyb. 6.19.1; Augustus, *Res Gestae Divi Augusti*, 17; Cass. Dio 55.23.1. On non-Italian soldiers: Yann Le Bohec, *The Imperial Roman Army* (New York : Hippocrene Books ; London : Batsford, 1994), 76, http://archive.org/details/imperialromanarm0000lebo; Richard Alston, *Soldier and Society in Roman Egypt : A Social History* (London ; New York : Routledge, 1995), 42–43, http://archive.org/details/soldiersocietyin0000alst.

[79] Korneel Van Lommel, "Heroes and Outcasts: Ambiguous Attitudes Towards Impaired and Disfigured Roman Veterans," *Classical World* 109, no. 1 (2015): 104.

Traumatic Brain Injury: A Review." *Journal of Neurosurgery* 119, no. 5 (November 1, 2013): 1235–45. https://doi.org/10.3171/2013.7.JNS121822.

Broshek, Donna K., Anthony P. De Marco, and Jason R. Freeman. "A Review of Post-Concussion Syndrome and Psychological Factors Associated with Concussion." *Brain Injury* 29, no. 2 (January 28, 2015): 228–37. https://doi.org/10.3109/02699052.2014.974674.

Capasso, Luigi, Lia Pierfelice, Elisabetta Michetti, Antonietta Di Fabrizio, and Ruggero D'Anastasio. "Lesions Linked to Athletic Activities in the Ancient Roman Population from Herculaneum (Italy, First Century Ad)." *Anthropologie* 42, no. 2 (2004): 181–88.

Castellani, Rudy J., and George Perry. "Tau Biology, Tauopathy, Traumatic Brain Injury, and Diagnostic Challenges." *Journal of Alzheimer's Disease* 67, no. 2 (January 1, 2019): 447–67. https://doi.org/10.3233/JAD-180721.

Collins, Michael W., Mark R. Lovell, Grant L. Iverson, Robert C. Cantu, Joseph C. Maroon, and Melvin Field. "Cumulative Effects of Concussion in High School Athletes." *Neurosurgery* 51, no. 5 (November 2002): 1175–79; discussion 1180-1181. https://doi.org/10.1097/00006123-200211000-00011.

D'Amato, Raffaele, and Graham Sumner. *Arms and Armour of the Imperial Roman Soldier from Marius to Commodus, 112 BC - AD 192*. London: Frontline Books, 2010.

Daneshvar, Daniel H., Christine M. Baugh, Christopher J. Nowinski, Ann C. McKee, Robert A. Stern, and Robert C. Cantu. "Helmets and Mouth Guards: The Role of Personal Equipment in Preventing Sport-Related Concussions." *Clinics in Sports Medicine* 30, no. 1 (January 1, 2011): 145–63. https://doi.org/10.1016/j.csm.2010.09.006.

Edwards, George, Jing Zhao, Pramod K. Dash, Claudio Soto, and Ines Moreno-Gonzalez. "Traumatic Brain Injury Induces Tau Aggregation and Spreading." *Journal of Neurotrauma* 37, no. 1 (January 2020): 80–92. https://doi.org/10.1089/neu.2018.6348.

Erfan, Moushira, Aly El-Sawaf, Muhammad Al-Tohamy, Azza Soliman, El-Din Sarry, Wafaa Kandeel, Rokia Abd, El-Shafy El-Banna, and Ayman Azab. "Cranial Trauma in Ancient Egyptians from the Bahriyah Oasis, Greco-Roman Period." *Research Journal of Medicine and Medical Sciences* 4 (January 1, 2009): 78–84.

Gabriel, Richard A. *Man and Wound in the Ancient World: A History of Military Medicine from Sumer to the Fall of Constantinople*. Vol. 1st ed. Washington, D.C.: Potomac Books, 2012.

Gilliver, Catherine. "Battle." In *The Cambridge History of Greek and Roman Warfare*, edited by Philip Sabin, Hans Van Wees, and Michael Whitby, II: Rome from the Late Republic to the Late Empire, 122–57. Cambridge: Cambridge University Press, 2007.

Goldsworthy, Adrian Keith. *The Roman Army at War, 100 BC-AD 200*. Oxford: Clarendon Press, 2009.

Golubovic, Snezana, Nemanja Mrdic, and C. Scott Speal. "Killed by the Arrow: Grave No. 152 from Viminacium." In *Waffen in Aktion: Xantener Berichte 16. Akten Der 16. Internationalen Roman Military Equipment Conference (ROMEC) Xanten, 13.-16. Juni 2007*, edited by Alexandra W. Busch and Hans-Joachim Schalles, 55–64. Mainz: Verlag Philipp Von Zabern, 2009.

Gorgoraptis, Nikos, Lucia M. Li, Alex Whittington, Karl A. Zimmerman, Linda M. Maclean, Claire McLeod, Ewan Ross, et al. "In Vivo Detection of Cerebral Tau Pathology in Long-Term Survivors of Traumatic Brain Injury." *Science Translational Medicine* 11, no. 508 (September 4, 2019): eaaw1993. https://doi.org/10.1126/scitranslmed.aaw1993.

Kanz, Fabian, and Karl Grossschmidt. "Head Injuries of Roman Gladiators." *Forensic Science International* 160, no. 2 (2006): 207–16.

Kleiven, Svein. "Why Most Traumatic Brain Injuries Are Not Caused by Linear Acceleration but Skull Fractures Are." *Frontiers in Bioengineering and Biotechnology* 1, no. 15 (November 7, 2013). https://doi.org/10.3389/fbioe.2013.00015.

Konsolaki, Eleni, Elisabeth Astyrakaki, George Stefanakis, Panos Agouridakis, and Helen Askitopoulou. "Cranial Trauma in Ancient Greece: From Homer to Classical Authors." *Journal of Cranio-Maxillo-Facial Surgery* 38, no. 8 (December 2010): 549–53. https://doi.org/10.1016/j.jcms.2010.02.009.

Le Bohec, Yann. *The Imperial Roman Army*. New York : Hippocrene Books ; London : Batsford, 1994. http://archive.org/details/imperialromanarm0000lebo.

Ligt, L. de. "Poverty and Demography: The Case of the Gracchan Land Reforms." *Mnemosyne* 57, no. 6 (2004): 725–57.

Mac Donald, Christine L., Ann M. Johnson, Linda Wierzechowski, Elizabeth Kassner, Theresa Stewart, Elliot C. Nelson, Nicole J. Werner, et al. "Prospectively Assessed Clinical Outcomes in Concussive Blast vs Nonblast Traumatic Brain Injury Among Evacuated US Military Personnel." *JAMA Neurology* 71, no. 8 (August 1, 2014): 994–1002. https://doi.org/10.1001/jamaneurol.2014.1114.

MacGregor, Andrew J., Amber L. Dougherty, Rosemary H. Morrison, Kimberly H. Quinn, and Michael R. Galarneau. "Repeated Concussion among U.S. Military Personnel during Operation Iraqi Freedom." *Journal of Rehabilitation Research & Development* 48, no. 10 (December 15, 2011): 1269–77. https://doi.org/10.1682/JRRD.2011.01.0013.

Marar, Mallika, Natalie M. McIlvain, Sarah K. Fields, and R. Dawn Comstock. "Epidemiology of Concussions Among United States High School Athletes in 20 Sports." *The American Journal of Sports Medicine* 40, no. 4 (April 1, 2012): 747–55. https://doi.org/10.1177/0363546511435626.

Melchior, Aislinn. "Caesar in Vietnam: Did Roman Soldiers Suffer from Post-Traumatic Stress Disorder?" *Greece and Rome* Ser. 2, 58, no. 2 (2011): 209–23. https://doi.org/10.1017/S0017383511000052.

Military Health System. "DOD TBI Worldwide Numbers." Accessed June 15, 2022. https://armymedicine.health.mil/MHSHome/Military-Health-Topics/Centers-of-Excellence/Traumatic-Brain-Injury-Center-of-Excellence/DOD-TBI-Worldwide-Numbers.

Mouzon, Benoit, Corbin Bachmeier, Joseph Ojo, Christopher Acker, Scott Ferguson, Gogce Crynen, Peter Davies, Michael Mullan, William Stewart, and Fiona Crawford. "Chronic White Matter Degeneration, but No Tau Pathology at One-Year Post-Repetitive Mild Traumatic Brain Injury in a Tau Transgenic Model." *Journal of Neurotrauma* 36, no. 4 (February 15, 2019): 576–88. https://doi.org/10.1089/neu.2018.5720.

NCAA.org. "Concussion Safety Protocol Management." Accessed October 17, 2022. https://www.ncaa.org/sports/2016/7/20/concussion-safety-protocol-management.aspx.

NCAA.org. "NCAA, DoD Launch Concussion Study." Accessed October 17, 2022. https://www.ncaa.org/news/2014/5/29/ncaa-dod-launch-concussion-study.aspx.

NFL Head, Neck, and Spine Committee. "NFL Concussion Diagnosis and Management Protocol," 2022. https://static.www.nfl.com/image/upload/v1665264248/league/moexjmdzy2kvjtqsdpbx.pdf.

"NIH Chronic Traumatic Encephalopathy Diagnosis Conference | National Institute of Neurological Disorders and Stroke." Accessed November 9, 2022. https://www.ninds.nih.gov/current-research/focus-disorders/focus-traumatic-brain-injury-research/nih-chronic-traumatic-encephalopathy-diagnosis-conference.

Nomikos, Nikitas. "Injuries in the Greek Epics of Homer." *Chinese Journal of Traumatology* 21, no. 2 (April 1, 2018): 109–12. https://doi.org/10.1016/j.cjtee.2017.09.005.

Nowinski, Christopher J., Samantha C. Bureau, Michael E. Buckland, Maurice A. Curtis, Daniel H. Daneshvar, Richard L. M. Faull, Lea T. Grinberg, et al. "Applying the Bradford Hill Criteria for Causation to Repetitive Head Impacts and Chronic Traumatic Encephalopathy." *Frontiers in Neurology* 13 (2022). https://doi.org/10.3389/fneur.2022.938163.

Ojo, Joseph O., Benoit Mouzon, Moustafa Algamal, Paige Leary, Cillian Lynch, Laila Abdullah, James Evans, et al. "Chronic Repetitive Mild Traumatic Brain Injury Results in Reduced Cerebral Blood Flow, Axonal Injury, Gliosis, and Increased T-Tau and Tau Oligomers." *Journal of Neuropathology & Experimental Neurology* 75, no. 7 (July 1, 2016): 636–55. https://doi.org/10.1093/jnen/nlw035.

Pischiutta, Francesca, Edoardo Micotti, Jennifer R. Hay, Ines Marongiu, Eliana Sammali, Daniele Tolomeo, Gloria Vegliante, et al. "Single Severe Traumatic Brain Injury Produces Progressive Pathology with Ongoing Contralateral White Matter Damage One Year after Injury." *Experimental Neurology* 300 (February 1, 2018): 167–78. https://doi.org/10.1016/j.expneurol.2017.11.003.

Poole, Victoria N., Evan L. Breedlove, Trey E. Shenk, Kausar Abbas, Meghan E. Robinson, Larry J. Leverenz, Eric A. Nauman, Ulrike Dydak, and Thomas M. Talavage. "Sub-Concussive Hit Characteristics Predict Deviant Brain Metabolism in Football Athletes." *Developmental Neuropsychology* 40, no. 1 (January 2, 2015): 12–17. https://doi.org/10.1080/87565641.2014.984810.

Querzola, Giacomo, Carlo Lovati, Claudio Mariani, and Leonardo Pantoni. "A Semi-Quantitative Sport-Specific Assessment of Recurrent Traumatic Brain Injury: The TraQ Questionnaire and Its Application in American Football." *Neurological Sciences* 40, no. 9 (September 1, 2019): 1909–15. https://doi.org/10.1007/s10072-019-03853-z.

Rees, Owen. "We Need to Talk about Epizelus: 'PTSD' and the Ancient World." *Medical Humanities* 46, no. 1 (March 1, 2020): 46–54. https://doi.org/10.1136/medhum-2018-011557.

Rocca, Julius. *Galen on the Brain: Anatomical Knowledge and Physiological Speculation in the Second Century AD*. Studies in Ancient Medicine. Leiden: Brill, 2003.

Sabin, Philip. "The Face of Roman Battle." *The Journal of Roman Studies* 90 (2000): 1–17. https://doi.org/10.2307/300198.

Schertz, Peter, Pamela Hatchfield, Richard Newman, and Reno Pisano. "Made from Life: A Roman Terracotta Portrait in the Museum of Fine Arts, Boston." *Memoirs of the American Academy in Rome* 61 (2016): 22–39.

Shay, Jonathan. *Achilles in Vietnam: Combat Trauma and the Undoing of Character*. New York: Scribner, 2003.

Sone, Je Yeong, Douglas Kondziolka, Jason H. Huang, and Uzma Samadani. "Helmet Efficacy against Concussion and Traumatic Brain Injury: A Review." *Journal of Neurosurgery* 126, no. 3 (March 1, 2017): 768–81. https://doi.org/10.3171/2016.2.JNS151972.

Tagge, Chad A., Andrew M. Fisher, Olga V. Minaeva, Amanda Gaudreau-Balderrama, Juliet A. Moncaster, Xiao-Lei Zhang, Mark W. Wojnarowicz, et al. "Concussion, Microvascular Injury, and Early Tauopathy in Young Athletes after Impact Head Injury and an Impact Concussion Mouse Model." *Brain* 141, no. 2 (February 1, 2018): 422–58. https://doi.org/10.1093/brain/awx350.

Trittle, Lawrence A. *From Melos to My Lai: War and Survival*. London: Routledge, 2000. http://www.crcnetbase.com/isbn/9780203462485.

—. "Xenophon's Portrait of Clearchus: A Study in Post-Traumatic Stress Disorder." In *Xenophon and His World: Papers from a Conference Held in Liverpool in July 1999*, edited by Christopher Tuplin, 325–39. Stuttgart: Franz Steiner Verlag, 2004. https://public.ebookcentral.proquest.com/choice/publicfullrecord.aspx?p=5806270.

Van Lommel, Korneel. "The Recognition of Roman Soldiers' Mental Impairment." *Acta Classica* 56 (January 1, 2013): 155–85.

—. "Terminology of the Medical Discharge." *Ancient History Bulletin* 27 (2013): 65–74.

—. "Heroes and Outcasts: Ambiguous Attitudes Towards Impaired and Disfigured Roman Veterans." *Classical World* 109, no. 1 (2015): 91–117.

WITH THE COURAGE OF AN EMPEROR: *VIRTUS* AND THE EMPRESS IN ROMAN PROPAGANDA

JULIA WETZEL

Introduction

To the Romans, war was a man's world and *virtus* the embodiment of male pride. Whereas v*irtus* is the most important male virtue, *pudicitia* is the defining virtue for women. In "Hadrian's Chastity," Carlos Noreña breaks this idea of traditional Roman virtues and scholarly assumptions that virtues were assigned to specific genders.[1] Whereas Noreña focuses on *pudicitia*, this paper centers on the versatility of *virtus*. Traditionally *virtus* is reserved only for males, but it is evident that there were rare instances when women could possess *virtus*. Second- and third-century empresses such as Faustina the Elder and Younger, and Julia Domna, are examples of powerful women who aligned themselves with their husband's virtues. Literary evidence often cites mythical women when mentioned in war, and most of the scholarship surrounding *virtus* looks at the late Republic through the Julio-Claudians. Therefore, coins are used to connect women to martial courage. Coins offer a unique link between imperial portraits and the reverse image, often of a god, goddess, or virtue. Images of goddesses such as Diana, Minerva, and Venus Victrix, who were associated with martial and masculine qualities, point to instances in which women could hold male qualities. A comparison of these same divinities in the emperor's coinage sheds light on their relationship with the imperial couple. Drawing attention to the empress's relation to masculine qualities as well as feminine ones orients the empress's image in Roman propaganda, revealing that the qualities of strength and courage associated with *virtus* were seen in conjunction with the empress to aid her husband's image and the empire.

[1] Carlos Noreña, "Hadrian's Chastity," *Phoenix* 61, no. 3/4 (2007): 296-317.

The multifaceted role of *Virtus*

Virtus was portrayed in multiple ways in Roman propaganda and thus requires a multifaceted definition that is important to its understanding. A full understanding of *virtus* also explains why women could not possess this virtue. Whereas *virtus* was a military virtue, it was also a word and abstract concept that became personified as a god in monumental propaganda. We understand traditional *virtus* as "manliness" and "courage" in battle, as well as "control over one's body and physical impulses."[2] *Virtus* was coupled with *disciplina militaris* which was the strict discipline that comes with military training to withstand the trials of war. *Virtus* was produced through military discipline and training, and when left undisciplined produced violence and anger.[3] All these things women were not allowed to do or be. Indeed, the Romans associated *virtus*, training, and discipline with their continued victory over "those less in control of their bodies, emotions, or political affairs," meaning women, slaves, and foreigners.[4]

Virtus can also be seen as a personification of an abstract quality and word. *Virtus* could not apply to women because it derived from the Latin word for man, or *vir*. When referring to courageous women, Romans used the term *fortitudo*, or strength, when referring to courageous women.[5] Although a different word was used, it was still rare for women to be seen as courageous.[6] Despite its etymology, *virtus* is a feminine noun with masculine construction. Personifications such as *Virtus* were the Romans' attempt to organize their identity externally.[7] Each personification represented a part of Roman identity that became symbolic and separate from the person. Once a virtue was personified, it fell into the social hierarchy. Personifications then took on a feminine gender, not because abstract nouns took a feminine gender, but because the Romans were dominant in the social hierarchy.[8] The Romans must take the masculine

---

[2] Julie Langford, *Maternal Megalomania: Julia Domna and the Imperial Politics of Motherhood* (Baltimore, Maryland: The John Hopkins University Press, 2013), 25.
[3] Sara Elise Phang, *Roman Military Service: Ideologies of Discipline in the Late Republic and Early Principate* (Cambridge Univ. Press, 2008), 3.
[4]Phang, *Roman Military Service*, 3.
[5] Myles Anthony McDonnell, *Roman Manliness: Virtus and the Roman Republic* (Cambridge: Cambridge University Press, 2009), 161.
[6] McDonnell, *Roman Manliness,* 161.
[7] Alex Dressler, *Personification and the Feminine in Roman Philosophy* (Cambridge: Cambridge University Press, 2016), 63.
[8] Dressler, *Personification,* 82.

subject so then personifications took the submissive feminine gender because taking the neuter gender would "dispersonify" them.[9] This gendered construction allowed the Romans to subjugate even virtues to their will and to "split the feminine from the female, so that the attribute was used to convey the good qualities" without needing the actual woman.[10] These different perspectives demonstrate the gendered dynamic in which personified virtues often find themselves. Where personifications and virtues represented an important part of Roman identity, they also expressed the complexity of social norms. *Virtus* was an important male virtue but also took a feminine representation and role in the social hierarchy. This context allowed the virtue to take on feminine characteristics even before *virtus* was applied to an empress, extending the realm of Roman identity to women.

A carefully crafted Image

Individual *virtus* began as a way to consolidate Augustus' charisma into a tangible and stable system of authority.[11] Once he consolidated his reign, Augustus figured out a way to pass his image to his successors. Unlike military maneuvers, which were easily formulated in training and easy to repeat and observe, charismatic qualities were unstable, in the sense that they were not easy to duplicate and invoked emotion.[12] *Virtus* became a stable way to produce great leaders while protecting Rome. Augustus is one example, but Rome's history is littered with hundreds of similar instances. Ancient historians formed and sculpted the definition of *virtus* through the stories they decided to include in their rendition of history. Roman historians were seen as "guides" who used historic and mythological examples to give a moral education.[13] If a reader wished for a particular outcome, then they only needed to follow the examples written in history.[14] Where history is unkind to women, mythology is not. In addition to the goddesses mention here, there are a few other examples of strong and courageous women. Virgil's Camilla possessed great leadership skills that won her honor like men who fight in war, while Ovid's Atalanta takes on the Calydonian boar.[15] Although rare, women could be seen in martial

---

[9] Dressler, *Personification*, 91.
[10] Dressler, *Personification*, 140.
[11] Phang, *Roman Military Service*, 22.
[12] Phang, *Roman Military Service,* 22.
[13] Catalina Balmaceda, *Virtus Romana: Politics and Morality in the Roman Historians* (Chapel Hill, NC: University of North Carolina Press, 2017), 242.
[14] Balmaceda, *Virtus Romana,* 246.
[15] Verg. *Aen.* 11.532-543; Ov. *Met.* 8.262-425.

settings. Historical writing created gendered expectations and traditions that hardly changed. Men and women knew their social expectations and societal beliefs from an early age, and who could argue with Rome's historic success. Yet in the case of the royal family, the charisma of the emperor expanded to cover the empress, allowing her to participate in the same propaganda campaign and identifying her image with her husband.[16]

A Transition of Images

The Julio-Claudians set the stage not only for imperial policy but for the imagery that communicated this ideology to the masses. Every dynasty that followed would pick and choose how it continued this legacy to support its own reign. Dynastic imagery was a mix of Greek and Roman ideas that communicated the dynasty's ideology to the public. Augustus' charismatic qualities were an accessory to the imperial image that helped communicate this ideology and were broken down into individual traits like *virtus.*[17] Women were originally not connected to *virtus*, but as their image and role changed over time, it moved into the sphere of their husband's image.[18] The use of similar strategies in the empress's image as her husband reflects his choices.[19] Looking at the transition of images through the first, second, and third centuries will set the context for feminine masculinity and the growth of the empress's image in Roman propaganda.

Women possessed little power in a militaristic culture they could not participate in. Numismatic and monumental evidence highlighted militaristic achievement to enhance the authority, glory, and honor of men.[20] Arches and columns boasted these achievements so fathers could pass these merits to their male children.[21] Women were links to male relatives for children to inherit honor and authority.[22] The Julio-Claudian women thus became links to the emperor for inheritance and dynastic continuity. Augustus's wife, Livia, was the fulcrum for the dynasty and her husband's

---

[16] Mary T. Boatwright, "Women and Gender in the Roman Forum," *Transactions and Proceedings of the American Philological Association* 141, no. 1 (2011): 134.
[17] Mary Taliaferro Boatwright, *The Imperial Women of Rome: Power, Gender, Context* (Oxford: Oxford University Press, 2021), 34.
[18] Boatwright, *The Imperial Women of Rome,* 34.
[19] Boatwright, *The Imperial Women of Rome*, 127.
[20] Charles Brian Rose, *Dynastic Commemoration and Imperial Portraiture in the Julio-Claudian Period* (Cambridge: Cambridge University Press, 1997), 7.
[21] Rose, *Dynastic Commemoration*, 7.
[22] Rose, *Dynastic Commemoration*, 16.

imperial policy. Her role as a mother became her image and identity as her children made their way into dynastic propaganda.[23] Children became a way to continue and commemorate a dynasty and were seen as the future of the empire.[24] Deities and personifications often held children to emphasize their association and continuation with the new regime. The "Tellus" panel on the Ara Pacis is the first and most famous example of children used in this manner, but another example exists in Faustina the Youngers coinage with the personification of piety, *Pietas,* holding a cornucopia and flower with a child at her feet.[25] Augustus' search for an heir serves as a reminder that when a direct heir could not be secured, adoption became a way to ensure the dynasty. This strategy is also seen with the Trajanic and Hadrianic dynasties. Imperial women were also used in imagery surrounding adopted heirs to connect the new heir to the dynasty and secure legitimacy. Augustus' death marked the adoption of Livia into the Julian family as his daughter, linking the clans and giving the Claudians access to the Julian divine ancestry.[26] A connection to Livia represented those divine connections. It was a way to communicate the "position of the emperor within the imperial family" by manipulating the women around him to suit his needs.[27] This connection was stressed in several ways but commemoration to an imperial mother was one of the more common ways to highlight a link to the imperial house. Where a mother or wife on the reverse of a coin was one way to commemorate imperial women, another can be seen through a *consecratio.* Antoninus Pius honors his wife with a golden *aureus* with the inscription, *CONSECRATIO,* that shows Faustina holding a scepter, next to a quadriga driven by a charioteer.[28] Reissued coinage was another way to highlight lineage and connection to predecessors. Honors like a *consecratio,* titles, deification and reissued coins were all ways to commemorate past family members.

---

[23] Elizabeth Bartmann, *Portraits of Livia: Imagining the Imperial Woman in Augustan Rome* (Cambridge: Cambridge University Press, 1999), 84.
[24] Charles Brian Rose, *Dynastic Commemoration and Imperial Portraiture in the Julio-Claudian Period* (Cambridge: Cambridge University Press, 1997), 15.
[25] *RIC* 1379, Sear 4715. Rev. PIETAS S C, Pietas standing, holding a flower and cornucopia, with a child at her feet. Rome, 154-56 AD.
[26] Marleen B. Flory, "Dynastic Ideology, the Domus Augusta, and Imperial Women: A Lost Statuary Group in the Circus Flaminius," *Transactions of the American Philological Association* 126 (1996): 297-8, https://doi.org/10.2307/370180, http://www.jstor.org/stable/370180.
[27] Flory, "Dynastic Ideology," 40-41.
[28] *RIC* 383. Sears 4558. Gold *Aureus*. Rev. *CONSECRATIO,* with a galloping quadriga beside Faustina holding a scepter. AD 146. Rome.

The image of imperial women changed in the second and third century during the Antonine and Severan dynasties. The age of adoption emphasized virtues along with dynastic and religious roles. Piety and modesty were exhibited for the health and wellbeing of the family and the state.[29] Even as empress, imperial women were expected to continue their domestic roles on an empire wide basis and their virtues reflected that expectation. The line between the domestic and militaristic spheres was drawn to separate the gendered roles of men and women and was not to be crossed no matter your level of status. Women were often celebrated in their roles as wives, mothers, and priestesses, and for their unwillingness for power.[30] The empress became an example, like her husband, for other women to follow. The emulation of her household virtues acted as a type of social control. We also see the empress highlighted in religious activities as well since those where the only activities women were allowed to partake in outside of the home.[31] The Severan women utilized the strategies of both centuries to promote their imperial women, creating a unique propaganda campaign.

The Severans utilized motherhood, religion, and virtues to support their rule but what highlights their individuality was their additions to the pantheon. As the Roman empire grew, new cultural centers added to the number of gods. While the Severans utilized the second century virtues, their propaganda also took advantage of the plethora of the provincial pantheon. The Severans placed the gods and goddesses that they had a connection to on their coinage to highlight their experiences. The personifications of locations are often connected to a particular campaign or visits, and examples of such experiences. A silver *denarius* highlights the personification of Africa and can be identified by its inscription and symbols of a scorpion and lion at its feet.[32] The *denarius* reflects Severus' visit home in 207 C.E. They used many of their local deities to emphasize their connection to Africa, but they also utilized the Roman pantheon.[33] The

---

[29] Eve D'Ambra, *Private Lives, Imperial Virtues: The Frieze of the Forum Transitorium in Rome* (Princeton: Princeton University Press, 1993), 85.

[30] Boatwright, *The Imperial Women of Rome*, 17.

[31] Mary Gilmore Williams, "Studies in the Lives of Roman Empresses," *American Journal of Archaeology* 6, no. 3 (1902): 263, https://doi.org/10.2307/496790, http://www.jstor.org/stable/496790.

[32] *RIC* 253. Sears 6260. Rev. *AFRICA,* Africa standing right while holding a scorpion with a lion at the bottom. Rome. AD 207.

[33] Clare Rowan, *Under Divine Auspices: Divine Ideology and the Visualisation of Imperial Power in the Severan Period* (Cambridge: Cambridge Univ. Press, 2013), 1.

Severan period constitutes a change in imperial ideology between the Antonine virtues that helped create the "Golden age" and the divine connections the Severans' possessed. Severans displayed themselves as *comes* or companions of the gods and these connections became increasingly seen as the third century begins.[34] The divine companionship of the Severans is demonstrated through the reverse side of the coin and reflects the emperor's power, ideology, and connections.[35] The visual connection between the bust on the obverse and the divinity on the obverse creates a visual link between the images.[36] This relationship is emphasized in the coinage and examples to come with both the emperor and empress and can be seen in several examples. The connection between Septimus Severus and *Mars Pater* can be seen in figure 1 and the connection between Julia Domna and Vesta in figure 2.

Fig. 1: Silver *Denarius*, Obv. Laurate bust of Septimius Severus, Rev. Mars advancing right with spear and trophy. *ANS* 1995.11.1490. Images courtesy of the American Numismatic Society
http://numismatics.org/collection/1995.11.1490

[34] Rowan, *Under Divine Auspices,* 1-3.
[35] Rowan, *Under Divine Auspices,* 1-3.
[36]Carlos F. Noreña, "The Communication of the Emperor's Virtues," *Journal of Roman Studies* 91 (2001): 153, https://doi.org/10.2307/3184774, www.jstor.org/stable/3184774.

Fig. 2: Silver *Denarius*, Obv. Draped bust of Julia Domna, Rev. Vesta seated holding a palladium in right hand and a scepter in the left. *ANS* 1997.10.9. Images courtesy of the American Numismatic Society
http://numismatics.org/collection/1997.10.9

The support of the deities and imperial virtues created an image of a family's right to rule. Emperors chose virtues and deities based on their experiences as each emphasizes different characteristics and qualities. The goddesses chosen such as Diana, Minerva, and Venus Victrix, each emphasized different qualities the emperor wanted associated with his reign. Gods were the guarantor of certain aspects of Roman power and pietal acts showcased on coins, monuments, and as a part of state religion were meant to keep the gods in favor of Rome and assure the dynasty's success and continuation.[37]

The imperial men and women featured in propaganda campaigns of each dynasty were seen as *exempla* to the masses of what to emulate. As part of the imperial *domus,* women reflect imperial ideology and dynastic continuity. Women of the imperial family are seen in conjunction with the emperor and share in his propaganda strategies. The connection of divinities and personifications on the reverse side of male and female coinage is one such strategy. This form of social control reinforced gender roles and established restraints. Numismatic evidence allowed the imperial image to reach outside of Rome. Women did not control their image in propaganda which participated in the same iconographic strategies as her husband. Coin and monumental evidence mirrors the image of the emperor and his court as well as an empress's relationship with her husband.[38] Women were not

[37] David Shotter, "Gods, Emperors, and Coins," *Greece & Rome* 26, no. 1 (1979): 50.
[38] Mary Taliaferro Boatwright, *The Imperial Women of Rome*, 127.

seen in coinage in Rome with the exception of Agrippina the Elder, and even then only a single occurrence until Hadrian.[39] The Empress' place remained on the reverse of her husband's coinage until 128 C.E when Hadrian gave his wife, Sabina, her own coinage.[40] There is a surge in women's coinage under Sabina and Hadrian where women became the focal point of imagery.

Personifying virtues not only externalized Roman identity but also allowed virtues to be worshipped as gods. The Romans believed that praying to deified virtues was how one manifested that trait within oneself. Personifications along with traditional deities show up in monumental and numismatic evidence in connection with an emperor or empress to highlight this connection to certain virtues such as marital harmony, abundance, and military courage that we will see in the following examples. Where virtues emphasized one characteristic, a traditional deity represented multiple virtues. The numismatic examples given highlight the multiple virtues represented and connected to the empress through goddesses such as Diana, Minerva, and Venus, *virtus* being one of them.

Figure 3: Silver *Denarius*, Rev. DIANA LVCIFERA, Diana holding a torch. *ANS* 1948.19.1490. Images courtesy of the American Numismatic Society http://numismatics.org/collection/1948.19.1490

[39]Clare Rowan, "The Public Image of the Severan Women," *Papers of the British School at Rome* 79 (2011): 243.
[40]Boatwright, *The Imperial Women of Rome*, 125.

The Goddesses

Diana or Artemis was the goddess of maidenhood, virginity, and the hunt. She watched over children as they became adults and was often depicted or mentioned on funeral monuments and in epitaphs.[41] Despite her association with purity and youth, she was also a huntress and was depicted with masculine undertones. Diana could be described as wild, fierce, and predatory as a huntress.[42] She was a model for bravery as well as virginity and was depicted on altars in "a running pose and brandishing her bow with her hunting dog by her side."[43] While Diana was a maiden and protector of youth, she was also a hunter with a masculine toughness. Steven Tuck argues that *virtus* also extended into the realm of hunting. He connects the virtue to the *tondi* on the Arch of Constantine which depicted the hunting of some of the most dangerous animals.[44] Tuck explains that the image of hunting brought the "benefits of [Hadrian's] rule to the Roman world by removing dangerous animals and thus ensuring peace and stability."[45] Tuck demonstrates that hunting was a way that emperors extended the realm of *virtus* outside of military engagement, allowing for Diana to partake in a man's sport and virtue. Diana's prowess as a hunter allowed her to enter the male sphere of *virtus* which was strictly a male and military virtue.[46] Her association with hunting tied not only her purity but her masculinity to the empress, giving her an attachment to virtue and bravery. An example can be seen on a gold *aureus* that depicts Julia Domna, the wife of Septimius Severus, with Diana on the reverse. The obverse contains a veiled bust of Julia with the title *IVLIA AVGVSTA*, and *DIANA LVCIFERA* with Diana holding a torch.[47] As the sister to Apollo the sun god, Diana was occasionally depicted with a crescent moon. She is depicted here with the epithet *Lucifera* which means "light-bearer" and refers to her association with the moon.

---

[41] Eve D'Ambra, "Daughters as Diana: Mythological Models in Roman Portraiture," *Memoirs of the American Academy in Rome. Supplementary Volumes* 7 (2008): 174, http://www.jstor.org/stable/40379353. CIL 610958.
[42] D'Ambra, "Daughters as Diana,"171.
[43] D'Ambra, "Daughters as Diana," 174.
[44] Steven L. Tuck, "The Origins of Roman Imperial Hunting Imagery: Domitian and the Redefinition of Virtus under the Principate," *Greece & Rome* 52, no. 2 (2005): 238, www.jstor.org/stable/3567870.
[45] Tuck, "The Origins of Roman Imperial Hunting Imagery," 238.
[46] D'Ambra, "Daughters as Diana," 174.
[47] *RIC* 548. Sears 6553. Gold *aureus*, IVLIA AVGVSTA, veiled bust of Julia Domna, Rev. DIANA LVCIFERA, Diana holding a torch. AD 196, Rome.

Fig. 4: Silver *denarius*, Obv. IMP CAESAR TRAIAN HADRIANVS AVG with laurate bust of Hadrian, Rev.℗ M TR P COS III Minerva, helmeted, levelling javelin, and holding spear. *ANS* 1916.192.209. Images courtesy of the American Numismatic Society. http://numismatics.org/collection/1916.192.209

Minerva or Athena was also seen attached to the emperor for her masculine qualities. A connection to Minerva "endowed [the emperor] with all of her virtues including wisdom and military strength" alluding to the fact that Minerva also holds masculine traits.[48] Diana and Minerva were *exempla* for women to model themselves after, allowing them to take on the same masculine traits.[49] Minerva can be seen in Hadrianic coinage where she is depicted on the reverse of a silver *cistophorus* holding a patera with a shield in front of her and a spear leaning on her left arm.[50] Like Diana, Minerva was also a masculine woman associated with male roles. What makes Minerva special is her role as a goddess of war. Since *virtus* was reserved only for men, Minerva's role and association to war is important. She is often depicted "armed, with a gleaming strudel, spear, and helmet, leading the fray."[51]

Minerva also protected the domestic sphere as well as the war front and is also associated with marital duties such as textile weaving.[52] Minerva combines male and female attributes, the "totality of sexual difference," due

---

[48] Eric R. Varner, "Transcending Gender: Assimilation, Identity, and Roman Imperial Portraits," *Memoirs of the American Academy in Rome. Supplementary Volumes* 7 (2008): 187.

[49] Varner, "Transcending Gender," 187.

[50] *RIC* 503, Sears 3442. Silver *cistophorus*, HADRIANVS AVGVSTVS P P with a bare bust of Hadrian, Rev. Minerva standing while holding a patera in one hand and a shield in front of her, and a spear leaning up against her left arm, AD 129-32, Asia.

[51] Eve D'Ambra, *Private Lives, Imperial Virtues: The Frieze of the Forum Transitorium in Rome* (Princeton: Princeton University Press, 1993), 10.

[52] D'Ambra, *Private Lives, Imperial Virtues*, 11.

to her origin and mythology.[53] Minerva was born by unconventional means, from Zeus's head splitting headache rather than from a woman.[54] Minerva takes on an androgynous appearance on coins like Diana but was also connected to emperors as a military goddess.[55] Minerva was also a feminine version of the male virtue of martial valor like Diana and Venus Victrix.

Venus was viewed as the ancestral goddess of the Julio-Claudian dynasty, the original Caesars.[56] She was given the epithet Genetrix and Victrix to symbolize her different virtues and different roles.[57] Venus's role "as the bestower of transcendence" in the *Aeneid* created her multifaceted image.[58] Venus was depicted as the mother of Aeneas, who was one of the originators of the Julian line. As a mother, Venus shared in her son's victory as well as the consecutive victories of the dynasty he created. Her image as a mother was then represented as Victrix, the "victor," and Genetrix, the "bearer." Julius Caesar viewed Venus as his own ancestor and built a temple to Venus as "Genetrix."[59] Where Appian notes her as "Victrix."[60] The two examples demonstrate the split usage and meaning of the temple. Different epithets highlight the different connection to Venus in each example. The empress's association with Venus Victrix highlighted the empress as the "bringer of victory" with the attendant deified personifications of peace, security, and welling.[61] Her image associated with Venus Victrix highlights her mirrored role as a mother who shares in the victories of her sons and dynasty. An *aureus* under Faustina the Younger demonstrates the connection of women to Venus Victrix. A bust of Faustina is placed on the obverse side with the title *FAVSTINA AVGVSTA*, and the reverse with Venus with a shield

---

[53] Marina Warner, *Monuments and Maidens: The Allegory of the Female Form* (Berkeley: University of California Press, 2014), 122.
[54] Warner, *Monuments and Maidens*, 122.
[55] Eric R. Varner, "Transcending Gender," 196.
[56] Martin Beckmann, *Faustina the Younger: Coinage, Portraits, and Public Image* (New York: American Numismatics Society, 2021), 28.
[57] Eleanor Winsor Leach, "Venus, Thetis and the Social Construction of Maternal Behavior," *The Classical Journal* 92, no. 4 (1997): 353, http://www.jstor.org/stable/3298407.
[58] Leach, "Venus, Thetis and the Social Construction of Maternal Behavior," 352.
[59] Martin Beckmann, *Faustina the Younger: Coinage, Portraits, and Public Image*, 28.
[60] App. *B Civ*. 2.68.
[61] Diliana Angelova, "The Ivories of Ariadne and Ideas About Female Imperial Authority in Rome and Early Byzantium," *Gesta* 43, no. 1 (2004): 6.

displaying a wolf and twins and a helmet. *VENVS VICTRIX* is inscribed in the margins of the coin.[62]

Fig. 5: Gold *aureus*, FAVSTINA AVGVSTA, bust of Faustina the Younger, Rev. VENVS VICTRX, Venus with a victory in one hand and a shield. *ANS* 1955.191.17. Images courtesy of the American Numismatic Society. http://numismatics.org/collection/1955.191.17

Each deity represented multiple virtues that assisted the empress in her roles as a wife and mother. Purity, fertility, and piety were shown in conjunction with courage and strength to help her rule alongside her husband. It is important to place these examples back into the context of their husband's image to show that the empress's association with *virtus* was part of her husband's overall image and propaganda campaign. His connection with these goddesses also demonstrates that these women were an asset to his image and connecting his wife to each was an extension of his image. Scholars debate who controlled the imagery on the coinage, whether the emperor or the *monetales*.[63] Clare Rowan argues that regardless of who controlled the mint, they needed to be sensitive to the emperor's preferences

[62] *RIC* 736. Sears 5246.Gold *aureus,* FAVSTINA AVGVSTA, bust of Faustina, Rev. VENVS VICTRX, Venus with a shield displaying a wolf with twins and a helmet. AD 161-75, Rome.

[63] For scholarly debates on who controlled the imperial image see: Rowan, *Under Divine Auspices: Divine Ideology and the Visualisation of Imperial Power in the Severan Period*, 20; Barbara Levick, "Propaganda and the Imperial Coinage," *Antichthon* 16 (1982): 106, https://doi.org/10.1017/S0066477400002999, https://www.cambridge.org/core/article/propaganda-and-the-imperial-coinage/9B67B0F14C0C4854A8E1FE073559192A; Boatwright, *The Imperial Women of Rome*, 122.

to remain in business.[64] The *monetales* who created the imagery, began using the empress as an extension of the emperor's image. Her image uses the same strategies as seen in the emperor's coinage. Where scholars previously assumed virtues stuck to specific genders, new research highlights their fluidity along with gender.[65] Divinities along with personifications of virtues moved between the imperial couple. The evidence of the same goddesses in the emperor's coinage demonstrates not only the versatility of the virtues each goddess possessed but also her established tradition in male oriented iconography.

Fig. 6: Silver *cistophorus*, Rev. COS II distyle temp of Diana of Perge with a cult statue, and inscribed DIANA PERG. *ANS* 1944.100.44674. Images courtesy of the American Numismatic Society.
http://numismatics.org/collection/1944.100.44674

Perhaps the more interesting of the three goddesses attached to the emperor is Diana. As the goddess of virginity and maidenhood, her attachment to a man must signify different virtues or an event. The image of a temple rather than the goddess herself could refer to an imperial trip to Ephesus or a connection to the local gods of the area. The temple of Diana of Perge shows up on a Trajanic *cistophorus* in 98 A.D. The reverse contains the outline of the temple with a cult statue inside with *DIANA PERG* at the top.[66] A Hadrianic cistophorus also highlights Diana as a cult object between two

[64] Rowan, *Under Divine Auspices: Divine Ideology and the Visualisation of Imperial Power in the Severan Period*, 20.
[65] see Noreña, "Hadrian's Chastity"; Varner, "Transcending Gender: Assimilation, Identity, and Roman Imperial Portraits."
[66] *RIC* 721. Silver *cistophorus*, IMP CAES NERVA TRAIAN AVG GERM P M TR P P P Rev. COS II distyle temp of Diana of Perge with a cult statue and inscribed DIANA PERG. Ephesus or Pergamum, AD 98-9.

stags.[67] She is referred to as *DIANA EPHESIA* rather than *Lucifera* as seen in the empress's coinage. Her epithet, cult image, and mint location reference her patron city of Ephesus, making her connection to the emperor different than the one to the empress and most likely functioned as the patron goddess of the city. The image of a temple rather than the goddess herself could refer to an imperial trip to Ephesus or a connection to the local gods of the area. The imperial mints rest in Rome and Lugdunum, allowing for provincial mints to mint different iconography.[68]

Diana's inclusion in the emperor's coinage was quite common and can be seen in multiple versions, allowing for her to communicate and associate different qualities. She can also be seen on a Hadrianic silver *denarius* holding a bow and arrow with no epithet.[69]

Fig. 7: Silver *Antioninianus*, Rev. VENERI VICTRICI, Venus standing left, holding a Victory and scepter while resting a shield on a helmet. *ANS* 1944.100.51519. Images courtesy of the American Numismatic Society http://numismatics.org/collection/1944.100.51519

The coin was minted in Rome rather than Ephesus like the empress's coinage, referencing a different connection. The mint location could

[67] *RIC* 474, Sears 3448. Silver *cistophorus*, IMP TRA HADRIANO AVG P P, Rev. DIANA EPHESIA cult image of Diana of Ephesus, center with two deer facing outward. Ephesus, AD 132-4.

[68] Tracene Harvey, *Julia Augusta: Images of Rome's First Empress on the Coins of the Roman Empire* (London: Routledge, 2020), 107.

[69] *RIC* 17, Sears 3466. Silver *denarius*, HADRIANVS AVGVSTVS P P with a laurel bust of Hadrian draped on far shoulder, Rev. COS III, Diana center, facing right, holding a bow and arrow. Rome, AD 126.

reference Diana's connection to Rome and the Roman pantheon rather than Ephesus. Diana's role in imperial coinage covers multiple roles. Her connection to the emperor and empress captures her versatility as a goddess and demonstrates that despite having connections to husband and wife, holds different meanings for the pair.

Venus Victrix was the rarest of the three and possesses the most variations, which makes assessing her use in the emperor's coinage a bit difficult. Her role as a mother and the originator of the Julian clan also makes her appearance in the emperor's coinage a bit bizarre. Where Venus has been shown as a "Genetrix" and "Victrix" she is also highlighted as "Felix" in coinage (figure 8).[70]

Fig. 8: Silver *Denarius*, Obv. FAVSTINA AVGVSTA - Bust of Faustina Younger, draped, right, head bare, hair waved and elaborately coiled, fastened in a bun on back of head, Rev. VENVS FELIX - Venus, draped, seated, left on throne, holding Victory on extended right hand and vertical scepter in left, *ANS* 1956.127.1020. Images courtesy of the American Numismatic Society.
http://numismatics.org/collection/1956.127.1020

She can also be seen without an epithet but for the sake of this study, she will be evaluated in the emperor's image as she was in the empress's coinage. The emperor's coinage highlights his *virtus* as an actual personification and did not need Venus Victrix or her variants for this

[70] *RIC* 1.3, *ANS* 1956.127.1020, Silver *Denarius*, Obv. FAVSTINA AVGVSTA - Bust of Faustina II, draped, right, head bare, hair waved and elaborately coiled, fastened in a bun on back of head, Rev. VENVS FELIX - Venus, draped, seated, left on throne, holding Victory on extended right hand and vertical scepter in left, Rome, AD.161-176.

purpose. *Virtus* as a personification can be seen in her own multitude of variants that change through the years and monumental evidence as well. Although extremely rare, there is evidence of Venus Victrix in connection to the emperor, Caracalla. Another connection to an emperor can be seen in the coinage of Gordian III. Venus Victrix shows up in the emperor's coinage when linked to a female family member. The rare examples of Venus Victrix connected to a man specifically is very rare. One example is a gold *binio* with *VENVS VICTRIX* as the inscription and Venus holding a Victory and scepter set on a helmet.[71] Another exists as a silver *antoninianus* with the inscription as *VENERI VICTRICI* in the dative case, and Venus still holding a Victory and scepter.[72] The reference of Venus Victrix in Caracalla's coinage may be to his mother, Julia Domna, with her role as a mother and victor. But without her name or bust, it is hard to be certain.

The Imperial House

The empress's role as a wife and mother fell under the idea of the *domus* or household and incorporated more than just a husband and wife. Richard Bauman explains that as Rome moved from a republic to an empire, the leading figures transformed as well. Instead of a senate, Rome was governed by one man who became emperor. This man and his family grew in status to make up the *Domus Caesarum.*[73] As the empress's image transformed over time, she became more important than a sole link to male members of the family. Her role in the *domus* became a wife and mother to her family and her image was used to support the house. The family rather than the emperor became the protector of the empire and secured its future. Gender roles only allowed women to exist outside of the house in a religious setting, making the empress's most fitting roles a wife, mother, and priestess.[74] These roles were utilized to the benefit of her image and the family's. The

---

[71] *RIC* 311c, Sears 6693. Gold *Binio*, ANTONINVS PIVS AVG GERM and a bust of Caracalla, Rev. VENVS VICTRIX, with Venus standing left and holding a Victory and spectre, resting on a shield set on a helmet, Rome, AD 216.
[72] *RIC* 310a, Sears 6783. Silver *Antioninianus*, ANTONINVS AVG PIVS P P, bust of Caracalla with laurel crown, Rev. VENERI VICTRICI, Venus standing left, holding a Victory and scepter while resting a shield on a helmet. Rome, AD 215.
[73] Richard A. Bauman, *Women and Politics in Ancient Rome* (London: Routledge, 2016), 6.
[74] Boatwright, *The Imperial Women of Rome*, 90, 145-46.

domus became a political unit where everyone within worked for the preservation of the empire.[75]

This same household also functioned to preserve the prosperity and safety of the empire and its citizens. The image of the imperial *domus* reflected the traditional gender norms and elevation based on a set of criteria that allowed the imperial family to lead effectively. The empress's image as a successful wife and mother supported her elevation as an example with her husband for other to follow.[76] This image not only benefited her own station but that of her husband and the rest of the household. While *virtus* helped men be victorious in battle, it helped women at home. Miles McDonnell cites Cicero's wife and daughter as women who exhibited bravery while the head of the household was absent. He concludes and demonstrates that *virtus* was possible for women when connected to men.[77] He also states that it was also acceptable for women to display strength as a widow.[78] Adding numismatic evidence to the discussion, provides a different picture of women and the military virtue.

This idea follows traditional domestic and gender norms. Wives were expected to obey the head of the household, the *paterfamilias.*[79] If her husband was a *paterfamilias,* she was then his equal as a *materfamilias.* Women were legally head of the house along with their husbands.[80] The *paterfamilias* referred to the head of the house not fatherhood where he managed the house and kept control of its inhabitants. Mothers were defined as a wife, who was also an inhabitant but shared in the management of the house. She possessed power as his equal but was submissive due to traditional gendered roles. *Mater* and *paterfamilias* were legal roles where one role was more often used than the other. Women possessed power, they were just expected not to use it, and when coupled with the image of the

---

[75] Richard P. Saller, ""Familia, Domus", and the Roman Conception of the Family," *Phoenix* 38, no. 4 (1984): 355, https://doi.org/10.2307/1088380, http://www.jstor.org/stable/1088380.
[76] Boatwright, *The Imperial Women of Rome*, 220.
[77] McDonnell, *Roman Manliness,* 165.
[78] McDonnell, *Roman Manliness*, 169.
[79] Barbara Levick, *Faustina I and II: Imperial Women of the Golden Age* (Oxford: Oxford University Press, 2014), 20.
[80] Richard P. Saller, "Pater Familias, Mater Familias, and the Gendered Semantics of the Roman Household," *Classical Philology* 94, no. 2 (1999).

imperial *domus,* gave the empress authority over the empire in her husband's absence.[81]

The wife's power was also dependent on her relationship with her husband, and harmony was important. The balance between "partners" is important to preserving the marriage.[82] Preserving harmony was dependent on the wife and husbands' personality and their standing with the public. It was up to the husband to divide household tasks evenly between spouses. The division of labor often favored the husband, and it was up to the wife to submit to his interests.[83] This division rests on the wife's role as the caretaker of the house and children, and developed the idea that the woman's place was at home.[84] Women were branded "useless" if they did not exhibit the correct virtues such as *pudicitia* and *fides.*[85] Men were expected to keep their wives beneath them in hopes of containing them. Just as the husband granted his wife power and authority, emperors granted their wives power as well.[86] Penelope Davies refers to it as a "mirror of power" that established that the empress as an equal to her husband.[87]

This idea also rested on the *concordia*, or harmony, between the couple. Both rulers were often connected to the personification of harmony to demonstrate that "all was well in the palace and the succession was secure."[88] Issues with succession created civil wars and risked the stability of the family and empire. Coins depict the virtue, *concordia,* with both the emperor and empress. A silver *denarius* of Antoninus Pius depicts his wife, Faustina the Elder, on one side with the personification of *concordia* on the reverse.[89] Similar coins can be seen under other dynasties like the Severans.

---

[81] McDonnell, *Roman Manliness: Virtus and the Roman Republic*, 169.
[82] Robert Brown, "Livy's Sabine Women and the Ideal of Concordia," *Transactions of the American Philological Association* 125 (1995): 314.
[83] Susan Treggiari, *Roman Marriage* (Clarendon Press, 1993), 202. https://public.ebookcentral.proquest.com/choice/publicfullrecord.aspx?p=4964651.
[84] Treggiari, *Roman Marriage*, 203.
[85] Treggiari, *Roman Marriage*, 237.
[86] Beckmann, Faustina the Younger: Coinage, Portraits, and Public Image, 27.
[87] Penelope J. E. Davies, *Death and the Emperor: Roman Imperial Funerary Monuments from Augustus to Marcus Aurelius* (Austin: University of Texas Press, 2004), 109.
[88] Suzanne Dixon, *The Roman Mother* (London: Routledge, 2015), 83.
[89] *RIC*.1.3, *ANS* 1905.57.401, Silver *denarius*, Obv. FAVSTINA AVG ANTONINI AVG PII FIL Bust of Faustina, draped, right, head bare, hair is elaborately waved and coiled on back of head, Rev. CONCORDIA - Concordia, draped, standing, with

Allowing the empress to hold the same level of power as the emperor provided a new level of stability for the empire. Despite the equal power given to the *materfamilias*, women were traditionally bound to their gendered roles as mothers and expected to produce an heir. In addition to her image as a loyal spouse, the empress was depicted as a "strong, virtuous, self-sacrificing, and devoted" mother to her children.[90] Roman mothers were responsible for the education of young children and the marriage of older daughters. The Romans viewed mothers as "shaping great leaders."[91] It was tradition to Roman households that the role of mother succeeds the role of wife. Empresses were viewed not just as mothers to their own children but to all of the empire.

The empress's role in the *Domus* can also be seen in the title *Mater Castrorum*, or Mother of the Camps.[92] Julia Domna was revered as the Mother of the Camps to emphasize the dynastic connection of the empress to the army and promote harmony with the emperor. The troops were "under the protection" of the empress, and they in turn, defended her.[93] This title and honor also demonstrated the unity of the army as well as the ruling family. The title was even extended under Julia Domna to include the *senatus et patriae*, or the senate and state.[94] The title also afforded a connection to those who previously held it. Julia Domna's possession of the honor also held by Antonine empresses, like Faustina the Younger, helped

right hand gathering up fold of skirt and holding cornucopia in left. Rome, AD 145-161.

[90] Lauren Hackworth Petersen and Patricia B. Salzman-Mitchell (eds.), *Mothering and Motherhood in Ancient Greece and Rome* (Austin, TX: University of Texas Press, 2013), 173.

[91] Hackworth Petersen and Salzman-Mitchell (eds.), *Mothering and Motherhood,* 173.

[92] Barbara Levick, *Julia Domna, Syrian Empress* (London: Routledge, 2007), 41-2. *RIC*.4, *ANS* 1948.19.1505, Silver *Denarius*, Obv. IVLIA AVGVSTA, draped bust, Rev. MATRI CASTRORVM, Julia Domna standing left with serpent rising from altar with right and holding scepter in left, to left, three standards set in the background. Rome. AD 196-211. *RIC* 1.3. *ANS* 1985.140.139. Silver *Denarius*, Obv. DIVAE FAVSTINAE PIAE Bust of Faustina, veiled and draped, hair is waved and fastened in a bun on back of head, Rev. MATRI CASTRORVM, Faustina seated left, holding phoenix, with nimbus on globe and a scepter, in front of her, two legionary eagles set on base to the left. Rome. AD 176-180.

[93] Barbara Levick, *Julia Domna,* 42.

[94] Anthony Barrett, *Agrippina: Sex, Power, and Politics in the Early Empire* (London: Routledge, 2005), 64.

establish a new connection to the previous dynasty.[95] The dynastic element highlighted in this title allows for the empress's children to honor her as well. These honors extended Domna's role as a mother to the empire and allowed her to receive acts of piety as well as present them.

Conclusion

Roman propaganda was constructed through a complex set of images meant to communicate a multifaceted image and ideology to the masses. While the emperor was the main subject for this system of communication, his family also played a vital role. As the empire grew, it moved further away from Rome and was harder for one person to cover the entire span of the empire. The empress began to serve a role in Roman propaganda to extend the emperor's ideology further. Her image began to grow in influence to aid her husband in ruling the vast empire. Since women were confined to the domestic sphere, her image needed to be set in gendered traditional settings. Motherhood became her way of serving the empire and *virtus* assisted her in that role. The use of Diana, Minerva, and Venus Victrix helped communicate the empress's *virtus* within the confines of Roman tradition and allowed her to be strong and courageous. These carefully selected goddesses were able to gently communicate the empress's strength without altering the ways of communicating to the masses. Her image became an extension of the emperor's image while becoming a tool for him to use, and an example for the masses to emulate.

Bibliography

Angelova, Diliana. "The Ivories of Ariadne and Ideas About Female Imperial Authority in Rome and Early Byzantium." *Gesta* 43, no. 1 (2004): 1-15.

Balmaceda, Catalina. *Virtus Romana: Politics and Morality in the Roman Historians.* Chapel Hill, NC: University of North Carolina Press, 2017.

Barrett, Anthony. *Agrippina: Sex, Power, and Politics in the Early Empire.* London: Routledge, 2005.

Bartmann, Elizabeth. *Portraits of Livia: Imagining the Imperial Woman in Augustan Rome.* Cambridge: Cambridge University Press, 1999.

Bauman, Richard A. *Women and Politics in Ancient Rome.* London: Routledge, 2016.

---

[95] Levick, *Julia Domna, Syrian Empress*, 43.

Beckmann, Martin. *Faustina the Younger: Coinage, Portraits, and Public Image.* New York: American Numismatics Society, 2021.

Boatwright, Mary Taliaferro. "Women and Gender in the Forum Romanum." *Transactions and Proceedings of the American Philological Association* 141, no. 1 (2011): 105-141.

—. *The Imperial Women of Rome: Power, Gender, Context.* Oxford: Oxford University Press, 2021.

Brown, Robert. "Livy's Sabine Women and the Ideal of Concordia." *Transactions of the American Philological Association (1974-)* 125 (1995): 291-319.

D'Ambra, Eve. *Private Lives, Imperial Virtues: The Frieze of the Forum Transitorium in Rome.* Princenton: Princeton University Press, 1993

—. "Daughters as Diana: Mythological Models in Roman Portraiture." *Memoirs of the American Academy in Rome. Supplementary Volumes* 7 (2008): 171-83.

Davies, Penelope J. E. *Death and the Emperor: Roman Imperial Funerary Monuments from Augustus to Marcus Aurelius.* Austin: University of Texas Press, 2004.

Dixon, Suzanne. *The Roman Mother.* London: Routledge, 2015.

Dressler, Alex. *Personification and the Feminine in Roman Philosophy.* Cambridge, MA: Cambridge University Press, 2016.

Flory, Marleen B. "Dynastic Ideology, the Domus Augusta, and Imperial Women: A Lost Statuary Group in the Circus Flaminius." *Transactions of the American Philological Association* 126 (1996): 287-306.

Hackworth Petersen, Lauren and Patricia B. Salzman-Mitchell (eds.). *Mothering and Motherhood in Ancient Greece and Rome.* Austin, TX: University of Texas Press, 2013.

Harvey, Tracene. *Julia Augusta: Images of Rome's First Empress on the Coins of the Roman Empire.* London: Routledge, 2020.

Langford, Julie. *Maternal Megalomania: Julia Domna and the Imperial Politics of Motherhood.* Baltimore, Maryland: The John Hopkins University Press, 2013.

Leach, Eleanor Winsor. "Venus, Thetis and the Social Construction of Maternal Behavior." *The Classical Journal* 92, no. 4 (1997): 347-71.

Levick, Barbara. *Julia Domna, Syrian Empress.* London: Routledge, 2007.

—. *Faustina I and II: Imperial Women of the Golden Age.* Oxford: Oxford University Press, 2014.

McDonnell, Myles Anthony. *Roman Manliness: Virtus and the Roman Republic.* Cambridge, New York: Cambridge University Press, 2009.

Noreña, Carlos F. "The Communication of the Emperor's Virtues." *Journal of Roman Studies* 91 (2001): 146-68.

—. "Hadrian's Chastity." *Phoenix* 61, no. 3/4 (2007): 296-317.
Phang, Sara Elise. *Roman Military Service: Ideologies of Discipline in the Late Republic and Early Principate.* Cambridge Univ. Press., 2008.
Rose, Charles Brian. *Dynastic Commemoration and Imperial Portraiture in the Julio-Claudian Period.* Cambridge: Cambridge University Press, 1997.
Rowan, Clare. "The Public Image of the Severan Women." *Papers of the British School at Rome* 79 (2011): 241-73.
—. *Under Divine Auspices: Divine Ideology and the Visualisation of Imperial Power in the Severan Period.* Cambridge: Cambridge Univ. Press, 2013.
Saller, Richard P. ""Familia, Domus", and the Roman Conception of the Family." *Phoenix* 38, no. 4 (1984): 336-55. https://doi.org/10.2307/1088380. http://www.jstor.org/stable/1088380.
—. "Pater Familias, Mater Familias, and the Gendered Semantics of the Roman Household." *Classical Philology* 94, no. 2 (1999): 182-97.
Sear, David Ronald. *Roman Coins and Their Values.* 5 vols. Vol. 2, London: Spink, 2000.
Shotter, David. "Gods, Emperors, and Coins." *Greece & Rome* 26, no. 1 (1979): 48-57.
Treggiari, Susan. *Roman Marriage.* Clarendon Press, 1993.
Tuck, Steven L. "The Origins of Roman Imperial Hunting Imagery: Domitian and the Redefinition of Virtus under the Principate." *Greece & Rome* 52, no. 2 (2005): 221-45.
Varner, Eric R. "Transcending Gender: Assimilation, Identity, and Roman Imperial Portraits." *Memoirs of the American Academy in Rome. Supplementary Volumes* 7 (2008): 185-205.
Warner, Marina. *Monuments and Maidens: The Allegory of the Female Form.* Berkeley: University of California Press, 2014.
Williams, Mary Gilmore. "Studies in the Lives of Roman Empresses." *American Journal of Archaeology* 6, no. 3 (1902): 259-305.

WARFARE AND *CANONES SACRORUM CONTEMPTORES*

DESPINA IOSIF

Did the Christian Church lose its soul once Constantine became emperor? The view that the Church after the conversion of emperor Constantine turned from a stage of pacifism to non-pacifism is still strong and it ought to be challenged.[1] One could argue in favor of a different development: from a stage of acceptance or toleration of wars, the Church moved after the conversion of Constantine to a stage when wars became acceptable only under certain circumstances. There was a change after the conversion of Constantine, but his ascension to the throne did not terminate the pacifist period, as alleged, simply because there never was such a period. The most influential voices in Christianity continued to accept wars as a legitimate option but simultaneously started to set requirements for the appropriate conduct of wars. For Christians, it was the arrival of an era of responsibility for policy and action. An unprecedented number of thinkers started to question the morality and efficacy of wars for the first time in the fourth century. This phenomenon continued in the following centuries. Ambrose in the fourth century, Augustine in the late fourth and early fifth century, Gratian in the twelfth century, Thomas Aquinas in the thirteenth century, Francisco de Vitoria. and Francisco Suarez in the sixteenth century, Hugo

[1] The idea that the recognition of Christianity as the official religion of the empire signified the end of the early Christian anti-militarism is supported by numerous scholars. See e.g.: Aldous Huxley, *An Encyclopaedia of Pacifism* (London: Chatto and Windus, 1937), 22; Bainton H. Ronald, "The Early Church and War," *Harv.Theol.Rev.* 39 (1946): 189-215; John Ferguson, *War and Peace in the World's Religions* (London: Open University Press, 1977), 104; Enda McDonagh, *The Demands of Simple Justice* (Dublin: Gill and MacMillan, 1980), 60; Rosemary Ruether Radford, "War and Peace in the Christian Tradition," in *A Council of Peace*, ed. Hans Küng and Jurgen Moltman (Edinburgh: Humns Ancient and Modern, 1988), 17–24, (20). See also George Kalanztis, *Caesar and the Lamb* (Eugene, Oregon: Cascade Books 2012); Nico Huttunen, *Early Christians Adapting to the Roman Empire: Mutual Recognition* Novum Testamentum, Supplements, vol. 179 (Leiden/Boston: Brill, 2020), 138-228.

Gorth in the seventeenth century, John Locke in the eighteenth century and recently other scholars engaged in determining all the more strict factors for characterizing a war as legitimate. These factors were: the declaration of the war by a competent authority, for a just cause, and with a reasonable hope of success.[2] Twentieth century intellectuals, being stunned by the horror of the 'nuclear age', added to the list of the requirements the use of acceptable weapons. We still are adherents and defenders of the just war theory.

If the early Christians had indeed been preoccupied about the legitimacy of having military careers and participating in warfare, then we would expect the Church to have helped them to be freed from their dilemmas by officially announcing its position. The first attested 'public appearance' of the Church, as we will see, belongs to as late as the beginning of the fourth century. It looks as if the Church had not intervened before simply because it felt there was no reason to; the issue did not constitute a major concern in the previous centuries.

In my paper I will examine the regulations that the Church announced concerning the appropriate behavior of its members towards war, violence and military service.[3] Some regulations were set by local clergy who had in mind solely their communities; some were composed by ambitious individuals with the desire that all Christians worldwide follow them; and some were the result of ecumenical synods and were addressed to all Christians worldwide. It is imperative to keep this distinction in mind when reading the sources.

The regulations will be examined, where possible, in chronological order since they were responses to specific situations and reactions to imperial measures. In late antiquity the Church and the State were engaged in a dialogue concerning the responsibilities towards each other. This paper contends that the Church initially did not have a unanimous position concerning Christian responsibilities. No pattern emerges showing clearly the early Christian official attitude to war, violence and military service. Different Christian communities perceived the issue in a different light and consequently proposed different solutions. It was only in the end of the fourth century when Christian religious leaders agreed that only clerics and

---

[2] Arthur Nussbaum, "Just War- A Legal Concept?" *Michigan Law Review* 42 (1943): 453–479.

3 The present paper is based on the chapter 'Councils or Those who throw away their arms' of my book *Early Christian Attitudes to War, Violence and Military Service* (Piscataway, NJ: Gorgias Press, 2013), 287-303.

monks were to live a 'pure' and 'higher' Christian life and to keep themselves apart from military affairs and expected the laity to get actively involved. This position was happily acknowledged by the State:

> A soldier who is in authority must not kill men; if he should be ordered (to do it), he must not do it. He must not take the military oath. If he will not agree, let him be rejected. A military governor or a magistrate of a city who wears the purple (toga) either let him resign or be rejected. If a catechumen or a baptized Christian wishes to become a soldier let him be rejected for he has despised God (*Traditio Apostolica* 16).

The author of the *Traditio Apostolica*,[4] who might have been Hippolytus, advised Christians not to seek a military career. He also provided for those who had started a military career prior to their conversion to Christianity. If they were ordinary soldiers they should refuse the order to kill, take the military oath, and shun the positions of superior officers or magistrates who inflict or carry out the death penalty. If they were superior officers and magistrates then they were expected to change careers. The author of the *Traditio Apostolica* made clear distinctions and put different limitations on what a Christian could/should do according to his position in the army and the time of his conversion. There was no universal rejection of the notion of military duty as such. A military career was seen as a possibility, but the limitations put on it (not to kill and not to take the military oath) made it virtually impossible for someone to continue to serve without attracting the attention and the fury of his superiors. The *Traditio Apostolica* is the earliest surviving collection of Church regulations for the appropriate Christian conduct of laity and clergy, composed at the end of the second[5] or the beginning of the third century,[6] and one of the earliest pieces of evidence for the existence of anti-military sentiments in the Christian Church. We do not know whether these regulations were ever followed or even accepted, nor if they were authorized as generally binding by the Christian

[4] Contemporary scholarship expresses skepticism whether the *Traditio Apostolica* has rightly been attributed to Hippolytus. The discussion is complicated but interesting to follow. See Jacques Fontaine, "Christians and Military Service in the Early Church," *Consilium* 7 (1965): 58–64; Philip Rousseau, *The Early Christian Centuries* (London: Routledge, 2002), 174f.

[5] B. Steimer, "Traditio Apostolica" in Dopp Siegmar (ed.), *Dictionary of Early Christian Literature* (New York: Herder and Herder, 2000), 580–583, (582).

[6] Gregory Dix and Henry Chadwick, *The Apostolic Tradition*, (London: Routledge, 1995): xi, xxxv-xxxviii, argued in favor of 215 CE.

congregation in which their author belonged, though he would certainly have wished the latter to be the case.

Those who throw away their arms in times of peace should be kept from communion, (*de his qui arma projiciunt in pace placuit abstineri eos a communion*). This is what Canon 3 of the Synod of Arles,[7] summoned at Emperor Constantine's behest,[8] promulgated in 314 CE. This Canon has greatly vexed scholars interested in early Christian attitudes to warfare. Several interpretations have been proposed to the phrases *arma projiciunt* and *in pace*. Some scholars replaced the *arma projiciunt* with *arma conjiciunt* assuming that the Canon prohibited Christians from carrying weapons in peacetime and referred either to gladiatorial games or the carrying of weapons in everyday life (contrary to Jesus' advice in Lk 22:35–38).[9] Other scholars asserted that the early Church was pacifist and understood either that Christians were only allowed to pursue military careers as long as they did not shed blood but were only engaged in police work, and if they abandoned their police duties then they were in trouble.[10] Similarly, others understood that Christians helped the Roman army only in periods when there were no wars and kept themselves apart when there were. The Church expected its members not to get involved in warfare but punished them if they refused to help the State when there were no wars.[11]

---

[7] James Stevenson (trans.), *A New Eusebius. Documents Illustrative of the History of the Church to AD 337* (London: Society for Promoting Christian, 1957), 144–147.
[8] Donatus, bishop of Carthage, was against the readmitting of clergy who lapsed during persecution and appealed to Constantine who referred the matter to a meeting of bishops in Rome and then at Arles. See Averil Cameron, *The Later Roman Empire* (London/Cambridge: Harvard University Press, 1993), 59.
[9] Andreas Bigelmair, *Die Beteiligung der Christen am öffentlichen Leben in vorkonstantinischer Zeit. Ein Beitrag zur ältesten Kirchengeschichte* (Munich: University of California Libraries, 1902), 182; K.H.E. De Jong, *Dienstweigering bij de oude Christenen* (Leiden: Brill, 1905), 28ff.; G.S. Windass, "The Early Christian Attitude to War," *Irish Theological Quarterly* 29, (1962): 235–245, (244).
[10] Stephen Gero, "Miles Gloriosus: the Christians and Military Service According to Tertullian," *ChHist* 39 (1970): 285–298; Eileen Egan, "The Beatitudes. The Works of Mercy and Pacifism," in Shannon A. Thomas, ed., *War or Peace? The Search for New Answers* (New York: Orbis Books, 1980): 169–187, (176).
[11] Henri F. Secrétan, "Le Christianisme des premiers siècles et le service militaire," *Revue de Théologie et de Philosophie* 2 (1914): 324–366, (364); Roland H. Bainton, *Actitudes cristianas ante la guerra y la paz. Examen histórico y nueva valoración crítica*, tr. Angel Alcalá (Madrid: Tecnos, 1963), 77; John Driver, *How Christians Made Peace with War. Early Christian Understandings of War* (Scottdale, Pennsylvania, Kitchener, Ontario: Wipf and Stock, 1988), 75–76. The idea that early Christians served in the Roman army but carried out only police duties is still very

A few scholars understood the term peace in quite a different way, as the peace that prevailed between the Empire and the Church with the termination of persecutions against Christians.[12] Finally, some replaced the *in pace* with *in bello*, discerning a Christian effort to please the State by prohibiting Christians from feeling an aversion to warfare and subsequently being uncomfortable in the army.[13] We can maintain that the Church was not pacifist but tried to please the State without having to replace any words if we assume that the Church prohibited its members from finding the military profession incompatible with their faith even in periods when there was no warfare and no bloodshed, and punished all those who dared to desert their posts. And the Council of Arles made no mention of throwing away arms in time of war (to which the Church also objected) because military laws had already covered this possibility and had assigned the death penalty.[14]

Admittedly, none of these interpretations is entirely convincing. However, some are less convincing than others. First of all, it seems unlikely that the distinction between police work and military work was a meaningful one in late antiquity. There was physical violence involved in both activities.[15] And the Church could not have expected its members to serve only in periods of

---

popular among modern theologians; see for example "War" in Ferguson B. Sinclair and Wright David, eds., *New Dictionary of Theology* (Leicester, Illinois: IVPAcademic 1988), 714–716, (715); Ronald F. Musto, *Catholic Peacemakers. A Documentary History. From the Bible to the Era of the Crusades* (New York and London: Garland, 1993), 158.

[12] Edmond Le Blant, *Les inscriptions chrétiennes de la Gaulle romaine*, vol. 1 (Paris: Imprimerie nationale 1892), 82, n.2 ; Adolf Harnack, *Militia Christi. The Christian Religion and the Military in the First Three Centuries*, tr. David McInnes (Philadelphia: Fortress, 1981[1], 1905), 100. See also: Tert. *De Fuga in Persecutione*, 3; Origen, *C. Cels.* 3.15, 8.70; Euseb. *Hist. eccl.* 5.16.19; Lactant. *De mort. pers.* 4, where the term peace has this meaning.

[13] For example, L. Surius, *Concilia omnia tam generalia quam provincialia atque particularia*, Colonia, 1576, cited by Charles Joseph Hefele, *A History of the Christian Councils from the Original Documents to the Close of the Council of Nicaea A.D. 325*, William R. Clark (tr.) (London: T&T Clark, 1894), 185.

[14] Alfred William Winterslow Dale, *The Synod of Elvira and Christian Life in the Fourth Century* (London: MacMillan 1882), 238f., 281; John Helgeland, "Christians and the Roman Army from Marcus Aurelius to Constantine," *ANRW* II.23.I.1979, 724–834, (806).

[15] Louis J. Swift, *The Early Fathers on War and Military Service* (Wilmington, Delaware: M. Glazier 1983), 91; Christopher J. Fuhrmann, *Policing the Roman Empire: Soldiers, Administration, and Public Order* (Oxford: Oxford University Press 2011).

relative calm and desert when war was threatened. No army would have tolerated such an arrangement, and no emperor. And we know that the emperor was present at the proceedings; it was explicitly stated so every time the Canons of the Council were published.[16] Furthermore, the military laws had already covered the possibility and had assigned penalties for desertion not only in times of war but also of peace.[17] This leaves us with the following possibilities: either to take *in pace* to mean the end of Christian persecution or to replace words (*arma projiciunt* with *arma conjiciunt* or *in pace* with *in bello*). We have no reason to believe that the author(s) of the Canon did not have the end of persecutions in mind. However, there is no strong evidence to support such an interpretation either since there is no evidence to justify lexical replacement.

Canon 3 is far too brief. The rest of the Canons of the Council were concerned about donatism, celibacy, the consecration of bishops, rebaptism and various other issues that unfortunately do not clarify the obscurity of the meaning of Canon 3. It is only helpful to keep in mind that since the Emperor Constantine was present at the Council meetings it is highly unlikely that the Council would have decided something that might displease him and jeopardize his loyalty to the Church. Condemning Christian participation in warfare would certainly not be to any emperor's liking. Thus, it seems reasonable to assume that service in the army could create problems for the early Christian conscience (and that was the reason a relevant Canon was included in the Council) but the Church's decision was in favor of Christian participation in warfare.

This Canon could be a Christian response to the generosity of Constantine who had a few years earlier obliged the Church by commanding the following:

> Once more, with respect to those who had previously been preferred to any military distinction, of which they were afterwards deprived, for the cruel and unjust reason that they chose rather to acknowledge their allegiance to God than to retain the rank they held; we leave them perfect liberty of choice, either to occupy their former stations, should they be content again to engage in military service, or after an honourable discharge, to live in undisturbed tranquility. For it is fair and reasonable that men who have displayed such

[16] Hefele, *Christian Councils*, 180-181.

[17] J.B. Campbell, *The Emperor and the Roman Army 31BC–AD235* (Oxford: Oxford Clarendon Press, 1984), 303f.; Dick Whittaker, "Landlords and Warlords in the Later Roman Empire," in John Rich and Graham Shipley, eds., *War and Society in the Roman World* (London and New York: Routledge, 1993), 277–302, (287).

> magnanimity and fortitude in meeting the perils to which they have been exposed, should be allowed the choice either of enjoying peaceful leisure, or resuming their former rank (Euseb. *Vit. Const.* 2.33).

Constantine with this rescript, preserved by Eusebius in the emperor's biography, the *Vit. Const.*,[18] permitted soldiers who had deserted, were demoted, or expelled from their posts on account of their faith to have the option of resuming their former ranks in the army or to obtain honorable discharge.[19] Modern scholarship has correctly understood the emperor as referring to the desertion of Christians from the Roman army under the anti-Christian measures of the emperors Diocletian and Licinius[20] or Diocletian and Galerius.[21] The rescript started and concluded with Constantine expressing his conviction that he enjoyed the unlimited favor of the Christian God he honored.[22] It announced measures manifesting the emperor's support to the Church: the return of Christian exiles to their homes, the release of the clergy from curial duties, the restoration of Christian property, the release of Christians from confinement, prison and hard labor, the restoration of Christian soldiers to military rank, the release of Christians from service in State factories, and the restoration of their noble status where applicable.[23] The document was sent to all eastern provinces.

The Church and the State had begun a dialogue in good faith to define the terms of their relationship. It would be interesting to follow the dialogue, as much as we can, by examining Canons, decrees and laws. The State needed

---

[18] This document is considered authentic by Timothy D. Barnes, "The Constantinian Settlement," in Harold W. Attridge and Gohei Hata Gohei, eds., *Eusebius, Christianity and Judaism* (Leiden, New York and Köln: Wayne State University Press, 1992), 653–657, (650); Averil Cameron Stuart G. Hall in their edition of *Eusebius, Life of Constantine*, (Oxford and New York: Oxford University Press, 1999), 5, 18, 239. Chapters 27 and 28, the end of 26 and beginning of 29, have been identified as the text written on the back of *P.Lond.* 878 (319–320), thus leaving no doubt about the authenticity of the document. B.H. Warmington, in "The Sources of some Constantinian Documents in Eusebius' Ecclesiastical History and the Life of Constantine," *St Patr.* 18/1 (1986): 93–98, (94–97), suggested that the document was given to Eusebius by an imperial notary called Marianus whose work is described in iv.44.

[19] See also Sozom. *Hist.* 1.1.

[20] Giovanni Crescenti, *Obiettori di coscienza e martiri militari nei primi cinque secoli del cristianesimo*, (Palermo: Flaccovvio, 1966), 320.

[21] Swift, *Early Fathers on War and Military Service*, 91.

[22] This idea pertains to Eusebius' *Vit. Const.*

[23] Euseb. *Vit. Const.* 3.24–42.

men to serve as soldiers. In 318 CE, a law was published that obliged the sons of veterans to serve in the army.[24] The State did not want the Church to avert its members from pursuing military careers.[25] This is also evident in 325 in the Council of Nicaea[26] (summoned by Constantine when the emperor discovered that theological disputes had broken out in the East concerning the nature of Jesus and His relationship to His Father)[27] which provided the following in Canon 12:

> Those who responded to the call of grace and initially expressed their faith by putting off the military belt, but who subsequently acted like dogs returning to their vomit when they offered money and gifts to get back to the army must remain among the hearers for three years and then among the supplicants for ten more. In all such cases, however, it is proper to examine into the purpose and nature of their repentance; for as many as manifest their conversion in deed, and not in appearance only, by their fear, and tears, and patience and good works, these having completed the prescribed time as hearers, may properly communicate in the prayers, and the bishop may be allowed to determine yet more favourably respecting them. But those who hear their sentence with indifference, and think the form of entering into the church sufficient for their conversion, must complete the whole time.

Did the Christians in Nicaea have a different position from the Christians in Arles? Was Canon 12 of the Council of Nicaea an expression of the incompatibility of the Christian faith with the military profession and a declaration towards the State of its decision not to help the State by supplying Christians to serve? This possibility is highly unlikely since Emperor Constantine presided at the proceedings and gave the Canons of

---

[24] *Cod. Theod.* 16.22.2, *Cod. Theod.* 7.1.5, 12.1.15, 35; Crescenti, *Obiettori*, 269.

[25] *Cod. Theod.* 16.2; Zos. *Hist.* 3.

[26] The Council's primary aim was the condemnation of Arius (who regarded Jesus as inferior to His Father), the agreement for the correct date of the celebration of Easter and the establishment of concord in the Christian communities. See: William G. Rusch, *The Trinitarian Controversy* (Philadelphia: Fortress Press, 1980); Colm Luibhéid, *The Council of Nicea* (Galway: Galway University Press, 1982); Arthur Stanley, *The Council of Nicea. Lectures on the History of the Eastern Church* (Piscataway, NJ: Gorgias Press, 2009); Young Richard Kim (ed.), *The Cambridge Companion to the Council of Nicaea*, (Cambridge: Cambridge University Press, 2021).

[27] Constantine vehemently desired the Church to be or to seem unanimous in religious matters. See Euseb. *Vit. Const.* 1.44.3, 2.65.2, 3.12.2, 3.21–22, 3.59–60, 4.42.

the Council the force of imperial laws.[28] Moreover, the Church would not have insulted her most precious ally, the Emperor, in such an obvious and careless way. Perhaps the Church and the emperor imagined that it would be possible to have a completely pagan army with no Christians serving in the ranks? There is no evidence to support such a hypothesis. One could argue that the Canon does not concern the military profession in general and its relationship to the Christian faith. It concerns rather only those Christian soldiers who under the emperor Licinius preferred to resign their posts instead of succumbing to the emperor's pressures to take part in pagan sacrifice.[29] However, they subsequently returned to their ranks, out of love for (or in need of) money, and paid the required homage to the traditional gods.[30] Whether the Canon referred only to a particular case or had more general character, what we ought to keep in mind is that for some early fourth century Christians returning to the army once they had left it was inconceivable.

The author(s) of the Canons 13 and 14 of the Church of Alexandria,[31] dated between 336–340,[32] probably did not have a specific case in mind when they declared the following:

> A man who has accepted the authority to kill, even if he is a soldier, he should on no account do it. And those who are soldiers and are ordered to

---

[28] See: Norman P. Tanner, ed., *Decrees of the Ecumenical Councils,* vol. I, *Nicea I to Lateran V*, (Washington: Georgetown University Press, 1999), 4.

[29] Euseb. *Vit. Const.* 1.54.1; Sulpicius Severus, *Historia Sacra*, 2.7.

[30] Canon 12 must have referred to the desertion of soldiers under Licinius, since the previous Canon, number 11, dealt with Licinius and allowed those Christians who performed pagan ceremonies during the emperor's persecution to return to their communities so that unity be restored. See: Harnack, *Militia Christi*, 103 and *The Mission and Expansion of Christianity in the First Three Centuries,* tr. James Moffatt (London and New York: Williams and Norgate, 1908), 63; Crescenti, *Obiettori*, 267; Helgeland, "Christians and the Roman Army," 807; idem. "Christians and the Roman Army AD 173–337," *ChHist* 43.2 (1974): 149–163, (163); Swift, *Early Fathers on War and Military Service*, 92; José Fernández Ubiña, *Cristianos y militares. La iglesia antigua ante el ejército y la guerra* (Granada: Universidad de Granada, 2000), 480.

[31] The *Canons of the Church of Alexandria* were edited by Oscar von Gebhardt and Adolf Harnack, *Texte und Untersuchungen zur Geschichte der altchristlichen Literatur* (Leipzig: J. C. Hinrichs, 1891). Their purpose was to explain the duties of bishops and the duties of Christian congregations. It is highly significant that the only two professions excluded from the Christian community are that of the soldier and that of the creator of idols.

[32] Swift, *Early Fathers on War and Military Service*, 93.

> fight should also refrain from slandering and neither put on crowns on their heads, nor obtain any military standard. ... Every man who has been elevated with the ornament of justice to the position of prefect or excellence or authority, which is in accordance to the gospel, and does not comply, he is to be separated from the congregation and a bishop should not pray in his presence (Canon 13).

> A Christian should not voluntarily become a soldier unless compelled to by someone in authority. He should have a sword, but he should not be commanded to shed blood. If it is ascertained that he has done so, he should stay away from the sacraments at least until he has been purified through tears and lamentation. He should fulfill his obligation without deceit and in fear of God (Canon 14).

Christians ought preferably to stay away from the army in order to avoid shedding blood. However, if someone in authority compelled them to join, then they would have little choice but to accept. Not upsetting the authorities was always a top priority for the Church. The Canons of the Church of Alexandria, also known as *Canones Hippolyti*, were based on the *Traditio Apostolica*.[33] The author(s) of the Canons of Alexandria was not as strict as the author(s) of the *Traditio Apostolica*; he was willing to pardon Christian involvement in the military if under pressure from the authorities. Finally, I must mention that the Canons in all probability had only local force; there is no evidence whatsoever to suggest that they were considered binding by other communities outside Alexandria.

For some Christians it was crucial to differentiate whose blood one sheds or was called to shed. If the killing was instigated from private initiative and for private gain then it was unacceptable, but if one was instructed by the authorities to do so on the battlefield and in order to defend an ideal and to protect the nation's interests then it was a completely different matter.[34] In that case, killing might even be praiseworthy as Athanasius, one of the most important Christian theologians of the fourth century, stated:

> One is not supposed to kill, but killing the enemy in battle is both lawful and praiseworthy. For this reason individuals who have distinguished themselves in war are considered worthy of great honours and monuments are erected to celebrate their accomplishments. Thus, at one particular time,

---

[33] Oscar von Gebhardt and Adolf Harnack, eds., *Die Canones Hippolyti*, (Leipzig: J. C. Hinrichs, 1891).

[34] See also Sen. *Ep. ad Lucilium* 95.30; August. *De civ. D.* 1.21, 1.26, *Contra Faustus* 22.70, *De Gratia et Libero Arbitrio*, 1.4.9.25, 1.5–11.32–13.41, and *Ep.* 47.5.

> and under one set of circumstances, an act is not permitted, but when the time and conditions are right, it is both allowed and condoned (Athanasius, *Ep. ad Amunem monachum* 68–69).

Basil was another prominent Christian theologian who was aware of this simplistic distinction between 'private and thus objectionable murder' and 'public and thus praiseworthy killing' but felt a little uncomfortable with it in 374,[35] and approximately twenty years after Athanasius sent his letter to Amun:[36]

> Our predecessors did not consider killing in war as murder but, as I understand it, made allowances for those who fought on the side of moderation and piety. Nonetheless, it is good to admonish those whose hands are unclean to abstain just from communion for three years (Basil, *Ep.* 188.13).

Soldiers who shed blood were to be simply excluded from the sacrament of Communion for three years. We do not know whether Basil was conveying an expression of opinion or an express injunction. However, it is interesting that he was not annoyed as much as to order something drastic like the excommunication of soldiers who shed blood. Basil did not detest soldiers; let us keep in mind that he had addressed *Ep.* 106 to a soldier who, he admitted, had perfect love for God.

Athanasius and Basil had slightly different views. For Athanasius service in the army was completely unproblematic; for Basil there was a problem, but it could easily be overcome. However, the Council in Trullo in 691,[37] did not discern any difference of opinion between the two passages from Athanasius and Basil and gave to both precepts the status of a Canon.[38]

It seems that by the end of the fourth century Christians were beginning to reach a consensus as far as the issue of shedding blood in warfare was concerned. The principle eventually applied only to the Christian clergy. *Constitutiones Apostolorum*, a voluminous compilation of earlier Church

---

[35] Swift, *Early Fathers on War and Military Service*, 95.

[36] Georgios Poulis, *Η άσκηση βίας στην άμυνα και στον πόλεμο κατά το εκκλησιαστικό δίκαιο* (Thessaloniki: Sakkoula, 1990), 58.

[37] George Nedungatt and Michael Featherstone, eds., *The Council in Trullo Revisited* (Roma: Pontifico Instituto Orientale, 1995); James C. Skedros, "Canons of the Council in Trullo," in Richard Valantasis, ed., *Religions of Late Antiquity in Practice* (Princeton and Oxford: Princeton University Press, 2000), 289–300.

[38] Athena Kolia-Dermitzaki, *Ο Βυζαντινός «ιερός πόλεμος». Η έννοια και η προβολή του θρησκευτικού πολέμου στο Βυζάντιο*, (Athens: Vassilopoulos, 1991), 129.

orders, created in Syria between 375 and 400,[39] ended with the *Canones Apostolorum.* These consist in part of synodal Canons from different sources that had perhaps been collected in an earlier Corpus Canonum.[40] Most of the Canons dealt with the ordination, the official responsibilities, and the moral conduct of the clergy, and only a few were concerned with the duties of Christians in general. Canon 83 required that any bishop, priest, or deacon devoting himself to military service and aiming at combining it with the duties of his office shall be forthwith degraded from his ecclesiastical rank on the principle of failing to give to Caesar what is his and to God what is God's. The clergy was allowed if necessary to get involved in a battle but not to kill a human being.[41] The clergy was preferably to avoid violence under all circumstances.[42] As far as the laity was concerned, they were not prevented from entering the army. And those already in the army prior to their conversion were simply encouraged to obey the injunctions given to soldiers by John the Baptist in Lk 3:10-18: to be unfair to no man, to accuse no man falsely, and to be content with their hire. If Christian soldiers gave that promise they were to be admitted to the Christian community and if they refused to do so they were to be rejected.[43] There was no real obstacle for a soldier to become a Christian or a Christian to become a soldier, unless he had a clerical office. Eusebius was probably the first theologian to draw an elaborate comparison between the avocations permissible to the clergy and to the layman. He specified as among those that belong solely to the latter the carrying on of 'just warfare': τοῖς τε κατά τό δίκαιον στρατευομένοις.[44]

From the end of the fourth century onwards more and more Christian Councils under the influence of industrious theologians started discouraging much more strongly the clergy from getting involved in warfare.[45] The 8th Canon of the Council of Toledo in 398 forbade anyone who after baptism has put on the military belt to be raised to the office of a deacon. The 7th

---

39 Franz Xavier Funk, *Die Apostolischen Konstitutionen* (Frankfurt: Verlag von Wilhelm Bader, 1970, 1891[1]); Richard Hugh Connolly, *Didascalia Apostolorum* (Oxford; Oxford Clarendon Press, 1929), xx.

40 B. Steimer, "Apostolic Canons," in *Dictionary of Early Christian Literature*, 43–44.

41 *Canones Apostolorum*, Canon 66.

42 *Canones Apostolorum*, Canon 27.

43 *Canones Apostolorum*, Canon 8.32.

44 Euseb. *Demonstratio Evangelica*, 1.8.

45 "Military Service" in William Smith and Samuel Cheetham, eds., *A Dictionary of Christian Antiquities*, vol. 2, (London: John Murray, 1880): 1181–1184.

Canon of the Council of Chalcedon in 451 anathematized all who having been once enrolled among the clergy or were monks returned either to warfare or to secular employment. Canon 5 of the First Council of Tours in 460 excommunicated all clergy who engaged in warfare. The 1st Canon of the Council of Lerida in 523, speaking of the case of clergy who might be in a besieged city, provided that all who minister at the altar should abstain from shedding human blood; those who had done so, even in the case of an enemy, should be removed for two years not only from their office but from communion. The two years were to be spent in fasting, prayers and alms-giving. At the end of the two years, they might be restored, but never promoted to higher stations. The penance might be protracted at the will of the bishop if not performed to his satisfaction. Canon 5 of the first Council of Mâcon in 581 provided that any clergy wearing arms shall be kept for thirty days on bread and water. Canon 19 of the fourth Council of Toledo in 633 forbade that any employed in secular warfare or pursuit should be ordained while Canon 44 directed that those of the clergy who ventured to take up arms *in quacumque seditione* (= in any unrest) shall be similarly treated and sent *in monasterium poenitentiae contradantur* (= to do penance in a monastery). The 2nd Canon of the Council of Lestines in 743 prevented any of the clergy from wearing arms and from accompanying armies, except one or two bishops with their chaplains in attendance on the prince and one presbyter attached to each division of the army. Canon 3 of the first Council of Soissons in 744 kept abbots from bearing arms; even those who by their feudal tenure were obliged to send soldiers from their lands. Canon 37 of the Council of Meaux in 845 forbade the clergy to engage in warfare and even carry arms under pain of deprivation of office as *sacrorum canonum contemptores et ecclesiasticae sanctitatis profanatores* (= despisers of the sacred canons and profaners of ecclesiastical sanctity).

The fact that numerous ecclesiastical Councils systematically condemned any involvement of the Christian clergy in military affairs requires some explanation. Was it a custom devoid of meaning? Were the Christian Councils simply repeating earlier injunctions? Or did the clergy tend to forget that war was not their business and the Councils had to remind them? There is ample evidence that the latter was true.[46] It seems that the clergy often preferred the excitement of the camp to the seclusion of the cloisters

[46] Paul Lemerle, *The Agrarian History of Byzantium from the Origins to the 12th Century* (Galway: Galway University Press, 1979), 148, n.1; H. Poulis, *Η άσκηση βίας*, 79, 196, discusses cases of the clergy's involvement in military matters.

or the monotony of their clerical duties. The Canons were frequently transgressed, especially during medieval times.

The State found the decision of the Church to keep only its religious officials away from warfare and politics convenient. Emperors endorsed the idea that the clergy and the monks[47] ought to avoid violence,[48] keep themselves apart from political affairs,[49] and be exempt from public duties.[50] However, emperors were willing, in collaboration with the Church authorities, to allow exceptions. In 398, according to *Cod. Theod.* 16.2.31, if someone who held a clerical office got involved in violence, they might consider forgiving him. The accepted norm was that the clergy would not mingle in military affairs or indeed any kind of violence.[51] Of course, Christians were not to take advantage of this 'agreement' and no one was to hold an ecclesiastical office in order to excuse himself from serving in the army or fulfilling his civic duties.[52] At the same time, the State expected only Christians to serve in the military; in 416, Theodosius II issued a decree stipulating that only

---

[47] *Cod. Theod.* 16.3.1. However, Christians were not to enter the monastic life merely to avoid civic duties (*Cod. Theod.* 12.1.63).

[48] See Justinian, *Novella*, 123.11.1, while in 545 CE, the emperor in *Novella* 131, endowed the Canons of the first ecumenical Councils with the validity of imperial legislation.

[49] *Cod. Just.* 1.3.17 of 416 was explicit: "it pleases our majesty that the clergy have nothing to do with public acts or political affairs."

[50] Euseb., *Hist. eccl.* X.vii.1–2; *Cod. Theod.* 16.2.11, *Cod. Theod.* 16.2.9, *Cod. Theod.* 16.2.14–16, *Cod. Theod.* 16.2.24; *Cod. Just.* 1.3.6; *Cod. Theod.* 11.16.15, *Cod. Theod.* 11.16.22, *Cod. Theod.* 16.8.13, *Cod. Theod.* 16.2.46; *Cod. Just.* 13.22, *Cod. Just.* 10.49.2, *Cod. Just.* 1.2.12. However, some emperors tried to cancel or control clerical exemption from public duties. For example, in 383, Gratian, Valentinian II and, Theodosius I, prohibited Jews and Christians from entering the clergy before they had performed their public services (*Cod. Theod.* 12.1.99); in 408, Arcadius and Honorius prescribed that ex-clergymen were expected to perform their civil duties (*CS* 9); in 423, Honorius and Theodosius imposed upon churches a share in the performance of some public services (*Cod. Theod.* 15.3.6); in 440–1, Theodosius II and Valentinian III ordered the Church to help the imperial tours by providing transport animals (*Cod. Just.* 12.50.21) and finally cancelled the clerical exemption from public duties (*LNV* 10).

[51] The norm became deeply rooted to the conscience of the emperors to such an extent that we find in the twelfth century the princess Anna Comnene filled with anger about the involvement of the Christian clergy in warfare, (Anna Comnene, *Alexiad*, 2.218.18f).

[52] *Cod. Theod.* 7.20.12, *Cod. Theod.* 16.2.3, *Cod. Theod.* 12.1.49, *Cod. Theod.* 8.4.7, *Cod. Theod.* 12.1.50, and Julian, *Ep.* 380D–IA.

Christians were allowed to serve in the army,[53] and in 519 or 520 it was only orthodox Christians, and not heretics, that were permitted to enlist.[54]

In conclusion, it seems that Christian involvement in military affairs started creating problems for the Christian conscience from the beginning of the fourth century. It was rarely generally forbidden. Normally several factors and conditions had to be taken into account before a decision was reached. It was examined whether a Christian had to kill or had already killed a human being, had to give or was given the order to kill on the battlefield, took the initiative to kill for his own reasons and profit, whether he had to take or taken the pagan military oath, whether he was a soldier or a high-ranking official, whether he was in the army prior, during or after his conversion to Christianity, whether he had left the army and later returned to his post, and, most importantly perhaps, whether he was a cleric, a monk, or an ordinary Christian. Christian involvement in military affairs was a complicated issue that could afford no easy solutions.

Bibliography

Bainton, Ronald H. "The Early Church and War." *HThR* 39 (1946): 189-215.

—. *Actitudes cristianas ante la guerra y la paz. Examen histórico y nueva valoración crítica*, tr. Angel Alcalá. Madrid: Tecnos, 1963.

Barnes, Timothy D. "The Constantinian Settlement." In *Eusebius, Christianity and Judaism,* edited by Harold W. Attridge and Gohei Hata, 653–657. Leiden, New York and Köln: Wayne State University Press, 1992.

Bigelmair, Andreas. *Die Beteiligung der Christen am öffentlichen Leben in vorkonstantinischer Zeit. Ein Beitrag zur ältesten Kirchengeschichte.* Munich: University of California Libraries, 1902.

Cameron, Averil. *The Later Roman Empire.* London/Cambridge: Harvard University Press, 1993.

Cameron, Averil and Stuart G. Hall. *Eusebius, Life of Constantine.* Oxford and New York: Oxford University Press, 1999.

Campbell, J. B. *The Emperor and the Roman Army 31BC–AD235.* Oxford: Oxford Clarendon Press, 1984.

---

[53] *Cod. Theod.* 16.10.21.

[54] *Cod. Just.* 1.4.20. See also *Cod. Theod.* 16.5.29, *Cod. Theod.* 16.5.42, *Cod. Theod.* 16.5.48, *Cod. Theod.* 16.5.61; *Cod. Just.* 1.5.22.

Connolly, Richard. *Didascalia Apostolorum*. Oxford: Oxford Clarendon Press, 1929.

Crescenti, Giovanni. *Obiettori di coscienza e martiri militari nei primi cinque secoli del cristianesimo*. Palermo: Flaccovvio, 1966.

Dale, Alfred William Winterslow. *The Synod of Elvira and Christian Life in the Fourth Century*. London: MacMillan, 1882.

De Jong, K. H. E.. *Dienstweigering bij de oude Christenen*. Leiden: Brill, 1905.

Dix, Gregory and Henry Chadwick. *The Apostolic Tradition*. London and New York: Routledge, 1995.

Driver, John. *How Christians Made Peace with War. Early Christian Understandings of War*. Scottdale, Pennsylvania and Kitchener, Ontario: Wipf and Stock, 1988.

Egan, Eileen, "The Beatitudes. The Works of Mercy and Pacifism" in Shannon A. Thomas, ed., *War or Peace? The Search for New Answers* (New York: Orbis Books, 1980), 169–187.

Ferguson, John. *War and Peace in the World's Religions*. London: Open University Press, 1977.

Ferguson, Sinclair B. and David F. Wright, eds. *New Dictionary of Theology*. Leicester, Illinois: IVP Academic, 1988.

Fontaine, Jacques. "Christians and Military Service in the Early Church." *Consilium* 7 (1965): 58–64.

Frend, William. *The Donatist Church: a Movement of Protest in Roman North Africa*. Oxford: Oxford Clarendon Press, 1952.

—. *Orthodoxy, Paganism and Dissent in the Early Christian Centuries, Variorum Collected Studies Series*. Hampshire and Vermont: Routledge, 2002.

Fuhrmann, Christopher. *Policing the Roman Empire: Soldiers, Administration, and Public Order*. Oxford: Oxford University Press, 2011.

Funk, Franz Xavier. *Die Apostolischen Konstitutionen*. Frankfurt: Verlag von Wilhelm Bader, 1970, 1891[1].

Gero, Stephen. "Miles Gloriosus: the Christians and Military Service according to Tertullian," *Church History* 39 (1970): 285–298.

Harnack, Adolf. *Militia Christi. The Christian Religion and the Military in the First Three Centuries*, tr. McInnes David. Philadelphia: Fortress, 1981[1], 1905.

—. *The Mission and Expansion of Christianity in the First Three Centuries*, tr. James Moffatt. London and New York: Williams and Norgate, 1908.

Hefele, Charles, *A History of the Christian Councils from the Original Documents to the Close of the Council of Nicaea A.D. 325*, tr. Clark R. William, London: T&T Clark, 1894.

Helgeland, John. "Christians and the Roman Army AD 173–337." *Church History* 43.2 (1974): 149–163.

—. "Christians and the Roman Army from Marcus Aurelius to Constantine." *ANRW* II.23.I (1979): 724–834.

Huttunen, Nico. *Early Christians Adapting to the Roman Empire: Mutual Recognition.* Novum Testamentum, Supplements, vol. 179. Leiden/Boston: Brill, 2020.

Huxley, Aldous. *An Encyclopaedia of Pacifism.* London: Chatto and Windus, 1937.

Iosif, Despina. *Early Christian Attitudes to War, Violence and Military Service.* Piscataway, NJ: Gorgias Press, 2013.

Kalanztis, George. *Caesar and the Lamb.* Eugene, Oregon: Cascade Books, 2012.

Kim, Young Richard. *The Cambridge Companion to the Council of Nicaea.* Cambridge: Cambridge University Press, 2021.

Kolia-Dermitzaki, Athena. *Ο Βυζαντινός «ιερός πόλεμος». Η έννοια και η προβολή του θρησκευτικού πολέμου στο Βυζάντιο.* Athens: Vassilopoulos, 1991.

Le Blant, Edmond. *Les inscriptions chrétiennes de la Gaulle romaine*, vol. 1. Paris: Imprimerie nationale, 1892.

Lemerle, Paul. *The Agrarian History of Byzantium from the Origins to the 12th Century*, tr. Mac Niocaill G.. Galway: Galway University Press, 1979.

Luibhéid, Colm. *The Council of Nicea.* Galway: Galway University Press, 1982.

McDonagh, Enda. *The Demands of Simple Justice.* Dublin: Gill and MacMillan, 1980.

Musto, Ronald. *Catholic Peacemakers. A Documentary History. From the Bible to the Era of the Crusades.* New York and London: Garland, 1993.

Nedungatt, George and Michael Featherstone, eds. *The Council in Trullo Revisited.* Roma: Pontifico Instituto Orientale, 1995.

Nussbaum, Arthur. "Just War- A Legal Concept?" *Michigan Law Review* 42 (1943): 453–479.

Poulis, Georgios. *Η άσκηση βίας στην άμυνα και στον πόλεμο κατά το εκκλησιαστικό δίκαιο.* Thessaloniki: Sakkoula, 1990.

Rousseau, Philip. *The Early Christian Centuries.* London and New York: Routledge, 2002.

Ruether, Rosemary. "War and Peace in the Christian Tradition." In *A Council of Peace* edited by Hans Küng and Jurgen Moltman, 17–24. Edinburgh: Humns Ancient and Modern, 1988.

Rusch, William. *The Trinitarian Controversy.* Philadelphia: Fortress Press, 1980.

Secrétan, Henri. "Le Christianisme des premiers siècles et le service militaire." *Revue de Théologie et de Philosophie* 2 (1914): 324–366.

Skedros, James. "Canons of the Council in Trullo." In *Religions of Late Antiquity in Practice*, edited by Richard Valantasis, 289–300. Princeton and Oxford: Princeton University Press, 2000.

Smith, William, and Samuel Cheetham, ed. *A Dictionary of Christian Antiquities*, vol. 2. London: John Murray, 1880.

Stanley, Arthur. *The Council of Nicea. Lectures on the History of the Eastern Church.* Piscataway, NJ: Gorgias Press, 2009.

Steimer, B. "Traditio Apostolica." In *Dictionary of Early Christian Literature*, edited by Siegmar Dopp, 580–583. New York: Herder and Herder, 2000.

Stevenson, James. *A New Eusebius. Documents Illustrative of the History of the Church to AD 337.* London: Society for Promoting Christian, 1957.

Swift, Louis. *The Early Fathers on War and Military Service.* Wilmington, Delaware: M. Glazier, 1983.

Tanner, Norman, ed. *Decrees of the Ecumenical Councils,* vol. I: *Nicea I to Lateran V.* Washington: Georgetown University Press, 1999.

Ubiña Fernández, José. *Cristianos y militares. La iglesia antigua ante el ejército y la guerra.* Granada: Universidad de Granada, 2000.

von Gebhardt, Oscar and Adolf Harnack. *Texte und Untersuchungen zur Geschichte der altchristlichen Literatur.* Leipzig: J. C. Hinrichs, 1891.

—. eds. *Die Canones Hippolyti.* Leipzig: J. C. Hinrichs, 1891.

Warmington, H. "The Sources of Some Constantinian Documents in Eusebius' *Ecclesiastical History* and the *Life of Constantine.*" *St Patr.* 18/1 (1986): 93–98.

Whittaker, Dick. "Landlords and Warlords in the Later Roman Empire." In *War and Society in the Roman World*, edited by John Rich and Graham Shipley, 277–302. London and New York: Routledge, 1993.

Windass, S. "The Early Christian Attitude to War." *Irish Theological Quarterly* 29 (1962): 235–245.

CONCLUSION: LOOKING TO THE FUTURE OF EXAMINING THE PAST

GRAHAM WRIGHTSON

This volume presents a selection of new areas of research into ancient warfare. Ancient warfare is often considered to be an old-fashioned avenue of research and scholarship. However, as long as wars continue to be waged there will always be new ways to examine the past to better understand the present and look to the future. This volume, and the annual conference from which it stemmed, promotes using these new ideas to widen the appeal of ancient warfare and provide a selection of future ways of examining the past.

Three papers (Mitsios, Brighouse, and Wetzel) examine the roles of women in war as depicted in visual or mythological settings. These show that the study of war has an enduring appeal outside of traditional avenues of military history. Women in war is a topic in its own right, but with the relatively recent rise in equality among the sexes, as well as the ongoing debate around transgender rights, it has become even more significant. As women continually raise the proverbial 'glass ceiling,' study of ancient warfare reveals just how far some women in the past pierced it. As sex and gender equality continues to increase across society so too will examinations of women in all aspects of history including war.

Two other papers (Atkinson and Iosif) concern religion and war and the motivations behind soldiers signing up to fight. The authority of religion compelling soldiers and generals to wage wars is as old as history itself and shows no sign of slowing down with what is happening in Gaza in 2023. Religion causes as many rifts and violent reactions now as it did thousands of years ago. The specifics of the religions involved may have changed just as the weapons used have done but the motivations of the soldiers and policy makers remain the same. Unless something drastically changes on a global scale, religion and its motivations will continue to shed light on warfare both ancient and modern.

Two papers (Gauthier and Kreiner) examine the logistics of waging war in the Roman world, one before conflict, the other the consequences. Logistics is an often-neglected aspect in the research of warfare, but it has and always will be the most important part of the actual campaigning. Still now, just as in the ancient world, armies could not fight battles if the soldiers had no food, water, equipment, or travel routes. More research is now happening around the world on how states waged war even more than the consequences of those conflicts. When it comes to consequences scholars most often focus on long-term military political and sometimes cultural events but the importance of the immediate effects on everyday society are often overlooked yet simultaneously the most important.

Following along the themes of how wars were fought, three papers (Feil, Wrightson, and McCarthy) examine the more traditional aspects of warfare, equipment, soldiers, and tactics. All three papers show that a detailed examination of the evidence in light of modern theories or scientific knowledge, no matter how scarce or minimal the ancient sources might at first appear, can shed new light on otherwise misunderstood events, peoples, or armies. In the ancient world especially, equipment is most important for determining how soldiers fought, and survived (or not), and all three papers demonstrate that understanding exactly how units or soldiers fought is the most important step for understanding any events on the battlefield. Though these aspects of study in ancient warfare are not new, the methods used to analyse them are innovative and rely on modern theories and ideas and open up a whole new avenue for continuing future scholarship.

Perhaps the largest thematic collection in this volume are the five papers (Vertedor Ballesteros, Brusa, da Silva Ferreira, Waldron, and Wetzel) that examine how the memory, or depictions of wars and military ideas affect our understanding of events. Representation matters. Whether it is the language that records the events (Waldron), the literary genre of the sources (Brusa, and da Silva Ferreira), the artifacts themselves (Wetzel), or the contemporary methods of representing power (Vertedor Ballesteros), the how is more important than the what. This is true of all history, but is perhaps even more relevant in military history in relying on so few sources to determine what actually happened.

This theme also connects with every other paper in the volume. The two papers on war represented in mythology (Mitsios, and Brighouse) examine the power of memorializing women in war in the past and the present. Two papers (Brusa and Wetzel) examine the representation of the same Roman term, *Virtus*, in literature and material culture revealing the importance of words in the popular imagination. The papers on religion (Atkinson and Iosif) assess how war is viewed in religious terms with connections to enduring memory as well as initial motivations. Source problems or the lack of contemporary evidence is the crucial theme in the more traditional papers on armament, soldiers' experiences, and logistics (Feil, Wrightson, Gauthier and McCarthy), and Kreiner's paper even concludes with an aside on how the specific events discussed actually shaped the surviving historical account rather than the other way around.

Overall, this volume demonstrates the enduring interest in ancient warfare and the continual avenues for future work in bringing new light onto old topics. There are other areas too that advance our understanding of ancient warfare, such as experiential history, geophysical and satellite surveys of topography and individual sites, and the ideas of creating sensory examinations of the past. These topics appear with increasing regularity in other collections of papers or journals, but the papers in this volume show how current research is still shaping our understanding of the past and how much more there is still to discover. That's why studying ancient warfare is so exciting!